# HAPPY
LANDINGS

# HAPPY LANDINGS

## A Military Family's Odyssey

**LT COL JJ SCANLON AND RE SCANLON PE**

Copyright © 2024 Lt Col JJ Scanlon and RE Scanlon PE.

All rights reserved. This book is protected by copyright. No part of this book may be reproduced or transmitted in any form or by any means, including as photocopies or scanned-in or other electronic copies, or utilized by any information storage and retrieval system without written permission from the copyright owner.

ISBN: 979-8-9889337-2-4 (paperback)
ISBN: 979-8-9889337-0-0 (hardcover)
ISBN: 979-8-9889337-1-7 (ebook)

Typesetting by Wordzworth.com

Dedicated to military dependents everywhere,
past and present; they also serve.

# Contents

Introduction ix

About the Authors xi

**Chapter 1**  2009 Austin, Texas  1

**Chapter 2**  1924 Denver, Colorado  6

**Chapter 3**  1935 California  18

**Chapter 4**  1942 War  31

**Chapter 5**  1947 Ferry Command  86

**Chapter 6**  1948 Fassberg, Berlin Airlift  92

**Chapter 7**  1951 Lake Charles, Louisiana  109

**Chapter 8**  1953 Randolph, San Antonio, Texas  114

**Chapter 9**  1957 Johnson AFB, Japan  122

**Chapter 10**  1959 Hill AFB, Ogden, Utah  135

| | | |
|---|---|---|
| **Chapter 11** 1961 Maxwell, Montgomery, Al | | 175 |
| **Chapter 12** 1962 Ent AFB, Colorado Springs Co. | | 186 |
| **Chapter 13** 1965 Taiwan Defense Command | | 227 |
| **Chapter 14** 1967 Griffiss AFB Rome New York | | 278 |
| **Chapter 15** 1970 Kwan Ju Korea | | 330 |
| **Chapter 16** 1971 Dyess AFB Abilene Texas | | 377 |
| **Chapter 17** 1973 Austin | | 392 |
| **Chapter 18** 1974 The Oilfield | | 405 |
| **Chapter 19** 1975 Retirement | | 439 |
| **Chapter 20** 1976 Austin Brentwood | | 453 |
| **Chapter 21** 1977 Austin IRS | | 463 |
| **Chapter 22** 1979 Barrington Way | | 490 |
| **Chapter 23** 1981 San Francisco | | 507 |
| **Chapter 24** 1983 Cindie | | 519 |
| **Chapter 25** 2009 Abilene | | 542 |
| Commendations | | 565 |
| Addendum | | 567 |
| References | | 575 |

# Introduction

**"Happy Landings"** was how Grandpa used to sign off on his letters to my Dad. He crash landed three times; many of his fellow Pilots had some very unhappy landings…we dependents landed here and there too, with similar outcomes.

**Disclaimer:** Some names and events have been changed to protect the innocent; memory is also fragile and fallible so please excuse my shortcomings. This is also my Dad's memoirs that he wrote and compiled on a script manual typewriter that Mom insisted he get in Hong Kong on a C-47 junket in 65. Dad occasionally flew this route in order to get some flight time in when we were stationed in Taiwan. Dad, eighty-years-old and no typist, started writing his memoirs after he was diagnosed with Lewy Body Dementia a disease sometimes associated with Parkinson's and the second most common type of dementia after Alzheimer's. I have taken the liberty of editing his heroic efforts, humbly trying to make something of it. I think he would have liked that.

# About the Authors

## Robert E. Scanlon P.E. (RES.)

The author earned a BSME from the University of Texas at Austin graduating in 1979, while employed by the University as a draftsman. After graduating Mr. Scanlon continued employment with UT as a Professional Engineer for 38 years where he designed and produced all manner of mechanical equipment and technical documentation for various Navy DOD contracts.

## Lt. Col. Scanlon. (JJS.)

Served 32 years with the Air Forces of the United States. He was a natural pilot, and an excellent navigator. Col Scanlon spent most of his career as a Command Pilot (over 12,000 hours) who excelled as an Operations Officer. In the 50's. Col Scanlon perfected a lob bombing technique flying the B-57 for tactical nuclear weapons delivery without which it was a one way trip. Col Scanlon flew with the 453rd, during WWII flying a full combat tour in Heavy

Bombers then volunteered for a second tour in Fighters with the 4th fighter group, composed of pilots of the ex-Eagle Squadron, the highest scouring fighter group in theater, with over 1000 destroyed. Dad helped Governor Perry get his commercial pilot license when he retired from the Air Force. He became a.hunting buddy of the governor and Jim Baker. Dad flew Lt Governor Hobby plus many other notable Texans around as a charter pilot and flight instructor after retiring as Base Ops Officer, Dyess AFB Abilene Texas.

# 2009 Austin, Texas

**RES.** On the road again, it's a long boring ride up to Abilene to see Mom and the "Colonel", Dad to my seven siblings, Sir to me; one of the few men that I have ever addressed as Sir. It's a rare person that can measure up to him, as far as I'm concerned. These trips to west Texas are getting to be a pain. After you go down the same road, on cruise control, over and over again you sort of go on automatic, you drift from landmark to landmark habitually, not even looking at the signs on the highway, only the seasons seem to change, along with the type and number of dead critters on the road. I was on this road before cruise control. I've seen this road go from: two lane bar ditch, to three lane soft shoulder, it's now a proper four lane highway; forty years of TxDOT evolution. At least I'm not driving 55 anymore. That was quite a cross to bear in Texas, were all the major cities are at least 200 miles apart. The local radio is not to my taste and I'm too cheap to renew my satellite radio account, never owned a recording of any type, same reason, so that leaves daydreaming. What do I think about? Probably what everyone thinks about: how did I get here, where am I going, what am I doing, am I doing things the right way, could things have been different? Choices made, opportunities passed up, people left in your wake; you roll mental images of faces you've known and places you have been, back and forth in your head, some as transitory as the fence posts flicking by. Don't know what I'm looking for. There really are no answers; it's just the way it turned out, just like everything else, OBE, overtaken by events.

Mom and Dad should have moved down to Austin a long time ago, down to where the rest of us live and work. I don't really know what happened, they never said, plans were made, lake view lots were purchased; hell, Brother Steve's girlfriend Kerry and I even cut brush and surveyed the property with transit and chain one hot summer day. Not knowing any better, both of us were chewed up by chiggers. I drafted the contours for the plot, laid in a foundation for a 3000sq foot house, a driveway, drainage and a tennis court. It all looked good, at least on paper. That was twenty five years ago. Somewhere in there two thirds' of the property was sold off leaving one half acre lot that I have been saddled with, paying taxes on since the dissolution. We only used to see the parents at Thanksgiving, Christmas, and Easter maybe, if they cared to make the trip; now they can't even do that. They are afraid to go anywhere since they spun out the Grand Marquis in a light snowfall and put it in the bar ditch going home last Christmas. We told them early on that the shear cost and logistics of getting the eight of us with our collective spouses and kinder up for special occasions would work a hardship. Mom is nonplused, she insists that the mountain must come to Mohammed, so we all have to make the pilgrimage north to the museum or its guilt city. No one wants to stay at Mom's house, it's

## 2009 AUSTIN, TEXAS

not very guest friendly and any suggested potential upgrades are out of the question, more guilt. I call it the museum because there isn't square inch of wall space or a horizontal surface that doesn't contain some family historical memento of our time on the road. Everything is dear to Mom.

    I used to be upset, almost embarrassed that my parents seemed to have so little interest in my kids, compared to my wife's parents, who dote on my girls. I mean, I did everything the way I was expected to; I kept out of trouble, worked my way through college, graduated with an Engineering Degree and got a good job, bought a house, married a wonderful girl, had three beautiful children, so where's the love? I really wanted my girls to spend time with Mom so they would get to know and love her like I do and grow up to be proper ladies. Too late now, the girls are older, two are in college, they're not interested in hanging out with old folks. Lacking that early bonding, they don't have much of a relationship; Cindie's Mom took care of the baby the first two years of her life and she still calls nearly every other day to talk to one or another of the girls. When I bring it up to Mom it's: "you never call me, or why don't you take me with you when you go on vacation?" me, me, me…wonder why, Cindie's parents take US to the beach, that's how you do it if you want to spend time with someone; you make plans, talk to people, fund your plans, invite and get people on board, follow through; hard for someone who is budget, math and schedule challenged.

    I can also take a hint, but no hint was ever required, the Colonel told me straight up one day "I'll come visit you" I guess having eight kids' makes you yearn for that empty nest in a bad way. Alone time is good, but 200 miles away? Mom swears it wasn't her decision, but I know better, she may not have come right out and said it, but I'll bet she came up with a million reasons and excuses not to move, served up with a truckload of guilt. I know Dad was for it, I have letters ending with "your mother is inclined to beat the dead horse; pity me", he even had me come up to get him once, just to back out at the last minute; said it would be "splitting the blanket". Dad is loyal and faithful to his own peril and Mom knows this. Mom complains of her many moves in the service and the hardships she suffered, moving every one or two years, talking to me like I wasn't along for the ride. We were always over our weight allowance going overseas and we were twice as heavy coming back with all her booty.

The main reason Mom will not move, the elephant, the huge, white elephant, in the room, is that any move would involve a dumpster; stuff would have to be thrown away, death by a thousand trash bags. She can't even bring herself to think about it. Mom can't throw anything away, she still has Montgomery Ward carpet samples circa 1960 that she simply can't part with!

There's some history there though, Mom was a child of the depression, her Dad was a junk dealer; WWII on the home front was all about rationing and saving. The kicker was her sea trunks with all her wedding presents, her dress, all the things she had collected overseas in Germany and France being "lost" in transit on the boat ride home. That, and when she went to her Mother's house she found out that she had cleaned out her room. She'd thrown out all Mom's shoes and clothes, all the records she bought when she worked in a music store, along with her books, school papers and magazines. Mom was furious.

I swear over the past forty years Mom has been taking inventory every night; drinking coffee into the wee hours, making sure everything is there and in its proper place. From jewelry to foolery, all the treasure she collected overseas and in auction houses, just occupying space. Mom still has every toy I ever possessed, even the rotten genuine rabbit fur Davey Crocket Coonskin Hat I had when I was five; found that last trip.

The 'stuff' has been a running gunfight between Mom and Dad for years. Dad would slip something that needed throwing away in a paper sack when Mom was out and about, sneak out the back to the alley, and throw the sack in the trash. Mom would check the trash cans when she came home, fish the stuff out, and hide her find back in the house like a pack rat. This went on for a long time both of them improving on their stealth, tactics, dumps, hides, carries and raw audacity. When the trashcans were replaced by a huge dumpster, my five-foot-tall Mom got a ladder and a grabber. Well her dumpster diving days are over now, she and Dad are too feeble for the game. Mom's Spidey senses are still fully functional and always tingling; she is very suspicious of all comings and goings, insisting on knowing what is coming and going because she is neither coming nor going. We could probably cleanout an entire room without her knowing it now, but if she ever found out there would be hell to pay, so no one dares move a thing.

Maybe Grandma threw out Mom's stuff because she ran off and got married on the sly to that dammed cowboy Jim Scanlon. She did it while on vacation, in Europe with her cousin Sonia and some girlfriends, a vacation they paid for and sponsored through USC, it was the first time they had let her out of their sight. Brother Sal was to go as chaperone but he backed out at the last moment. Mom is a good Catholic girl from a high Spanish Los Angles family. It was just after the Berlin Airlift, Dad was stationed in Munich; they had been writing each other on the sly and scheming. In her letters she calls him Jimmy Boy so there is something there already. It's a long road to Munich from Los Angles for the both of them.

# 1924 Denver, Colorado

**JJS.** I was born at 11:25 PM in our home at 41 Cherokee, Denver, Colorado on 17th June 1924. Dr. George E. Osborn M.D. conducted the home delivery because Mother refused to go to a hospital with an Osteopath as the doctor in charge. Mother had a very difficult delivery with Brother Bob on 27th May 1923 and had made up her mind to stay clear of hospitals, so I was hatched at home.

My Father, James Emmett Scanlon, was a WW1 Veteran. He enlisted 29th April 1917, and received a Medical Discharge on January 10, 1919. Dad was an army cook.

Private Scanlon was standing outside the mess tent when a Captain rode up on a frisky horse and told Dad to hold the reins. Something spooked the horse, which reared up pawing and hit Dad knocking his teeth out and kicking him in the groin. Dad was in the hospital for months and almost died. The doctors performed several operations on him. They removed some of his small intestine and his appendix; then peritonitis set in. Dad was given up for dead. The year 1918 – 1919 was a bad time to get sick in the Army with the Spanish Flu raging! When Dad finally recovered, he was medically discharged from the Army at the Medical Detachment General Hospital #14, Fort Oglethorpe, Georgia. Dad was a bona fide Disabled Veteran. Dad took advantage of the Disabled Veteran Education Program and attended Dental College at the University Dental College in Denver. The course at that time was three years; Dad almost finished, but had to drop out to make a living after Brother Bob was born.

Dad was raised mostly in the Upper Peninsula of Michigan. He grew up in the small towns of the Iron Range and lumber mills: places like Iron Mountain and Ishpeming. My Grandfather, Jim Scanlon was an independent Diamond Driller – one who used diamond bit drills, prospecting for and proving up the locations of Iron ore and coal deposits.

Family history concerning my Grand Parents on both Dad's and Mother's sides is real slim. (That is why I am now putting down what I recall).

Grandfather Jim Scanlon was born in Ireland and migrated to Canada with his family. They later moved to the United States, and settled in Michigan. Grand Dad Jim was a tall man about 6' 2". He took up the trade of a Diamond Driller. He made several trips to Africa to buy diamonds. Stories were told of his time on sailing ships. In one, there were some monkeys jumping about in the rigging causing problems. One unhappy sailor caught the pests and put grease on their tails, this eventually caused the monkeys to "lose their grip" and fall overboard. Granddad brought back a parrot which lived to a ripe old age. It died at our house in Denver after an overdose of green grapes from the kind hands of little Bob and Jim!

Grandfather Jim married a beautiful redheaded Irish girl named

Catherine Kennedy around-about year 1890. They had four children as follows: Anastasia, James Emmett (my Dad), Gratton, and Joseph. James Emmett was born September 10, 1895 in Chicago, Illinois. He was baptized September 15, 1895 in St. Dennis Catholic Church, located in Lockport, Illinois.

Grandmother Catherine Scanlon died during World War One while her son Jim was overseas with the US Army. I have very little knowledge of Catherine.

We bought a house at 223 South Irving in Denver around 1927 or '28. It was small! There were two small bedrooms, a living room, and a large kitchen. There was no bathroom! We had an outhouse in the back of the lot next to the chicken coup and rabbit pens. There was a tiny dirt floor basement beneath the kitchen. Access to the basement was through a trap door in the middle of the linoleum-floored kitchen. We had a wood/coal kitchen range a large sink with a drain board and running water. There wasn't hot water; instead, we had to heat water on the cook stove for washing and bathing. I clearly remember Mother giving Bob and myself baths in a galvanized tub placed next to the cook stove in the kitchen. Water was heated in two tea kettles to a boil and then poured into the tub half full of cold water. As a young lad, I was kind of afraid of the dark; therefore, it was a real problem to use the outhouse on a cold dark winter night when there was a foot of snow on the ground. My usual solution was to wake up brother Bob, who would kindly accompany me. We would put on our bathrobes and goulashes then hike the thirty or forty yards to the two-holer. Bob hardly ever refused to go with me; he was a bed wetter and my getting him up in the wee hours often prevented a wet bed. The room we slept in was too small for two beds, so Bob and I slept together in a small twin bed. I often ended up at the foot of the bed to avoid the wet spot.

## 1924 DENVER, COLORADO

The Irving Street house, located halfway up a hill on a dirt road, was steep on top, giving one a fast start when riding our toys. Dad bought us a wagon and flexible flyer sleds; we each had a scooter too, so we had plenty of transportation to negotiate the hill summer or winter. Dad bought us all ice skates and taught us to skate: he was a fine athlete, a good skater, and a good softball and baseball player. Mother was a good sport but not too good at sports. Dad also taught us how to swim, dive, fish and hunt. Dad gave Bob a 16ga Winchester Mod 12 shotgun when he was eight. When I turned seven I bought a model 15 20gage Ithaca pump and a box of shells for $8.50, new in the box. I just walked into the general store, laid my money down and walked out with the shotgun. Dad taught us to swim at a public pool about a mile from home. He allowed us to play in the shallow end at first then he took us to the deep end and had Bob and I jump off the diving board while he treaded water below. Dad would snag us when we bobbed to the surface; then give us a push toward the pool's side. We had to swim or sink! Needless to say, Dad took us along in easy stages. After that Bob and I went to the pool weekly, walking both ways in the summer during the six years that we lived at S. Irving St. We fished in local lakes and a small nearby frog pond that was full of crayfish, fine bait for bigger fish. The pond was also great for gigging frogs to eat.

The first real toy I recall having was a tricycle, which was my "undoing." I was about four-years-old, sitting on the seat with Bob standing behind and pushing. My foot got tangled between the pedals and the front wheel, and I was pitched over the handlebars smack onto the sidewalk. The result was all four front uppers breaking *off!* My smile was quite unusual for the next few years. To ensure that we would experience future spills, we got ice skates. Since we were young and growing, not wealthy, we received the kind of skates that clamped on to regular shoes, so they frequently came off. We had snow sleds (Flexible Flyers), which lasted forever. Another item of our rolling stock was the *Red Wagon* which had lots of hard use. Our folks always bought *two* of each, so that we could play together and go rolling and sliding down South Irving hill, racing. Bob got a Violin, and I got a Banjo. We both took lessons while in Denver, and later in San Diego. We let the music slip on our move to Los Angeles in 1939.

Mother saved my two four-string Banjos and my Tenor Four-String Guitar, also Bob's Fiddle. She saved those instruments until 1959 when I returned from Japan, she asked me if I wanted them; and foolishly, since I was moving with six kids to Utah, in a Volkswagen bus, I told her, "Maybe later". Later came. Mom kept those instruments from 1930 to 1959 then pitched them. !@*# I sure goofed! @#*$!

We had: scooters, homemade racing chuggs along with our sleds and wagons, BB Guns, .22 Remington pump rifles besides our shotguns. We also both had fishing poles and tackle. We used our *Stuff*, which kept us occupied, exercised, and outside

We had a few different dogs; I don't recall how many, most were pups with no brains. One was a German Shepard that would bite our legs as we coasted downhill. This dumb dog got run over by a car on a street that seldom saw more than ten cars a day. We were not sad when "leg biter" died. Then along came Pepper. Pepper arrived at our house in pretty poor shape, with a few feet of chain around his neck. We could tell that Pepper had been mistreated. Mom felt sorry for him and took the chain off and fed him for a few days; from then on he was ours. Pepper had a long black and white coat, and a full flowing tail. Pepper was a family member for five years until we departed for California in the fall of 1935. We gave Pepper to good friends who sure spoiled him. When Pepper was with us he never stayed in the house, instead his doghouse was under our back porch. Neighbors hung small rugs to beat and clean. Our neighbors had no vacuum cleaners, and for that matter no cars or telephones; people were poor. Pepper would shop around the neighborhood for small rugs, then would drag them home to line his quarters. He was a fine watchdog and could whip any other dog near or far. If we said sic-em he would attack!!!! Pepper loved canned salmon, which we bought for ten cents a can. Bob and I attended P.T. Barnum Elementary School, down Irving St. three or four blocks. I attended this school Kindergarten through Fifth Grade, 1929 to 1935. Bob was a year ahead.

We didn't have to put up with the primitive plumbing for too long. After a year or so we were connected to a gas line, got a gas furnace, a hot water heater, and Dad had a lean-to built back of the kitchen where he had a bathtub, toilet, and wash tub installed. Man we were in high cotton. We were doing great; Dad was bootlegging dental work, had steady employment, we had a Model 'T' Ford, then the Great Depression hit in 1929.

Growing up in Denver during the Great Depression was not easy. Dad, having almost completed Dental School and being a Master Dental Technician, was the only one of all the relations to have steady employment. Dad provided all the dental needs for many, many folks, including false teeth, (uppers and lowers), for our relations. Dad would barter dental work for window drapes, house painting, and backyard cement work, whatever he needed that folks could provide. He even had most of the garage built through dental barter. About half the time Dad charged cash for his work, but there was little cash to be had during the depression and he very frequently ended up not collecting, or accepting small payments of

$1.00 per week. Since Dad never kept accounts of what was due and who owed what; hundreds of dollars owed were never collected. Dad always maintained a small but complete dental lab workshop, wherever we lived.

Dad still had full time employment with Painless Parker, so he junked the Model T Ford for a Model A. The economic depression did not affect us as it did others. With so many people unemployed and unable to pay for professional dental work, Dad's bootleg dental business flourished. My Dad was an expert in most phases of the dental profession; he was the best gold technician in town, partial dental plates were his specialty. I can remember him charging $25.00 for a full set of dentures uppers and lowers. About the only work he didn't do were tooth fillings and difficult molar extractions. For a fact all the people he knew in Denver did not call him Jim, he was known by one and all as Doc Scanlon.

The hill we lived on was steep enough to make it difficult for a car to climb when slick with ice and snow. Irving Street was an unpaved road; in the summer city sprinkler trucks came by to keep the dust down, and in winter the city would put coal cinders and sand on the frozen hill for traction. When we had the Model T, Dad would get stalled part way up the hill, so he would back up and if the clutch was slipping he would get out, grab a handful of coal cinders and throw it on the exposed clutch plate. Of course this procedure shortened the life of the clutch, but it did get one up the hill. We had a lot of activity, both summer and winter, on our hill. Bob and I must have slid down South Irving over a thousand times on our sleds; both Mom and Dad went sleigh riding with us on many a cold winter night. There was not too much traffic on the hill; but once I was racing down the same time that our neighbor backed out his Model T on to the road — I lucked out and steered my sled under the car between the front and back wheels. Fortunately for me the Model T was built pretty high off the ground. That incident scared the neighbor to a degree where he tried to stop our fun.

When the ponds and lakes froze over we went ice-skating. Dad was a fine skater; Mom put on skates and just stood around. It cost ten cents to ride the streetcar across town to Sloans Lake. The city would scrape the snow using a team of horses to smooth paths clear around the lake, and they provided a warm building where we could thaw out after hours on the ice. We wore long johns, two pair of pants, and a sheep skin/canvas coat; we still froze. Denver is cold in the winter!

In the days before antibiotics and modern medicine it was hard to stay healthy. Both Bob and I caught just about every disease — Chickenpox, three types of measles, mumps and Scarlet fever. When Bob, Mom, and I had scarlet fever we were quarantined with the sickness for about six weeks. Mom caught it from me, probably from sleeping with me cause I was so sick. Mom would rub my back with all kinds of stuff to soothe the terrible itch and to relieve the fever.

Dad worked six days a week, but found time to teach his boys to fish, shoot, hunt, play baseball, chop wood for the stove, haul coal, and feed the chickens; Dad was not one to sit around and waste time

Since we lived only two blocks from the city limits, we had ample time to hunt rabbits and pheasants before and after work/school. Dad liked to hunt out of the car. He would put Bob and me on the front fenders and drive through fields flushing up game with the both of us hanging on for dear life to our guns and our seats on the fender. Dad had many friends who owned land all over Colorado, so we always had lots of good places to hunt and fish. I don't know why, but Mom was always afraid that we were in mortal danger of getting shot or drowned; frequently she would be so concerned that she would have to go to bed with a sick headache. Mother did finally learn to like fishing.

On more than one occasion Dad would get a Saturday afternoon off. Bob and I would be at the Saturday Kids movie matinee. Dad would come to the movie house and shout out our names while we were watching a Tarzan serial. We knew right away that we were either going hunting, fishing, or to a baseball game, something special with our Dad!

By not having four front teeth for five years since my trike accident, my permanent teeth stuck almost straight out when they came in. Just to indicate how good a dentist Dad was, he rigged an upper palate plate with a gold wire then very slowly, over a three year period, pulled all four front teeth back into perfect alignment. Of course Dad was not always perfect in his dental decisions. When Bob was nine and I was eight Dad decided that it was time that our baby teeth should be coming out to be replaced with our permanent teeth. When Mother and Aunt Bertha were on a short vacation to Chicago, Dad proceeded to extract six of Bob's baby teeth; and, as long as he was at it he pulled four of mine. That night, immediately after the extractions, Dad treated us to a steak dinner and a baseball game. Both Bob and I were spitting blood and not in any shape to enjoy either the meal or the game. Dad made us fried bacon and eggs for breakfast, and gave us fifteen cents for school lunch, which bought a Coke and a hotdog. For supper he would fry something up. With all the fried food and teeth pulling, my poor little body reacted by busting out in a monstrous case of hives. When Mother arrived home there was hell to pay. Oh well, there was no permanent harm done, but I had to Gum my food for quite some time.

Dad loved to fish and hunt. He had a .30cal Remington Pump rifle, a 410 gauge shotgun, and access to several 12 gauge shotguns.

## 1924 DENVER, COLORADO

Dad once flushed a pheasant and knocked the bird down three times. Every time he would go to pick it up, it would take off. He was using the 410 at the time. That was the last time he hunted with a 410 other than shooting cottontails out of the car window. Dad came back from a hunting and fishing trip in Wyoming with enough meat and fish to feed the whole neighborhood, and that's exactly what he did. Those were the days before refrigerators and our icebox was small. It was the depths of the depression. Our relatives and neighbors were most happy for the handout.

Dad never hit us, but do recall the first and only spanking! It was in our Cherokee St. house in Denver. Dad and Mother had gone to church, leaving Brother Bob and me with a stupid babysitter who showed how much fun it was to strike kitchen matches and drop them down the hot air registers. Mom had stopped by to visit a neighbor, so Dad beat her home, smelled the sulfur matches and found out what we had been up to. I do not recall what Dad did to the stupid babysitter, but he whipped brother Bob and me with a stick. Mom discovered our bruised rear ends at bath time and lit into Dad under no uncertain terms. Dad never hit Bob or me again; however, Mother did, but she never used a stick or switch. The stupid babysitter was taken in to live with us for some reason unknown to me. At the time of the firebug episode we were babies aged three and four. Our parents were the kind who would help poor relatives and people in need.

We most always vacationed in the Colorado Mountains; twice we rented a cabin at Grand Lake. The outlet stream of Grand Lake is the start of the Colorado River; brook and rainbow trout were pretty thick in the shallow water but hard to catch in such clear water. I fished with worms without much luck until I ran out of worms. I tied a fly on that Dad had used and discarded. With little skill this seven-year old kid with a beat up rod and fly caught a dozen trout in jig time while three elder fishermen stood by in utter disbelief. They asked me what I was using as bait; when they saw the worn-out fly they all trimmed down their flies. They still were not having any luck when I left. We rode horses at 25 cents an hour. Bob and I were pretty lucky kids to have a father like ours who really took an interest in raising his boys!

Dad built us a swing, ring, and trapeze set which was a great exercise. Along with walking to school, running up, sliding and rolling down, South Irving Street hill, we had plenty of exercise.

We had a fair sized wall tent. Dad built a wood floor for the tent. Bob and I lived and slept in the tent starting in the spring, after the last snow, returning to the house in the fall, after the first or second snowstorm. We each had a steel cot, a real blessing for me to escape from the single indoor bed and my bed-wetting brother. We had our flashlights and an excellent Coleman lantern. We had the outhouse nearby and the roosters in the chicken coop to wake us at daybreak. I remember having to adjust the tension on the tent tie down ropes when it rained, and also hanging on to the tent poles during some pretty heavy thunderstorms when we thought it might fly away. We had our BB guns to ward off the varmints and any Indians or bad guys that our young minds might conger up in the dark of night. I don't think we could have made it without our guns!

We were always well cared for. We had plenty of good food; my Mother was a real fine cook. Mom was very traditional. She had certain days of the week to clean house, do the washing, and bake: all the family breads, sweet rolls, coffee cakes, pies and cakes. There was always a set time for: spring cleaning, beating the rugs, waxing the floors, washing the windows, etc.... etc. When Mom was not totally occupied with her scheduled chores she would do:

some very beautiful embroidery, tatting, and all kinds of sewing. Mom made most all of Dad's shirts. Dad dressed like a real spiffy dude sporting: double breasted suits, vests, spats, derby hat, and loud colored shirts — yellow, green, red, etc. Dad had a part in a minstrel show. He dressed up in black face and did a tap dance routine. His costumes were all made by Mom. Mother canned a bunch of food: jams, jellies, fruits, and veggies. We also had plenty of root beer that Mom made and bottled.

# 3 CHAPTER

# 1935 California

**JJS.** During the Great Depression, Dad found himself constantly having to help out our poor relations. Mom and Dad were generous. They always helped out when they saw real need; but real need soon became too great as the depression worsened. Mom wanted to fix up our house. Dad thought it was a waste of time and money to fix up the old frame four room house with the shed addition bathroom. Dad finally got fed up with the hangers on, the poor relations and our small house, so he decided to move to California. Everyone thought he was crazy to give up a home and a good paying job, especially when times were tough and 35% of the population were out of work. Dad was the chief Dental Technician for Painless Parker Dental in Denver. His good paying job plus his rather lucrative bootleg dental business was not easy to give up; however, Dad was a real go-getter and totally unafraid. We had a 1933 Plymouth with wood spoke wheels and suicide doors that opened from front to rear. Dad drove it to San Diego, California in 1935. He immediately landed the same job that he had in Denver with Painless Parker, accepting a position with Dr. Beauchamp who had offices in San Diego, Los Angeles and Long Beach.

When we moved to San Diego, we left most of our stuff in Denver. We sold it to a used furniture outfit; you can't imagine how little we got. Mom was not in favor of the move, when Dad told her to sell out because it was cheaper than shipping it. She did ship her good stuff: dishes, pots, pans, her clothing and things she treasured, but sold Dad's .30 Cal Remington Pump Rifle. That's how wives get even.

We took the train from Denver thru Salt Lake City to San Diego. In order to make sure that Mom didn't sell Bob's or my stuff we boys hand carried our shotguns and .22 cal rifles, plus Bob's violin, my banjo and guitar.

Leaving our home in Denver we experienced little remorse. Brother Bob and I seldom quarreled and we got along well with the neighbor kids. We really looked forward to California. The train took us thru the Rocky Mountains, via the Moffett Tunnel and into Salt Lake City where it was snowing and bitterly cold. We made the entire trip in a coach car. The train was not crowded so we nine and ten-year-old kids could stretch out and sleep on the seats. Our eyes sure bugged out entering Southern California; palm trees, orange groves, fields of grapes, tall eucalyptus trees, things that we had never seen before.

We rented a small home on Estrella Ave, between 48th and 49th St for $25.00 a month partially furnished. We lived on the east side of San Diego, almost on the edge of town. Bob entered Woodrow Wilson Jr. High School and I attended a grammar school that was real neat compared to old P.T. Barnum in Denver. We only had one house near us. Our house was on the edge of a big canyon, and when it rained hard we had a small lake in the canyon for several days after the rain stopped. We built a wood raft for the lake to play on and had a grand time on Estrella St. for the three and a half years that we lived there. I would tie my 20Ga shotgun on my back, hop on my bike and would be hunting rabbits and quail in less than twenty minutes. We had our BB guns and were always plinking at something or another almost every day.

My first bike cost $12.00 used but I needed a way to get to school go hunting. The bike was made of crowbars and railroad tracks it was heavy as hell! It would go downhill at great speed; uphill, ya pushed it. Flat tires were fixed with punched in rubber bands and a big squirt of condensed canned milk. We seldom got Toy, Toys; we got things we could *use!*

Dad was a fascinating person. He was a sharp card player; he taught many of our young friends how to play Poker for keeps. He was an expert pool player. He taught me how to play three-cushion Billiards and Snooker. Dad was always clean-shaved, and always wore a necktie. Dad had a Speech Impediment. He Stammered/ Stuttered all his life. His World War I Injury *greatly aggravated the condition . . .*

I had a real good friend, Carl 'Corkey' Warner, who lived close by. He was a big help in getting me into lots of activities. I went camping in the desert with his folks where quail and rabbits were pretty thick. Corkey's Mom cooked our game for us over a campfire. We thought we were real pioneers. Carl knew the territory all around San Diego and he and I hit most of the interesting spots. We had a wonderful swimming hole at Adobe Falls, located downstream of two small lakes, Little Murray and Big Murray where we hunted and went fishing. The lakes were probably private property but there were no fences or houses, just open land with many canyons and no farms. Years later, I discovered that most of the land belonged to The University of San Diego; therefore, I was swimming, hunting and fishing on private property, out of hunting/fishing season without permission, and without a license. We rode our bikes everywhere, as far as La Jolla, and to Mission Beach, one of our favorites.

During summers in San Diego we could get Transient Passes for $1.00, which were good for a week's travel on any bus, streetcar, or ferryboat. Mom would pack me and brother Bob a lunch and off we'd go. Some days we took the ferry to Coronado Island, ride it for six or eight trips across the bay, take time for a swim at one of the beautiful public beaches, and jump back on the ferry for lunch. Other times we took the inter Irving Red Cars to La Jolla and Mission Beach. At any rate we got to know San Diego by riding the ferryboats and traveling to the end of, every bus and streetcar rout.

I believe it was 1936 that San Diego had a World's Fair. The Fairgrounds combined with the famous San Diego Zoo kept us fully occupied and entertained. A buddy and I mowed lawns and did yard work to make just enough money to support our rather limited needs; we never paid where we could either wrangle a free ticket or sneak in. The area where we mowed lawns was populated mostly by elderly female retired teachers who were on a very limited budget. We mowed lawns for as little as $0.25 which included edging, trimming and sweeping; all mowing was done with an old reel-type pusher with grass catcher, trimming with hand clippers, and all carried from job to job on my little red wagon.

We camped out at quite a few locations. Once we spent one winter night about fifteen feet above ground huddled in a lifeguard stand at Mission Beach. We had spent the day rowing about in a derelict boat we had rescued from some unknown source along with some oars we had conveniently obtained from another unknown location. To make a long story short, we four boatmen got caught near the bay opening to the Pacific during the peak period of an outgoing tide. We had to abandon the boat in shallow water because we weren't strong enough to hold it against the current. We lost the boat, the oars, our food, and darn near our hides. We were wet and a long way from where we needed to be. So four wet cold kids spent the night trying to get warm in a lifeguard lookout station. We cycled home the next morning, starved, but much wiser.

Bob and I were hunting at Little Murray Lake on a very cold day. I had already bagged a cotton tail and a jack rabbit when some ducks came in to land on the shallow lake; naturally we blazed away knocking one down. It fell in the middle of the lake dead. We debated, leave it or wade out and pick it up. Dad always told us to keep what we shot so we were kinda obligated, besides ducks were good eating. Bob took his shoes, socks, and pants off then waded out a few yards. He came back to shore saying that the bottom was too rough on his bare feet. I told him to put on his shoes. We were pretty dam poor and the most we ever had was two pair of shoes. We knew that Mom would have lots to say about stupid boys ruining a good pair of shoes. Bob thought he could swim out quicker so he stripped down then swam out and retrieved the duck. By the time he hit the shore he was blue and shaking. Here we were, five miles from home, two great hunters on bikes, one frozen stiff. Bob was near shock, and I suspect we could have lost him if we had much farther to go. Bob was sick for a long time!!

San Diego, at that time, was the most pleasant spot on earth. Food was cheap, rent and utilities were had for a song. The climate was just grand. We had the ocean and beautiful backcountry hills only minutes away by auto. We made frequent trips to Tijuana Mexico, both to hunt and for Dad to buy his booze. Rock lobsters went for $1.00 a dozen. I paid $0.87 for a box of shotgun shells and only $0.12 for fifty .22caliber rifle shells. We made frequent trips to Warner Hot Springs to hunt, vacation and swim. Years later I returned to Warner Hot Springs with my good friend Lloyd Clark to hunt doves, quail, and rabbits. It was 1942 just before we both entered the Army Air Corps and become pilots.

San Diego was not all a bed of roses; the nation was still in a deep depression. We had one of Dad's ex coworkers from Painless Parker living with us. As I recall Larry made only $25.00 a week. Out of that small income he paid for his room and board, owned and operated his Ford Model A, and had enough left to see a movie once a week. Larry and I bunked together in the basement bedroom. This room was separate from the rest of the house. It had one door to the outside leading to stone steps up to the back door of the house. There was only one bathroom located in the main house and only one bedroom. It had a living room/dining room area and a real small kitchen. Mom and Dad had the bedroom, brother Bob slept on a foldaway cot in the dining room. Needless to say we didn't spend too much time hanging around the house. Larry was 25 years old, he clued me in on a few subjects including Sex, which neither Dad nor Mom ever mentioned. This liaison, rooming with Larry, put me a step ahead of Bob streetwise!

Mothers' health was a real problem. She suffered from migraine headaches. She would be bed-ridden for days at a time, unable to eat or function any way near normal. Part of her problem was worry; she could dream up problems that never happened. We did have some serious problems when our poor relations followed us to California and when old friends from Denver came to visit. Fortunately the visitors came by two's and four's. When it got really crowded; we almost had to eat and sleep in shifts. I do recall having to sleep on the floor and once I bedded down in the garage. Uncle Earl Kofford lost his umpteenth job in Denver, so he headed west with wife Bertha and kids, Barbara and Kent Next came Aunt Bess with husband Bill Kinney plus son Billy and daughter Joanne. Feeding and housing poor relations left little for us. The poor relations were all from mother's side of the family. Besides the Koffords and Kinnies, there were: Uncles Dick, Doc, George, Ed, and some never-do-well friends of theirs.

All these visitors eventually found employment in California. Not a single one of the free loaders ever repaid us from saving them from the poor farm and starvation. We never ever visited our Mom's relations; likewise Dads' folks, except for one trip to Chicago to visit Dad's sister Anastacia.

Bob and I played a lot of tennis. I played on the school's baseball team, and summer leagues every year. Our hunting and fishing, swimming, etc. came to an end in 1939. For some unknown reason that I was never told, we moved to Los Angeles in spring of 1939 where I enrolled in Bret Hart Junior High and Bob attended Washington High.

We boys were not too happy to depart San Diego, where we had girlfriends, swimming holes, the Pacific Ocean to fish and swim in, only a bike ride from where we lived. Now we had to start all over again from scratch. Our first home was one half of a duplex, but it was bigger than the house we left, and again, we were fairly close to schools. We soon moved from the duplex to a single house on 97th St. between Vermont and Hover in Southwest L.A. This second house was much closer to my two best friends, Bob White on 101st, and Lloyd Clark on 106th St. I went to Bret Hart Junior High for one semester and then joined Bob at George Washington High School

Soon after we moved from San Diego to Los Angeles, I was

invited to visit my buddy Carl Warner back in San Diego. I had saved up $10.00. I know this is hard to believe but that $10.00 bought me a round-trip train ride plus lunch. It was during this visit when the radio announced that Hitler's Germany had started World War Two. Mr. Warmer was a World War One Navy Veteran. He informed me that it was certain that the United States would have to go to war sooner or later. I was only fifteen-years-old and had thoughts of joining the Navy. The Navy would take recruits at seventeen with their parent's permission. I heard quite a few stories about World War One from Dad and other veterans which raised my interest

I had real mixed up thoughts and feelings for a fifteen-year-old. The Japs had invaded China, and had been at it since 1936. They had taken over Manchuria and most of China's seaports, all of Korea, and several South Sea Islands. I remembered when I was living in San Diego that there had been numerous Jap ships in port loading scrap metal. President Roosevelt had since cut off oil shipments, and other materials to Japan. This action deprived the Japs of the raw material they need to keep their war machine supplied. Now the Germans, under Hitler, were invading the Sudetenland of Czechoslovakia which Hitler claimed was part of Germany. So here I am a fifteen-year-old kid all worried about world events, that I had no idea of, and a gut feeling that sooner or later I might be involved in.

I thought of all the good times we had living in San Diego; how we tossed horseshoes with my Dad and the neighbors. Carl Warner and I spent a lot of time on the dock and noted all the various Navy ships sailing in and out, and the constant stream of aircraft operating off Lindberg Field and the Navy's field on North Island. Most all of the naval aircraft were single engine bi-planes. We used to be allowed to board the Navy launches and visit almost all the ships in port, but now all non-essential visits were cancelled. It sure made us wonder. Carl had two older sisters, both were married to sailors, and they were greatly concerned. So my imagination was conjuring up all kinds of situations which I could be involved in, in the not too distant future.

I went to see my girlfriend, Charolet Spainhower, whose father was a World War One Vet and a border control/Immigration Officer. We had a long conversation, in which he agreed with Mr. Warner, that the US was bound to get involved since our World War One allies, France and England, had already declared war against Hitler's Germany. He told me not to worry about it, saying I was much too young to ever be involved.

L.A. was not a segregated city per say, but the black folk lived on the East side, the Mexicans camped in the Northeast. There was very little mixing. Most people kept to themselves. There were no black kids in the schools that I attended, and very few Mexicans. Those of Spanish descent were well off high class types, and not the typical Tijuana South of the border people. The only noticeable minority were the Japanese, who, unfortunately were interned in camps after the Japs attacked Pearl Harbor, Dec 7th 1941.

We soon settled into the big city, found new friends, the beaches, places to hunt, and got along well in school. Summer jobs were easy to find, although the pay wasn't much. During 1939 and 1940 the grip of the depression began to slacken. In Los Angeles and the West Coast in general, the economy perked up. During school months I delivered local shopping news and had a small newspaper route. During summer vacation I worked full time at a photo processing plant, and stocking shelves at a large Sears's store. My last job, prior to entering the Army, was with the Los Angeles Times, processing Los Angeles Telephone books, working 12 hours a day 7 days a week which didn't leave much time to horse around.

Bob and I both enjoyed school; we both did quite well, but were split in our choice of friends and sports activities. I can remember Bob having only two close friends, both male, a bit odd, very studious, and no girlfriends. Bob was a serious student; he read extensively on many subjects and was a whiz at math, geometry, trig, chemistry, and science. We both took college level subjects, and were in the top 10% of our respective classes. Bob graduated the winter of 1941 and I finished high school in June 1942, just a half year after Bob, so I suppose there would naturally be a difference in our activities. I always had a job and a few dollars to date girls, and buy gas.

I was involved in basketball for three years where my position was mostly bench warmer. Since I was on the team I always had sixth period gym which gave me plenty of time to goof off during the baseball and track seasons. I had a newspaper route which got me up early and earned just enough money for dance dates at the Palladium, and other social events. Living in Southern California the beach was always the main attraction. We always lived close enough to ride a bike to Hermosa Beach, Manhattan Beach, or Mission Beach when we lived in San Diego. After body surfing and swimming all day it was a real effort to pedal home. Between pedaling a bike, walking a paper route, playing basketball and tennis, dancing and swimming, I kept in real good physical condition. Of course our family was never short on rations! Mother was a great cook, and all the chow was made from scratch! Mom still canned all kinds of vegetables, jellies, jams, pickles, and baked most of our bread. We were never out of cake and cookies. Mother made the world's best oatmeal cookies, full of raisins and walnuts, I can still taste them.

I dated over a dozen gals in my high school class. Dates were usually to dance halls at the beach, or the Palladium Ballroom in Hollywood. Admission to dance halls was $1.00 to dance all night to big bands like: Woody Herman, Harry James, Stan Kenton, Tommie and Jimmy Dorsey, Tinny Hill, Jack Teagarden, Jan Garber, and all the bands that were in Los Angeles during the war years. In the summer, when we were not working we hit the beaches, surfing and chasing the gals. The economy was picking up but food and entertainment was still cheap and affordable for a seventeen-year-old kid.

Dad bought a new 1940 Chevy and was quite liberal in letting me use it. Brother Bob drove it a few times until he had an accident and kind of shied from asking. I was sixteen years old and allowed to use the car to go hunting with my best friends, Bob White and Lloyd Clark. Dad found the Los Angeles streetcar and bus systems more handy than driving downtown in traffic and paying big bucks for gas then parking blocks away from his work. As long as I was careful, kept the gas tank full, and kept it clean, I could use the car.

We went hunting many times with Hillary White, Bob's Dad, his Mom Jessie, came along too. The Whites were like another Mom and Pop to me. I went camping with the White family to Big Bear Lake and hunting with them all over Southern California shooting mostly rabbits and quail. Bob White was quite small, about 5'3" but he grew rapidly during the war to a bit over 6'. Since he was so small, his Dad always wanted me to drive. Bob White's family moved to Blue Diamond, Nevada, just West of Las Vegas, where Mr. White was chief electrician and Sheriff. I was invited to spend a few days with the Whites and had a fine time hunting, swimming, and doing a bit of gambling in Vegas.

Mothers' health was a big problem! She had female disorders, headaches, and a bout with TB. This pile up of illnesses resulted in her being bedridden for about three months. Grandma Mary Hinds came to live with us to take care of Mom but she was not much help. She talked incessantly and drove Mom nuts. Grandma Hinds departed after two weeks just as school ended for summer vacation. Since Dad and brother Bob were working full time I was elected to play nurse, housekeeper, laundryman, cook, and whatever else was needed. The year 1940 was a very busy time for me. In the background the Selective Service Draft was putting the USA on a war footing, all men ages 18 to 36 were subject to military service. Eighteen-year-olds still in high school were being called up.

Dad and Bob chipped in to pay me $15.00 a week to run the place, cook all the meals, and watch after Mom. Dad let me have the 1940 Chevrolet to do the shopping, take Mom to the doctor, and go to the beach twice a week to get the hell out of the house and relax. This was quite a responsibility for a sixteen-year-old kid! I sure learned a whole lot real fast. I already knew how to clean house and take care of the yard. Mother answered any questions I had about cooking and what to buy at the grocery store.

The weeks flew by as the wars in Europe and the Far East grew into a worldwide conflict with the Axis powers winning on all fronts England stood alone and barely won the Battle of Britain, defeating the German Air Force and stopping a German invasion. Germany then turned east and attacked Russia! Japan, sensing the allied powers weakness, attacked Pearl Harbor December 7th 1941. Brother Robert Emmett Scanlon, just graduated from high school, joined the Army Air Forces shortly thereafter. Not too long after that, Harold Baker quit college at Long Beach Jr. College and joined up with Bob. Brother Bob was assigned to an experimental flying school; Class 42-X, where primary was skipped and the cadets were put in the BT-13 basic trainer in an effort to accelerate training. The idea was to make flight instructors of them in short order. The only students to pass the course were those who had previously flown. Bob washed back to Primary and joined class 43-D. He completed pilot training early 1943. He paid me a visit while I was in preflight February 1943 in San Antonio, Texas.

High School was enjoyable; I made A's and B's, graduating the same month that I turned eighteen. I was the only one from the George Washington High School class of June1942 with a red beard! I registered for the military draft as required. I knew that I would be called up soon, so along with my best friends Lloyd Clark and Bob White I took the aviation cadet written and physical exams. We all passed. All three of us eventually became Army Air Corps pilots.

Bob White was called up before Lloyd and myself. Bob was sent to a cadet training detachment at a University back east. Just prior to when Lloyd and I were called to active duty we went hunting at Warner Hot Springs for a last visit to Las Vegas with his family. While in Vegas Mom called and told me that I was to report to the Army; I caught the next bus to L.A. and was sworn in on the 6th of August 1942. I was told to stand by in L.A. to await active duty. During the period of August 1942 to February 1943, when not in Vegas, I worked for Sears as a stocking caddy six days a week. I asked for a raise when they made me the chief supply man. No deal Sears said, wages were frozen by the government, thanks Franklin, so I quit and went to Las Vegas. My last job was with the Los Angeles Times Newspaper where I worked 7 days a week 12 hours a day once again working on printing the L.A. Telephone Book.

# 1942 War

**JJS.** I was sworn into the United States Army Air Corps August 6 1942, at the Pacific Electric building in down town Los Angles California. February 1943 the Army put me along with a few hundred more men on a train which took about five days to get to San Antonio Texas There we get every shot known to medical science and a short arm inspection. We are issued coveralls because no uniforms are available. I learn how to pull KP, peel potatoes and wash pots and pans. I take a bunch of tests. I get shin-splints running in freezing weather half naked on concrete roads. I learn how to stand real straight, and in a straight line, with a hundred other guys, this is learned during roll call at 0430 in the dark winter morning. Cannon shot next morning, Revelry. I stay put. I'm sick and burning up. The 2nd lt. comes in and tells me to arise; he then feels my forehead and tells me to stay in bed, no KP today. The sergeant gives me an APC.

I'm washing windows after the next cannon shot. The spring latch holding the window up sprung and the window came crashing down smashing my hand. It was dark and cold. The Corporal of the quarters tells me to go to the hospital, I wander around, find the medics who tell me to return to barracks and to come back the next day if it still hurts. It's cold and dark walking back, I am stopped by a cadet guard holding an Enfield rifle with bayonet attached, who says "halt". I halt. He says "advanced and be recognized". Then with a most commanding voice he says "what am I supposed to say now?" "Beats me" I said. I went to bed that night with the assurance that the Army would always take care of me.

After a couple of weeks they take our coveralls, give us a very short haircut and issue us uniforms. They next classified us as Bombardiers, Navigators and Pilots based on the current demand for aircrew and the tests we took. We then are all shipped off to Preflight, where we are further tested, reevaluated, and exercised until we drop. We're then fattened up for the kill.

I was classified a Pilot and transferred to Coleman Texas, for Basic. There we flew the PT-19A Fairchild Trainer with the upside down ranger 17hp engine. I soloed on June 15, 1943, just two days before my 19th birthday. A real great instructor told me where to look when landing; after that problem was solved I had no more problems. I sailed through the rest of the course with ease. That one check ride with that most excellent instructor pilot who helped me solve my landing problem, will always be remembered; he was a guide for me throughout 12,000 hours and forty eight years as a pilot. His name was EO Young. Of course much credit is due to Mr. Roselle my primary instructor who taught me how to fly, except for landing, and soloed me off the grass field at Coleman. My logbook shows 65 hours primary trainer flight.

The next step was Basic at Perrin Field, Sherman Texas, where we flew BT 13's, the Vultee Vibrator. This was a big change; we had a radio and even an airspeed indicator. The PT-19 was an open cockpit, hand crank to start, no radio aircraft; now we had a closed cockpit interphone and radio, plus an electric engine starter. The big 450hp engine with a two speed prop put us in high cotton, we were now real aviators. I had a ball at Perrin, despite the East Texas heat; the chow was good, and the folks in the nearby towns and Dallas were real friendly. I sailed through basic without a hitch and logged another 81 hours, which included: aerobatics, spins, instrument flying, plus plenty of night flying. At the completion of basic we were given a choice of either single engine or multi engine advanced flying. I chose to fly the bigger planes for two reasons, the first was for better training which might lead to postwar opportunities flying commercial airlines, the second, brother Bob was flying B 24's and I didn't want him flying anything bigger than what I might end up flying.

Advanced multi-engine training was at Pampa Army Air Force Base, way up in the Texas Panhandle, North East of Amarillo. There we flew: Cessna Bobcats, AT-17s, Uc-78s, and Curtis AT-9s. The Cessnas were twin engine planes built primarily of wood, fabric, and dope held together with miles of kite string to dampen the vibrations. They were powered by two 225hp Jacobs radial engines which were very prone to carburetor icing. The wheel rims were also the brake drums; too much heavy braking resulted in a hot rim and a flat tire. The Cessnas were simple aircraft and easy to fly; a short field takeoff could be executed at 35 to 40 miles an hour if nothing went wrong. My pilot logbook shows 125 hours in advanced multi-engine planes, for a total of 271 hours as an aviation cadet. Pampa was a bleak cold and windy place the winter of 1943; however the people were most friendly. We were measured for new class A Officer Uniforms, we put them on, when we graduate as Second Lieutenants, December 5, 1943; we were then granted two weeks leave. My orders directed me to report to Liberal Kansas on December 19, 1943, about 80 miles north of the Oklahoma Panhandle in the Southwest corner of Kansas.

When I arrived at Liberal about three feet of snow covered the ground and only three B-24's were on the field. The weather was quite a contrast from my leave in Los Angeles. The Instructor Pilots at Liberal Army Air Force Base were all authorized Green Instrument Cards, which authorized them to fly almost anywhere in any and all kinds of weather. The B-24 instructors were assigned about five students each and told to train them anyway and anywhere they could find suitable weather. I went on trips to Oklahoma City, Tinker Field Oklahoma, Scott Army Air Force Base, Illinois, Denver Colorado Lowry Field, Laramie Kansas, up to Wyoming etc...etc. After 108 hours of intensive flying and ground school, I was awarded the rating of a four engine B-24 Pilot. In a short period I had gone from a student rating flying the two engine Cessna Bobcat trainer with two 450hp engines to being the Captain of the largest bomber in the inventory with four 1200hp engines.

At the completion of B-24 transition training, I was transferred to Hammer Field Fresno California. I thought I was to be stationed there, but no luck; it was at Hammer Field that I was assigned my bomber crew, which took only a day or so. We were issued orders to report to Tonopah Nevada. Tonopah Army Air Force Base is about halfway between Las Vegas and Reno exactly in the middle of nowhere. We went through Operational Training at Tonopah: dropping bombs, flying formation, extended navigation, gunnery missions, experiencing attacks by P-38's and P-47's, anything in general that would simulate what was to come in combat. I had a fine crew: Bill Young age 26 was Copilot, Bob Busby age 23 was Navigator, Emile Paulin age 23 was Bombardier, Harold Rohmer age 36 was Engineer/Top

Gunner, Charles Burnhan age 34 was Armorer/Nose Gunner, Chas Fonow age 19 was Waist Gunner, David Sullivan age 21 was Waist Gunner and later on Radio Operator substituting for Lonnie Click age 24 Radio Operator, when he was wounded, and James Gibson age 22 Tail Gunner. I am nineteen-years-old. We started together and finished 35 missions together except for Burnhan and Click, both were Purple Heart recipients having received flak wounds, they missed a mission while healing.

We were transferred to Hamilton Field San Francisco California on a rattling old troop train that was sidetracked at almost every crossroad village in route. At Hamilton Field we were: given a final physical and dental examinations, issued any clothing shortage that we had, and issued a brand-new B -24 serial number – 0576. On June 15, 1944 we test hopped our bird ran some fuel flow calculations, and went sightseeing over the Golden Gate Bridge and the San Francisco Bay Area.

We were issued parachutes with backpacks containing: mosquito nets, birdshot shells for our .45 sidearm, a large machete, and various other items one would need in a jungle. We thought for sure we were headed for the Pacific Area of Operations. The crew wanted to name and paint a logo on number 0576 but I discouraged that idea, because I knew that once we delivered the B-24, we would never see her again. The test hop at Hamilton Field was satisfactory, except for a slight tendency for the number three prop to over-speed.

The next day we loaded our personal gear and the bomb bay full of toolboxes which were lashed down to wood planks installed across the bomb bay catwalk. We were issued ball ammunition for our .45 caliber side arms and boxes of 50 Cal for our machine guns. We topped off with fuel to a full 2700 gallons of 100 octane which I calculated brought the aircraft way up to 64,000 pounds max gross weight. Next in order, I was briefed along with my crew, to proceed east along with our jungle kits to Harvard Nebraska; this gave is a clue that we were heading for the ETO, the European Theater of Operations. So on July 16, 1944 we sallied forth into the world of air combat. I can only suppose that hauling jungle kits East was designed to fool either Tojo or Hitler. We checked over the loaded plane, made a good preflight, set power at Turbo Eight with 48 inches of manifold pressure and took off with number three engine prop over-speeding but controllable with the feather button. Seven hours later we touched down at Harvard Army Air Force Base where we remained overnight. We pressed on without delay the next day, the 17th of June, my 20th birthday, to Grenier Army Air Force Base New Hampshire. I had the same problem with number three prop over speeding on takeoff so I wrote it up in the Aircraft Form 1. Maintenance at Grenier worked on the problem, ran up the engine and pronounced it fixed.

On 19 July we departed Grenier for Goose Bay, the Canadian Airbase in Labrador. Crossing over Maine, the St. Lawrence Seaway, Quebec, into Labrador. We encountered nothing but an unchartered wilderness of lakes, streams and trees; maybe that's why we were issued the mosquito nets.

Number three prop was again a problem, which the Copilot controlled with the feather button. Both Bill Young and I were getting fed up with the number three prop malfunction. We were taking off at maximum gross weight and a loss of power or a runaway prop spelled disaster at low speed during liftoff. I wrote up number three again and informed maintenance of the continued problems since departure from California.

Maintenance did their bit again but only ground checked it;

what the beast needed was a test flight, with a test maintenance type doing the flying. I suppose they thought that a Second Lieutenant was not qualified to doubt a maintenance job pronounced fixed.

After a one day delay we hauled ass in the dead of night on the 21st of July in route to Iceland. About an hour out, number three acted up. I stopped the engine, feathered the prop and returned to Goose Bay where I made my first really heavyweight landing and was congratulated by the crew for grease job. A Maintenance Officer met us and questioned my return. After my crew deplaned the maintenance people climbed aboard and much against our advice, started number three, ran it up to speed and blew it all to hell!

We had a short vacation, took things easy, drank a lot of beer, went bowling, visited an Eskimo camp, and sweated out the arrival of a new engine. Maintenance finally hung the new engine. Pappy Rohmer, Billy Young and I conducted a test flight on 25 July. The engine checked out okay; however the damn prop would still over speed and had to be controlled again with the feather button. Got a new engine installed behind the same old prop!

The next leg of our trip overseas was a long haul to Meeks Field Iceland, over the frigid North Atlantic and Greenland ice cap. It was about a nine hour trip with an alternate airfield BW-1 on the Southwest tip of Greenland. Due to our spooky number three prop situation I requested a daylight departure and arrival, which was granted. We took off from Goose Bay on 26 July 0850 hrs. We later landed at Meeks Field Iceland, still with a number three prop problem. Next day we flew the final leg from Meeks to Nuts Corner North Ireland and that was the last we ever saw of the B 24 Liberator serial number 0576.

We were shortly transported across the Irish Sea to Scotland in a real small wooden ship, we messed about England for a few days and then flew back to Northern Ireland for some useless ground training and indoctrination. Back to Scotland, I can't remember how, but we were put on a slow train south, to our final destination, the 453rd Bomb Group the 734th Squadron based at Old Buckingham. They lost our baggage in route but we were weary and sick of travel, glad to just stay put.

The food at the Officers Mess was consistently poor. I was informed by Capt. Wendell Jeske, a 734rd Squadron Bombardier, that several food service personnel were court-martialed for some crooked dealings. Each week we were supposed to have chicken; the meal would usually come out as Chicken Feather Stew Ala King weak soup. We always had a supply of Brussel sprouts and black bread. In order to supplement my diet, my folks mailed a five pound box of American cheese about every two weeks along with a one pound box of soda crackers plus frequent candy and cookie shipments. The mess hall had no screens so during warm weather thousands of honey bees swarmed all over anything sweet, dishes of orange marmalade would be covered with them. On a few rare occasions we were able to shoot rabbits and cook them on a coat hanger hanging down in our potbellied stove.

Our indoctrination to flying in the local area was conducted by a Capt. Sears. Frankly I'm certain that this orientation flight was the most dangerous, ill thought out, hazardous mission that we experienced during our stay with the 453rd, including all our combat missions. We took off in poor weather, low clouds, and limited visibility; with Sears in command in the left seat. After milling around without anything worthwhile in the way of information being gained, Sears landed at another base to pick up his laundry; we then departed in a worse weather condition and Sears got lost. Bob Busby tried to tell Sears to let him navigate us home but Sears started calling Darkie (local stations that are able to provide steering vectors) for help. The low clouds were now down to 300 feet and visibility was less than a mile. I asked Sears to please stop milling around and get us on the ground. He evidently had originally planned to go up to Bovington , an RAF base north of London. Darkie finally got us near Bovington; Sears spots the runway, banks on to a short final, dropped the gear, and called for landing flaps. We were high and hot to a very wet downhill runway. I, as a new guy and a Copilot, was hesitant to advise an Old Hand, but after two seconds I said "let's go-around". He ignored me and touched down about halfway down. I could see that we were in trouble but not too far down the runway to still make a safe go-around, again I advised a go-around, at which time the meathead applied wheel brakes. We slowed a bit, skidding down the runway, across the perimeter track, off the end and into a ditch, just missing a kid on a bike. When I saw we were not going to stop in time, I cut the mixtures and switches. There we sat with her nose down in a ditch with the tail high in the air! Capt. Sears disappeared; left us stranded, no explanation, no apology; some leader! We notified home base and they sent a UC-64 to pick us up the next day. I don't recall ever seeing Sears again.

To my recollection, we were never welcomed into the group; not by a Squadron CO or XO nor any group wheels. We got off the train onto a 6 x 6 truck and were dropped off at the living quarters led by a Staff Sergeant from the 734th Squadron. It appeared to me that favored Pilots from the original cadre flew part of a combat tour, then slacked off considerably to assume supervisory positions. This favoritism blocked the promotion of replacement personnel very effectively. The same very poor and ineffective supervisors stayed welded in same the position. We had a CO named Dowda for most of the period I was in the Group. He never knew me, never shook my hand, I can't recall his meeting us after a mission to give us a pat on the back, strictly impersonal. I didn't even know who he was until he walked into our hut one afternoon. Unannounced, he walked halfway in, looked around at nothing in particular, turned about and left without saying a word. There were six of us, all in attendance, no one called attention no one stood up, only one person present knew who he was, that was Jim McNew and Jim had been there for over ten months. We asked, "Who was that?" Jim said "that was T5 Doda", T5 was the rank of a technical rated corporal, Doda was a Major. How about that for respect! The only squadron supervisor I had any real meaningful conversations with was Major O'Dwyer, the Operations officer.

In late October 44 a Lt. Col. Davidson was assigned to our 734th squadron. I do not have a clue as to where he came from, it was very unusual to have Lt. Col. as a squadron commander; such high rank was reserved for commanders and executive officers. Lieut. Col. Davidson seem to be acquainted with fighter planes; but he was just a green bean when it came to flying heavy bombers and heavy bomber flying techniques. Davidson was the only, repeat only 734th commander that ever called an officers meeting while I was in the outfit. At his first meeting he instructed pilots to haul their heavy bomb laden B-24s off the ground as soon as possible and retract their landing gear ASAP. Of course his advice was deadly; the only safe way to take off was to accelerate to flying speed then very gently raise the nose and fly the bird off the ground. To yank one off the runway prior to attaining flying speed is good way to get killed. Col. Davidsons acknowledged his error when I clued him in.

We never had a formation to: present medals, announce promotions, discuss topics of mutual interest, or for any reason, prior to Davidson's arrival. When Lieut. Glass, carelessly rammed Lieut. Rollins, during final approach causing the death of Rollins and his crew, we should have had an immediate meeting to correct such gross negligence. Dowda nor anyone else called a meeting. Other squadrons got together regularly for meetings, get-togethers, to party, present metals, etc…etc. Possibly we had no awards/decorations people; hence no Distinguished Flying Crosses for crews completing 35 missions.

Another prime example of this permanent party original crew syndrome, is shown in the 453rd history edited by Andy Low, wherein he lists only original crews. What about the rest of the group? What about all the ground support personnel? What about the replacement bomb crews that far outnumbered original personnel?

# Flight Logs

**Mission 1:** My first mission was August 30, 1944 to the coast of France. We had about 2300 pounds of fuel and 8000 pounds of bombs, that's eight 1000 pound bombs. We're flying an airplane named *Paper Doll*. We took off at 0530. I flew this first mission as copilot without my crew, Lieut. Emerson was in the left seat to acquaint me with the procedures for: takeoff, climb out, form up, flying formation, bombing runs, and return landing procedure. We flew lead in the slot, positioned directly behind and below Squadron lead. Nearing the target chaff was dispensed; being very close to the lead aircraft bundles of chaff hit my windshield and I thought we were hit by enemy fire. There was no flak! The weather was bad we were bombing a V1 buzz bomb site.

**Mission 2:** My second mission and my first as pilot with my full crew, was September 5, 1944 we went to Karlsruhe Germany with about 2700 pounds of fuel and 6000 pounds of bombs. We were flying a plane called *Crow's Nest*\*. We left at 8 o'clock, this was our first mission as a complete crew. Flack was accurate, we got a few holes, nothing too serious, but I got a real good idea of things to come. We lost two aircraft on this mission, one ditched into the North Sea on the way home. (\* *Crow's Nest* ended the war with 101 missions to its credit)

**Mission 3:** The third mission was on September 8 to Karlsruhe Germany. We had 2700 pounds of fuel and 6000 pounds of bombs; were flying a plane called *Lucky Penny*. We hardly ever fly the same plane twice it seems. This was the same target as mission number two. I lost one engine near Mannheim Lubitsch. Our element leader got hit in his bomb bay; his firebomb clusters caught fire so he jettisoned. Upon seeing bombs fall, flight officer Paulin, our Bomb Aimer salvoed miles from the target. He should've dropped on Squadron lead not element Lead. Since the Nose Turret Gunner Charlie Burnham had a much better view, I had an extension of the Bomb Salvo Switch routed to Burnham for the remaining 32 missions Charlie always dropped on Lead. The squadron proceeded in a left turn to target while I eased out of formation, seeing no sense of proceeding over target with no bombs. Flack was quite accurate, three B 24's were shot up enough to cause them to the land in France; one ditched in the channel. Lieut. Whitehead my element lead landed in France, while I proceeded back to old Buckingham all alone. I never caught up with the group, I did some fancy evasive action, and looked for fighter escort. PS we never again broke formation.

**Mission 4:** The fourth mission was September 9, 1944 we were going to Maintz Germany we had 2700 pounds of fuel and 6000 pounds of bombs. We're flying a plane called *Hollywood and Vine*. We left at 0630 in the morning. The Maintz gunners were pretty active. We got a few flak holes, but no serious damage as with the previous mission, the weather was lousy, and I logged some actual weather/instrument time. After four sorties we were becoming old hands at this stuff.

**Mission 5:** Mission number five September 10 was to Ulm Germany. We were flying *Hollywood and Vine* again, we left at 0850, the third day in a row, early get up. Breakfast at Old Buck was not good. We had black bread French toast with simple syrup, oatmeal which was half hulls, and powdered milk with coffee. Our in-flight ration was either hard candy or a candy bar, both which froze and were almost impossible to eat. Our Squadron got into some terrific prop wash directly behind a closely packed Wing (three or four groups, thirty B-24s per group) the turbulence caused one ship ahead of us to lose control, flip inverted and split S out of sight. We lost an engine, probably due to oil breather problems. This was our second three engine return, sure made it difficult to stay in formation.

**Mission 6:** This mission was September 11, we went to Hanover. This bird had 2500 pounds of fuel and 6000 pounds of bombs. We were flying a plane called *Paper Doll*. A fourth day in a row of early up. We were introduced to a terrific display of flak! There were Kraut fighters in the area; however our escort of P-51 and P 47's keep the Huns busy, and we had none attack our Group. There is not a whole lot of good things to say about our 453rd Group, 734 Squadron, but one thing we excelled at was flying formation! We always held tight formation when the squadrons maneuvered within the group which presents to the enemy a very dangerous target to attack. A fourth day in secession was beginning to tell with the lousy breakfasts, long hours on oxygen in close formation, no lunch, lengthy debriefing, and absolutely stupid requirement that we had to dress in class A uniform for supper, resulted in a weakened skinny aircrews. The 453rd displayed a definite permanent party syndrome! All other flying units that I have been in during the 32 years and seven months and 28 days of my career gave first priority to the guys who put on parachutes not to the folks who strapped a chair to their ass!

**Mission 7:** My seventh mission September 13 was to Wissenhorn/Ulm. We took off at 0840 with 2700 pounds of fuel and 6000 pounds of bombs we are flying *Crow's Nest* again. We had no opposition in route to target, we bomb successfully, and head west for home. Charlie Burnham, Nose Turret Gunner, came on the interphone, saying how clear and beautiful the weather was and that this sortie was a milk run. Shortly after crossing over the Rhine River somewhere between Karlshure and Mannheim/Ludwigshafen all hell broke loose! We passed over an unknown flak concentration that had evidently been tracking us as we passed over and had the group zeroed in prior to firing a shot. That particular Kraut 88 gun crew certainly knew their business. The first few bursts of flak took out a B-24 and crippled our Squadron Lead Capt. Don Gillies, who had to make an emergency landing at the Manston Emergency Strip. He bailed out most of his crew due to the danger of landing without hydraulic pressure, engines out and some controls shot out. We caught some hits in Crow's Nest, Charlie Burnham got hit in the foot with a hot chunk of flak about ½ of an inch which tore up his boot and cut up his foot pretty bad. Charlie nor any of our crew ever used the term "milk run" while we were still airborne again. Jim Mack a new navigator on Gillie's crew, parachuted over Marston airfield. He bailed out through the camera hatch and was injured; seems like he could not turn loose so another crew member stepped on Jim's hand hard enough to make Jim let go, at the same time inflicting a small cut. Jim bailed out clutching his navigation log, and with grand flair entered end of mission in his bloodstained log! Charlie Burnham had a hard time convincing both Bob Busby the Navigator and Frenchie Paulin, Bombardier to let him out of his nose turret, seems like Bob and Frenchie were busy snapping on parachutes and flak vests, we were hit at 19,000 feet where oxygen is essential; I called for my crew to check in on intercom, to report: damage, oxygen check, and other systems checks, also to ensure no one became disconnected while attending to the wounded. We landed with only minor damage. For the next few missions Charlie Burnham arrived at the aircraft on crutches, a real loyal gunner! Of course Charlie was awarded the Purple Heart and so was Jim Mack!

On the 16th of September I flew to Southampton to pick up gasoline in 5 gallon cans. We loaded over 200 Jerry cans on board in the bomb bay, waist section, put-put (APU) well, and navigation/bomb sections. What a load. On September 17th we delivered gasoline to Gen. Patton's Army which had advanced beyond their supply line. We landed at Classtres France, a very recently liberated German airfield. We had minimum crew, two Pilots an Engineer and a Navigator. There were several wrecked German aircraft and a few dead Germans on the field, the area was also heavily mined. We had a continual top cover of P 38's and P 47's. I considered this mission to be just as dangerous as a bombing sortie. There were heavy gas fumes and it was almost impossible to bailout without first removing gas cans from the nose. Without a doubt the gasoline haul sortie should've counted as one of the 35 missions required to complete a combat tour. When my crew arrived in the ETO there were many many crews on station who were required to complete only 30 missions, many of which were completed in a relatively short time period and at less risk, bombing targets in France. To add insult to injury all crews completing 30 missions were awarded the Distinguished Flying Cross, while crews like mine who bomb 99% of their targets deep in Germany, were denied an automatic DFC. To my knowledge we were recommended for that metal three times and never heard a word yes or no, guess Squadron admin filed and forgot!

**Mission 8:** Mission eight September 25 is to Koblenz with 2500 pounds of fuel and 8000 pounds of bombs. We are flying *Crow's Nest* again. Take off is at 0630 Koblenz is on the Rhine River not too far south of the bridge at Remagen which became very important and famous March 1945, it is where the American forces first crossed the Rhine. This mission went real smooth, hitting the marshaling yards. I don't recall that we lost any aircraft; however as usual there was plenty of flak. The 88mm antiaircraft cannon was probably the finest quick firing and accurate weapon of WWII. We were always shot at every mission! Damage was inflicted on us relative to the weather and the proficiency of the German gunners. On this mission we dropped through a solid undercast.

**Mission 9:** Mission number nine was September 26, 1944 to Hamm Germany with 2500 pounds of fuel and 8000 pounds of bombs, flying *Crow's Nest* again. We left at 0630. Can't recall much about this sortie except that it was another early get up We had an extra crew member along, whose duty was to listen in on German radio frequencies for any intelligence regarding German fighters. The 8000 pound bomb load was four 2000 pound bombs to break apart reinforced concrete targets like canal locks.

**Mission 10:** Mission number 10 was September 27 the next, day we went to Kassel Germany this time with 2500 pounds of fuel and 6000 pounds of bombs. We were flying *You've Had It*, taking off at 0640. This was a very bad day for second air division B-24 groups. We lost one near the target, two were forced to land in France. 445$^{th}$ group made a serious navigation error and got out of the bomber stream where they were attacked by a mix of around 100 ME 109's and FW 190s. The Huns attacked in three or four waves of fifteen resulting in twenty B-24s shot down over and near Kassel, only four planes returned home to their base at Tibenham. That's two aircraft crashed in France one in Belgium, and two pranged in England; that's 29 out of 34 that did not get home! It was later found that 29 German planes were shot down in the air battle. I was assigned to fly with the Hethal 389th group and was monitoring C channel, the division fighter VHF frequency. We were headed west after bombing Kassel when I first heard the 445th call for help, I relayed to the P-51 and P-38 escorts that they were urgently needed east of Kassel. C channel was jammed with chatter, all too late to fully engage and drive off the Huns. September 27, 1944 was the day Tibenham was nearly cut in half.

Stand down. September 25, 26, and 27, three days in succession of little sleep early to rise, and choking down breakfast with no lunch, Red Cross brown bread orange marmalade sandwiches and a double shot of old overshoes rot gut booze definitely caused me to lose weight.

**Mission 11:** September 30, 1944 was to Hamm Germany we were flying *Iron Pants*. We left at 0625. Flak was moderate. German fighters were in the area, but our fighter escort was pretty thick, and as usual our formation was so tucked in that they passed us over. Evidently this large and very important marshaling railroad yard required lots of heavy duty bombing to keep it out of operation. The long hours of formation flying were very hard both mentally and physically. Fortunately I had a very good Copilot, William Young age 27. Bill and I traded off formation flying duties every 10 to 15 minutes. Close formations in the B 24 Liberator was tiring; one got a crick in the neck from constantly looking sideways. When tucked in real tight, were never more than 15 feet apart. I believe it was on this mission that the Squadron lead lost the group and dropped too soon. According to the records for 453rd group we lost forty B-24s over Germany, that's four hundred men, all in about eight months!

**Mission 12:** This was on October 5, 1944 bombing the Rhine marshalling yards. We were loaded with 2600 pounds of fuel and 6000 pounds of bombs. We were flying a plane called *Never Mrs*. We left at 0530. The marshaling yards were fairly close to the front lines and needed to be destroyed; the Rhine yard was a supply center for the German Army west of the river. The weather was good for a change and we bombed visually. My record shows that we were misaligned on the first run into the target so a 360° turn was made resulting in excellent bombing. It must be noted that a second run on any target was a very dangerous thing to do. Bombers were then twice in harm's way; 88mm Flack had another crack at them. Many crews were lost on second bomb runs. One of the main reasons to avoid another run was that it would cause mix-ups; following squadrons and groups and even following wings of three or four groups

600 aircraft or more would get out of sync and ball up causing fuel problems and disrupting the defense. The famous Ploesti Romanian oilfield raid is a prime example of Squadrons bombing out of turn, they ended up killing one another by bombing too early or too late on the wrong course. Weather was a factor causing confusion where squadrons got lost, and pressed on regardless; some even hit the wrong targets.

Slow timing. On occasion we had to do some additional flying, usually to slow time an engine after an aircraft was repaired. The engineering section would test fly most of the ships which had major battle damage. However aircraft with new engines were turned back over to the crews to test them and to put at least one hours' time on them to prove it airworthy. We usually always test flew our own ship; of course the ground crew chief went up with all test flights. Slow time flights were always conducted in good weather, which gave us a chance to see the country, and do a bit of low-level flying. All pilots like to buzz, not too low, but low enough to feel and see speed. You make people on the ground take notice, not enough to cause folks to dive in a ditch, just low enough to give them and us a bit of a thrill. If we got caught, what could they do to us, send us home? Not likely!

**Mission 13:** This was October 7 to Kassel Germany we were carrying 2500 pounds of fuel and 6000 pounds of bombs. We were flying *Crow's Nest* again. Eighth Air Force laid on a maximum effort and launched a mix of B 24's and B-17s which numbered about 1500 four engine bombers, and well over 450 fighter escorts. My friend Lieut. Parsons got hit but I can't remember the result. I do however know that Crow's Nest took enough flack hits to knock out one engine. This was my third engine loss; one over Manheim, one over Ulm, and number three over Kassel. The B-24 had probably the best radial engine of any aircraft of that time, the Pratt & Whitney 1200hp 14 cylinder two row radial. The engine was supercharged, it normally pulled 48 inches of manifold pressure on takeoff, but it could produce 60 inches of boost at Turbo 10. On a few very heavy takeoffs, with 2700 gallons of gas and an 8000 pound bomb load, I wound all four engines up to max power. The maximum recommended gross weight was 64000 pounds for this aircraft. When loaded with ten crewmen, heavy flak suits, one for each man, ammunition for ten .50 caliber guns, full gas and bomb loads, and don't forget the ten candy bars for lunch, we were way over maximum allowable that's for sure! And now get this, in order to increase bomb load capacity all belly ball turrets were removed that deleted two .50 caliber guns from the planes armament. Two guns doesn't sound like much, but when multiplied times thirty six, the number of planes in a group; that leaves the group about seventy two machineguns short! On this mission the 453rd launched ten planes from each of the four squadrons. A force of forty birds, short eighty machine guns.

Other flying duties. When not flying combat, or on an infrequent three day pass, we kept busy flying practice missions, slow timing new engines, and other maintenance duties. On one occasion we flew in review for the big brass, including Gen. Eisenhower. The Generals viewed the flyby at Hethel, home of the 389th Group. We cruised by in group formation at very low altitude and at slow speed. I had to use landing flaps on every turn to keep from stalling out. Just one more example of poor leadership. It sure would have been a fine display for the Generals to see a few B-24s stall, spin, and crash, while maneuvering at 500 feet! We also flew some night missions as targets for night fighters. These sorties could get pretty dicey, when one felt the prop wash and saw the exhaust plumes of our attacker. We conducted these night sorties with minimum crew, no sense in having ten pairs of eyes aboard when you couldn't see anything anyways. We also flew as targets for the new British and American jet fighters. The first American jet was the Bell 59 and the Brit was the Gloucester Meteor. The Germans had their Jet in combat long since. We evidently didn't use ours because the Krauts might shoot one down and find out that we too had jets. Some logic eh? Although the new jet aircraft had a short range and endurance, they still could've popped in and out of the battles, just to give the Krauts something to worry about. I'll say more when I get to my days as a fighter pilot!

**Mission 14:** This was October 15, it was to Reinholtz Germany. We were carrying 2500 pounds of fuel and 6000 pounds of bombs. We are flying *Sky Chief* today. Light flack came at us from this small oil target on the Rhine River. I believe we launched only two Squadrons and all planes returned safely.

Now that winter was upon us, weather became a great factor. It was always very cold at bombing altitude -22-40°f was normal in the fall, however temperatures as low as -60° were common over the continent at 22,000 to 28,000 feet in winter. Our breathing condensation formed sheets of ice on our chest and frequently we had to squeeze the oxygen mask to break up the ice in order to breathe. We had no heaters!

The normal heaters installed at the factory burned a fuel/air mixture piped from the engines carburetor; this uses precious fuel and was a fire hazard if damaged by enemy action; they were removed. Also removed was the prop anti-ice /de-ice system; it used alcohol, which was also a fire hazard. To my knowledge all deicer boots were removed, probably to save weight. At any rate, all we had was electric heated: suits, boots, gloves, and pitot, for the airspeed indicators. It is my firm belief that many aircraft and aircrews were lost due to lack of equipment to combat icing.

**Mission 15:** This was October 17, 1944 to Cologne. We were carrying 2500 pounds of fuel and 6000 pounds of gas. We were flying *Sky Chief* again. We took off at 0600. The marshaling yards were close to the Rhine River at Cologne we bombed through a solid undercast using radar. Big water features show up in definite contrast to land; radar bomb aiming becomes fairly accurate when you have those features. Cologne was a big important city protected by many big guns. My friend Bill Lofton's plane got hit and started leaking gas. His engineer stopped the leaks soon enough for them to make it to friendly territory where they all bailed out safely, it was a real close one. Lieut. Lofton later jumped out of a 6 x 6 truck and broke both arms. After I transferred to the 4th Fighter Group, I learned that Lofton had been killed while test hopping a B-24. I guess we only have just so much luck.

**Mission 16:** This was in *Sky Chief* for the second day in a row; we were going to Leverkusen. We took off with 2500 pounds of fuel and 6000 pounds of bombs at 0640. Leverhausen is just north of Koln on the Rhine and is the home of IG Farben. The weather was the same as the previous day, totally overcast. Bombing results were unknown.

Weather over England was becoming worse as fall edged towards winter. We were allocated only fifty pounds of coal per week to heat our Nissan huts. The huts were prefab corrugated steel half cylinders set on cement foundations with no insulation, in other words a bloody icebox; housing usually three officer crews, four officers per crew for a total of twelve men. Fifty pounds of coal fits in a sack about twelve inches in diameter and thirty-six inches tall, about enough to keep our small stove heating the hut for one day. To supplant our coal we did a lot of scrounging. Our main alternate source of fuel was bomb rings, big circles of impregnated paper used in shipping bombs. We all took turns preparing and lighting the morning fire; some methods were explosive! Hydraulic fluid, hundred octane gasoline, and lighter fluid were real good fire starters.

Weather during this mission was real bad! We formed up over England at 24,000 feet in and out of clouds and contrails. The flack was the most accurate I had seen coming up through a 10/10 under cast. It was not the usual barrage but accurate tracking. Radio Operator Lonnie Click was hit in the foot, which put him on crutches for a week or so. We got over 40 holes, the elevator trim was shot out, the radios were shot, out the number three prop governor was shot out, and the number four prop got a big dent in it. It was -40° at bombing altitude. I came home alone after entering clouds at 23,000 feet, and broke out at four or five thousand feet, we fired red flares on final to ensure the medics met us to take care Lonnie.

**Mission 17:** October 25 is to Newmunster Germany; we have 25,000 pounds of fuel and 6000 pounds of bombs on board. Newmunster is in Schiewig-Holstein south of Denmark. Our target was an airfield some six miles south of Keil, the German submarine center. Flack was heavy; it looked like a mix of 155, 105, and 88mm guns. The big guns burst with big red centers. It looked like they had our range so I did a bit of dodging at Charlie Burnham's insistence. The weather was solid undercast again so we bombed using radar with unknown results. We returned in *Sky Chief* with no holes.

# HAPPY LANDINGS

On October 29 and October 30 we flew local. I checked out copilot Bill Young in the left-hand seat. There is not many opportunities to fly local, and copilots had hardly any chance to practice landings. The B-24 was quite easy to land, but it was sure as hell no glider. The high-speed Davis wing lost lift in a big hurry; a partial power approach was best, not a drag-in high-power final which caused too much dangerous prop-wash for following aircraft. Upon recovering from combat we peeled off at thirty second intervals more or less and got on the ground as soon as possible, with one turning off the runway one in the middle rolling and one touching down. We were always low on gas, tired, had battle damage, etc.; missed approaches and go arounds messed up the sequence. Proper spacing in the pattern and landing on the right spot was essential.

October 31 we flew for two hours primarily for Navigator Bob Busby to practice using the GEE Navigation set. Bob was an excellent Navigator who always knew where we were. With the GEE Navigation equipment Bob could tell where we were: over the perimeter track, middle of the field, or on the approach end of the runway. All of the crew appreciated each other's work and got along well.

**Mission 18:** This was November 1 and it was to Gelsinkirchen. We took off with 2500 pounds of fuel and 6000 pounds of bombs at 0355. We're flying good old *Sky Chief* again. Our target was near Essen in the Ruhr complex which was always a hornet's nest of flack. We were well on our way when we lost an engine crossing into Holland. As usual the flack gunners at the Zuider Zee had us sighted in and put up a few accurate bursts. This was our first mission abort; got shot at while crossing enemy territory, lost an engine but received no mission credit.

Mission 18 repeated this mission was the next day November 2 to Castrop-Rauxel Germany. We're carrying 2500 pounds of fuel and 6000 pounds of bombs; we are flying **Sky Chief** once again. Floyd Evge, the best Ground Crew Chief in the Eighth Air Force had put a new engine in Sky Chief and had her ready to go in one day! This mission was almost a copy of the November 1 mission; same area on the Ruhr and same dangerous flack. Like Gelsenkirchen, Castrop-Rauxel was an oil target; great emphasis was placed on destroying oil, all oil, and the Germans put even greater effort into defending the oil. An aircrew on their first mission took a direct flack hit; their plane broke apart directly over the target. It's tough to go down, but doubly so on number one, without the opportunity to do even a part of what you were hoping to accomplish; what you trained for.

**Mission 19:** November 6 is to Minden Germany we have 2500 pounds of fuel and 8000 pounds of bombs today. We are flying *Sky Chief* again. My Pilot Log indicates that the target was the canal locks on the Wesser Elm Canal north of Minden. We dropped through an undercast; photo recon showed good hits with further damage to this vital water transport system. Weather over the Continent as well as in England was a serious problem; particularly since we had no anti-ice/deice systems on our B-24's. Freezing rain, fog and drizzle, are common in the area. The temperature over Minden was -40, the flack was pretty heavy, accurate and right at our altitude. Fighter escort was excellent, and kept the Kraut fighters at bay. The B-24 Groups ahead of us got hit pretty badly; we saw two go down with most of the crew parachuting safely. German ME-262 jet fighters were reported in the area; my gunners reported seeing many enemy fighters; none attacked our group.

**Mission 20:** November 8 is to the Rhine. We load 2500 pounds of fuel and 6000 pounds of bombs on *Sky Chief* and leave at 0530. The weather was very bad. My logbook records two hours of actual blind navigation. The group commander led the mission; lousy weather combined with poor leadership resulted in only one Squadron bombing. The rest of us haul our bombs back to old Buck, home base. Over five hours flying about in clouds, risking our necks for nothing.

The next two sorties were six hours night flying to support various air defense exercises named bull's-eye and spotlight for British radar equipped night fighters.

**Mission 21:** This was to Bielefeld, Germany. We had 2500 pounds of fuel and 6000 pounds of bombs. We are flying *Sky Chief* again. We left at 5030 in the morning. British ground forces were stalled on the left flank; in fact they didn't make it through the Netherlands and into Northwest Germany until the wars end. This situation caused us to fly routes over heavily defended occupied territory or longer routes over the North Sea whenever we hit targets north of Luxembourg. We always encountered accurate flack crossing into Holland, and I remember that the Germans launched their V-2 Rockets from the Zwolle just east of the Zuider Zee. A V-2 just missed our formation this day as we approach Zwolle.

We lost Capt. Conrad's crew on takeoff. The cause was never determined; however I strongly suspect icing and a center of gravity problem. There were times when I arrived at our hard stand to see Sky Chief down by the stern, sitting back on the tail skid; now that's aft CG! When you see a nose wheel aircraft, with the nose wheel high off of the ground, you know something is wrong. The aft CG was usually the result of an 8000 pound bomb load; four 2000 pound bombs, two loaded forward, and two loaded in the aft bomb bay. Aft

CG made the aircraft very very unstable; somewhat akin to a teeter totter with a big kid seated aft. In order to taxi I would move all crew except one crew member forward resulting in nine men forward of the front bomb bay, two in the nose and seven on the flight deck/cockpit and put-put-well area. Even with everyone forward, I would still have to tap the wheel brakes on the start of roll out to get the nose wheel on the ground. Takeoff technique was critical. If you raised the nose too early creating a high angle of attack and increased drag, the Beast would squat tail low, raise a few feet off the ground in ground effect, leaving the pilot between a rock and a hard place. Any attempt to raise the nose causes a stall crash, and the bird was too low to lower the nose to gain flying speed. We lost too many on takeoff: icing conditions, aft CG, and poor takeoff technique, especially icing, killing four crews, that's forty men!

**Mission 22:** This was November 29 to Attenbaken/ Paderborn. We took off with 2300 pounds of fuel and 8000 pounds bombs. It was another cold winter day. The group launched three squadrons, we were flying *Sky Chief,* and we left at 0530. We usually flew eleven aircraft in each Squadron, three in the lead element, three directly in trail of lead, one in the slot under and behind the second element, two in the high right abreast the leading element, and two in the low left below and abreast of lead. This formation was: compact, easy to maneuver, afforded good gun protection, and excellent bomb patterns. The target was a railroad line of communication chokepoint. Weather again was undercast bombing with GEE. Twenty five minutes prior to target we lost an engine. I used excess power on the remaining three, working hard to stay in formation to the target. I sure did like those Pratt & Whitney R-1830 engines. Turning west, now without an 8000 pound bomb load and half of the fuel burned, it was easy to stay in formation on three engines. Ground crew under Floyd Evje, installed a new engine and we slow timed it on a test hop the next day.

**Mission 23:** December 6 was to Minden Germany. *Sky Chief* again carrying 2300 pounds of fuel and 6000 pounds bombs. The weather on takeoff was really bad. The weather was even worse over Germany, so again we bombed through solid undercast with poor results. The target was the Mittelland Canal, a water target, which should have been easy for the lead Navigator Bombardier team to hit. So another potato patch bit the dust! This was the only mission that I can recall where we were not shot at; this was, a Milk Run!

**Mission 24:** This was December 11, it was to Hanau Germany. We carried 2700 pounds of fuel and 6000 pounds of bombs. Hanau is located east of Frankfurt, along the main rail line and was an important target. We were flying *Sky Chief*. We left at 8 o'clock. Weather again prevented visual bombing, so we salvo on lead and hoped for some hits. Eight hours round-trip, with little result, meant we had to play it again the next day.

**Mission 25:** December 12 is Hanau again. We're caring 2500 pounds of fuel and 6000 pounds of bombs, three tons. We're in *Sky Chief*. Take off at is at 7 o'clock. Second verse same as the first, except the weather in England was terrible. We made an instrument takeoff, visibility down the runway was about two lights or about 250 feet. I set the Directional Gyro, and kept my head down. Flight Engineer Pappy Rohmer called out airspeed, and Copilot Bill Young advised me to turn left or right two to three degrees to keep on track in the center of the runway. Near liftoff airspeed I eased the nose up a bit and we were flying. We practiced making instrument takeoffs; those who didn't crashed and burned, or scared themselves to death.

Weather cleared enough to bomb visually; we saw the bombs hit the marshaling yard a good lick. At least we didn't have to return soon. Recovering the several hundred bombers and fighters in lousy English weather was a real task when we returned. We broke up formation over our home radio beacon and flew a prescribed pattern. After turning on final, we would watch for flares continually fired at the approach end of the runway. It was amazing how few mishaps we had using such primitive navigation and approach procedures as landing aids.

**Mission 26:** This is December 24, Christmas Eve. We're flying to Mayen/Hillesheim Germany. We have 2500 pounds of fuel and 6000 pounds of GP 500 pound bombs. We are flying *Sky Chief*: my plane now after thirteen missions. We take off at 0530.

We had been stood down the past twelve days because of impossible weather. Ground crews had the time to put all our aircraft in commission, and we flew every bomber we had. It was during this period of terrible weather that the Germans launched the Battle of the Bulge, when Allied Air Forces were blind and grounded.

The weather finally cleared December 24. The Eighth Air Force put over 2000 heavy bombers in the target areas east of the bulge attacking lines of communications, troop concentrations, crossroads, all designed to break the German attack up and destroy German supply lines.

The 453rd group put sixty four B-24s, up and sixty two bombed their assigned targets; bombing visually with great success.

This maximum effort of the 453rd group established a record that has *Never* been exceeded. Most targets were attacked by six ship elements bombing at low altitude. Poor old Sky Chief took a heavy pounding. We got hit by flak several times near the target, but maintained formation, dropped on time, and plastered the target. We were severely crippled and just made it past the battle lines into France with: number one engine shot out and unable to feather the prop due to a hole shot through the prop spinner; number two engine had the turbo shot out and was leaking oil, number three was also leaking oil and was not thought to be reliable for much longer!

Bob Busby expertly navigated our shot up Sky Chief around known flak positions to a safe haven in France. Miraculously most of the aircraft systems still worked. We were losing altitude at a slow rate as we arrived at Cambrai/Miorginne, a British Airfield that Polish Airmen were operating Mosquito Bombers out of.

We fire red/red flares to let them know we were in emergency. I circled once and set up an approach. I had a the crew take crash landing positions, not knowing if the landing gear would work, or if the wheel, tires, or brakes were shot out. I lowered the landing gear and flaps, slipped to lose excess altitude, and made a smooth landing. I taxied clear of the runway and on to the frozen ground. Then I shut her down. We counted over fifty holes as big as your fist in the left-wing, and numerous other flak holes elsewhere. None of the crew were hit. We never saw Sky Chief again!

I had the crew disarm the guns, remove all the parachutes and all our flight related stuff. The Brits trucked us to a nearby American B -26 base where we spent the night. I'm a little foggy concerning all our activity. However, I do recall us wandering about at night trading cigarettes for wine, with Emile Pauline our French-speaking Bomb Aimer, to translate for us and his being unsuccessful in obtaining French fried potatoes.

While we were browsing around the small French towns at night looking for relaxation with Pauline speaking French to several Dutch Flemish displaced persons, the Germans were dropping paratroopers all around the area. We, with our green flight suits knee-high boots .45 pistols etc., could have been easily shot! I'd rather be lucky than smart any day. The Army guys told us of capturing and killing Krauts in the area dressed as Americans.

We had an extra crew member on this mission, an intelligence type that listened in on German radio frequencies; he insisted that he had to relay his data back to England right away. He was put in touch with a Major at the B-26 base.

As we were returning to our digs late at night we were challenged by an unseen guard. The escort Major had the wrong password and things got a little tense, but we were finally recognized. We milled around France trying to get a ride home for a while. We ended up at Leon where C-47's were flying in troops and about 200 double clutching black truckers were hauling the troops to the battle.

On 26 December we hopped on a C-47 but the Pilot could not get one engine to start; he depleted all battery power trying to start it. We even tried to hand crank it but no luck. That night German twin-engine planes came over, dropped flares and strafed a few disabled C-46s, a Goony Bird and a B-17, entirely missing about 30 C-47's parked on a perimeter track nose to tail, and the 200 6 x 6 trucks which shortly scattered to the four winds, some were not back until 1000 the next day.

We finally got a ride in a C-47 on 27 December, direct to home base, Old Buckingham, tired, dirty, and half starved! The squadron put us to work the next day slow timing a new engine and telling Floyd Evje how where when and what concerning old Sky Chief.

**Mission 27:** December 30, 1944 is to Euskirchen Germany we're flying in *Male Call\**. We have 2500 pounds of fuel and 6000 pounds of bombs onboard. We leave at 0510 in the morning. The target was just east of the Bulge. The weather again was solid clouds over the battle area. We dropped through a solid undercast using Radar/GEE. My pilot log says "easy" deal so I guess that we were not shot at too much and were not attacked by fighters. Having been through 26 missions, forced to land in France all shot to pieces, coming back to base with engines out, and contending with terrible weather/icing with no aircraft deice/anti-ice systems we were, so to speak, veterans who called flak light and inaccurate when New Guys called it, heavy, tracking, and so thick you could walk on it!

It was during the extreme weather in December and January that we lost several crews on takeoff for two reasons. The first was freezing drizzle, freezing rain and zero visibility; the second reason was absolutely extremely poor judgment of some real stupid commanders! I do not recall the exact mission but during this period I flew a mission where there was so much ice and freezing precipitation that one could not even stand on the taxiway without slipping and falling. We were trucked to the hardstand where we found the ground crews were running up the engines. They had the wheel chocks staked to the frozen ground. After they checked out the engines they sprayed all surfaces with deicer fluid. Surface ice was so bad that we could not taxi; we were towed to takeoff position with cleated track tractors, there we started our engines, and took off.

I remember that we could see about 200 feet down the runway, Bill watched the runway lights on the right, you couldn't see the runway lights on the left. I had my head down glued to the directional Gyro, Pappy Rohmer called out airspeed and we got the beast flying. I recall that the one ahead of us crashed. Visibility was so restricted by our ice coated windscreen that we were unable to see anything.

On that particular mission they only got a few airborne before the CO finally wised up and called a halt to it. We had to pull excess power during our climb out. We finally broke out on top at 24,000 feet. We eased back to cruise power settings and immediately lost speed as I pitched the nose up to stay on top. Back on climb power we headed for Buncher Radio Beacon to form up with those few able to make it up. We were covered with a layer of clear ice, which slowly abated off in the clear air. There was an occasional loud pop like a pistol shot as pieces of ice flew off the propellers and hit the fuselage. There was one B-24 slightly above us that seemed to stall then drop a wing and disappear into the undercast; I think they jettisoned bombs, recovered, then headed southwest to land at an ice free base. It appeared to me that the squadron COs were afraid to stand up for their aircrews and tell group COs to call a halt when flying conditions were deadly. Likewise group COs failed to cease flight operations because Air Division COs might think group COs gutless, and so on and so on up the line, one thing for certain, individual pilots had to go regardless! Many men died under a weak CO who failed to stand up for subordinates!

(Note* *Male Call* was Lt Col Jimmy Stewart's bird. Jimmy Stewart was the Operations officer for the 453$^{rd}$. Col Stewart transferred one month before Dad arrived at Old Buckingham. Col Stewart flew twenty combat missions, losing half his squadron, six planes and sixty men, on one of them, when he was a Sqd. Cmdr. with the 703rd. *Male Call* was a veteran of 95 missions. It was the sole surviving plane of the original complement of 61.)

So ended war year 1944. On January 5, 1945 I was Instructor Pilot for a new crew. The procedure now was to have an experienced Pilot fly with new crews instructing them through and briefing them on: engine start, taxi, take off, climb patterns, joining formation, fly to mid channel and return to the base, we also shoot a few landings. Some crews would go through this training twice if needed. Of course there were occasions when Copilots upgraded to First Pilot would need some work, but in general we only flew combat missions.

**Mission 28:** January 15, 1945. The target is Urich Germany. We're flying *Lucky Lucky* today. We take off at 8 o'clock with 2700 pounds of fuel and 6000 pounds of bombs. This mission was a real long haul in close formation, we climbed out and recovered on instruments. The flight log says we smashed two towns, bombing visual for a change. Lots of fighter escort to and from the targets. ME 262 German jet fighters were reported in the area.

**Mission 29:** January 21 the target is Heilbrunn, Germany. We have 2700 pounds of fuel and 6000 pounds of bombs. We're flying *Lucky Lucky* again; take off time is 0830. This is another long haul, but it's longer than the last! We had gas tanks in each wingtip called Tokyo tanks. This fuel had to be transferred into the tanks behind each engine in order to use it. I always had the Flight Engineer Pappy Rohmer transfer wingtip fuel as soon as these there was room for it. My reason was to make sure early in the mission that we have all the gas on board in tanks where we could use it. On this long mission we needed all the fuel we had to get home.

A few planes had to land in France; one crew had to bail out because they failed to transfer fuel from the Tokyo tanks due to a frozen transfer pump. It is a the fact that flak or enemy gunfire that hits an empty fuel tanks would more likely result in an explosion then hits in a tank with fuel, because vapor explodes and fuel burns. I made a wise choice to transfer early. Aside from extreme cold and limited fuel the mission was NORMAL!

We launched on a mission to Magdeburg on January 16 but had to abort when one of our engines blew apart. I've seen/had engines: stall, quit, vibrate, catch fire, but this one went to pieces. We salvo our bomb load over the channel, and do a bit of site seeing on the way home.

**Mission 30:** February 3 the target is Magdeburg Germany were carrying 2700 pounds of fuel and 6000 pounds of bombs. We are flying *Lucky Lucky*, and we take off at 7 AM. The exact target is Rothensee, a Synthetic Oil Plant near Magdeburg. The oil plant was a small target and had to be hit by visual bombing but no luck there, clouds obscured the target. So the entire 2nd Air Division unloaded on the marshaling yards in Magdeburg.

We were also briefed to strike Big B Berlin as a secondary target. Berlin was often designated as a secondary target if we were in the area. The Russians were just east of Berlin, so any hits we made there, would create confusion, and help the Russians take the city. I believe that fourteen Bomb Groups of B-24s, about 430 aircraft, struck Magdeburg that day.

**Mission 31:** February 6, back to Rothensee / Magdeburg again. We have 2700 pounds of fuel and 6000 pounds of bombs aboard. We take off flying *Lucky Lucky* at 0220. We had to abort this mission because of an oxygen leak. We had an extra crewmember aboard for this mission, manning some electrical countermeasure equipment.

We strongly suspect that he turned his oxygen regulator to the emergency constant flow position, and depleted the oxygen supply. We thought we might make it but the oxygen pressure got too low after two hours and twenty minutes time. We were briefed to hit Magdeburg or Berlin as an alternate target. Maybe the kid didn't wish to go that far into Hitler's lair.

We lost a crew during form up. Evidently he got caught in prop wash at slow speed while in a relatively steep bank, stalled the aircraft and spun in. None of them got out; everyone was killed. A fully loaded B-24, with aft center gravity, was a vicious totally unforgiving airplane! Unless there was sufficient altitude and bombs were salvoed, to shift the CG forward, spin recovery is almost impossible. The centrifugal force trapped the crew!

Mission 31 repeated on February 14 and mission 32 on February 15 were both to Magdeburg Germany again with 2500 pounds of fuel and 6000 pounds of bombs. On the 14th we flew *Iron Pants* and on the 15th we flew *E for Easy*. We takeoff between six or seven in the morning. Magdeburg was the target for several days. The 8th Air Force kept trying to demolish the oil target nearby with little success. About half the time the target was bombed through solid undercast using H-2X, the other times poor old Magdeburg received the bomb loads of several hundred 24s.

We were frequently briefed to hit Berlin as a secondary target. I believe we were briefed to strike Berlin five different times. We had excellent fighter cover, flak was moderate and not too accurate; however the flak was the most multicolored that I had seen so far.

Earlier in the month my good friend First Lieutenant Rawlins and his entire crew were killed when they crashed on final approach returning from a combat mission. Rollins was on short final approach when a big dumb stupid pilot named Glass overtook Rawlins and rammed into his tail assembly causing the crash. Glass was able to safely recover and land. I told Glass what I thought of him; he threatened to punch me out. I told him to wait until my back was turned, he got the point! To my knowledge Glass was never reprimanded, we never had a Squadron meeting to discuss such accidents. If I'd been the commander, I would've shipped Glass and his entire crew to the Infantry as Forward Air Controllers.

**Mission 33:** February 17, we go to Hanover Germany today. We have 2500 pounds of fuel and 6000 pounds of bombs, we are flying *E for Easy*. We take off for four hours and 20 minutes including an hour of solid weather flown to nowhere, just milling around. The group was recalled at the enemy coast. We lost one crew that ditched in the channel on the way home; reason unknown!

On February 19 we went to Burtonwood Depot to test hop a brand-new B-24L, and got to fly it back to Old Buckingham. The Engineering Officer advised me that the nose wheel shimmy damper had been written up for severe vibration, and that it been worked on, and signed off as repaired. However he suggested I make a fast taxi run to check it out.

We received clearance onto the runway and since this was just a fast taxi check we had not run through a Takeoff Checklist. The mixture was set to lean mix, cowl flaps are full open, the wing flaps were up, the bomb bay was open, my pilot side window was open; in other words, we were not configured for flying. As luck would have it, we had a: minimum fuel load, crew of only four, no guns or ammo, we were real light. I advanced power on all four engines to 30 inches of manifold pressure and released the brakes; at 40 or 50 miles an hour, there was no shimmy, so I increased speed further. Suddenly I realized, we were going too fast to stop on the runway. Bill Young and Pappy Rohmer recognized the situation; I shoved the mixtures to rich rich rich and called for takeoff flaps and advanced throttles to the stops. Bill lowered half flaps, eased the electronic turbos to #8, and with our hearts somewhere in our necks, we had the beast flying! Another lesson learned the hard way! The trip back to home base was uneventful.

We have to repeat number 33 February 21. We are headed for Nurnberg Germany were flying *Patient Pat*; we take off at 0830 with 2700 pounds of fuel and 6000 pounds of bombs. We were briefed for either of two targets, Nurnberg or Berlin. We were briefed for Berlin many times, this time Berlin was scrubbed prior to takeoff.

Eight and a half hours of close formation flying was probably the most tiring strenuous activity one could ever encounter. Bill and I took turns flying about every fifteen minutes, however the Pilot not at the controls had to remain at constant alert for any activity or movement within the formation. The Debriefing shot of Old Overshoes booze help some to unwind, but strapped into a seat sucking on oxygen for hours on end was something booze could not cure. Some tried that avenue of escape, and experienced great regret when they had to fly a mission the next day with a fat throbbing hangover!

I can't remember the exact mission, but it was on one of those long-haul sorties when we were riddled full of holes and our elevator trim was shot out just prior to bombs away. With a full bomb load we had to trim elevators nose down to counteract the aft CG, so when the bombs went we needed immediate nose up trim; with trim gone, we had to hold constant back pressure which became too much for both Bill and I to cope with. We eased a bit clear the formation and rigged my long neck scarf between an upright stanchion and the control yoke; this relieved us from having to manually hold the nose up.

Later when we were clear of enemy flak and fighter threats, I moved the Bomb Aimer aft to the tail position where he plugged his oxygen bottle into the system, both Waist Gunners moved as far aft as their oxygen systems would allow. Prior to landing I sent two more aft of the bomb bay, which helped balance the bird enough for me to round out at touchdown. The B-24 Liberator was heavy on the controls. With two engines out on the same side, and high power on the other two, pilots had put both feet on the rudder opposite the dead engines to counteract the yaw. In such a situation the copilot was essential. Full deflection of the aileron took 80 pounds of force. My left arm became so strong that I could beat most right-handers when arm wrestling left-handed.

**Mission 34:** February 22, 1945 is to Halberstadt Germany hauling 2500 pounds of fuel and 6000 pounds of ordinance. We are flying *E for Easy* and took off at 0650. Just about all the Allied Air Heavy Bomber Forces, consisting of 6000 planes of the: 8th,9th 15th and RAF bomber forces, plus some twin-engine bombers and fighter-bombers, struck German rail and crossroad chokepoints from the battle lines in the East to the West from Denmark in the north to southern Germany. Bombing was from medium to low altitude, 1500 to 12,000 feet. We hit the marshaling yards deep in Deutschland southwest of Magdeburg from 4000 feet. I told the gunners to fire at will, not that they could do much damage at that range, but to relieve frustration from sitting hours on end at 20000 to 25,000 feet with little activity. Of course Charlie Burnham, Nose Gunner, had the best seat, and I believe that he did do some damage. It is very unusual to be cruising along down low, taking in the sites, and not having to wear an oxygen mask. The mission was an unqualified success!

**Mission 35**: my last mission in bombers is February 25, 1945 and it's to Berlin, carrying 2700 pounds of fuel and 6000 pounds of bombs in *E for Easy*. We take off at 0740, we were briefed for Berlin and actually hit it. Our aim point was a marshaling yard in the Northwest Tenement District. Mission. The objective was to create confusion, to enable the Russians an easier entry into the city. Most of us at briefing voiced some complaint at bombing a tenement area. Up to that time we had always had a more specific military target. Over 2000 four engine bombers hit Berlin this day along with 800 escort fighters. The bombing was by Squadrons in trail, which resulted in bombs raining down on Berlin for over three hours. German fighters were observed with P 51's after them in droves, the flak was heavy.

**Mission 36:** February 27 is a mission to Halle Germany, the plane is *The Spirit of Notre Dame*. I felt no great relief at completing a bomber combat tour. I briefly considered flying this mission to finish up the required 35 for my crew. However, Operations told me to Stand Down, and assigned another old-timer as to pilot to finish up my crew's final mission.

We survived! The war was over for my crew, but not for me. My brother Robert Emmet Scanlon proceeded me into the Army Air Corps, he flew B-24s in the 15th Air Force in Italy. Bob was a twenty-one-year-old Major and I did not want him to fly any planes bigger than me; that's why got into bombers, I actually volunteered for bombers! I soon discovered that although heavies were offensive weapons, one did not get that feeling when flying over the valley of the Ruhr River where Kraut gunners took potshots at you; it was a constant flack barrage for thirty minutes.

I was very proud of my older brother Bob, Robert Emmet Scanlon, a major at twenty-one-years of age and a B-24 Group

Operations Officer in Italy. Bob completed a combat tour of 50 missions with the 15th Air Force; crews actually flew 35, but some of their most dangerous and long sorties counted as credit for two. This was not the case for many of the command elements of the 453rd. Brother Bob led the 15th Air Force on a mission to Munich Germany and also led that Air Force to the Ploesti oil fields for which he received the Distinguished Flying Cross and an additional Oak Leaf Cluster. The awards were for exceptional leadership whereby he had no, repeat no losses from his group on either mission, in contrast to the 453rd history which mentions a leader receiving the DFC for a mission he led while losing ten aircraft; that's one hundred men.

I had had enough of Bombers and ten man crews. I did not want to return stateside, and start training all over again for B-29s with an even bigger crew; so I set about finding a spot in one of our 2nd Air Division fighter Groups. There were five Fighter Groups in the Division and I went to each one using the 453rd UC-64 liaison plane. I was accepted by Col. Everett Stewart commanding officer of the 4th fighter group at Debden Airfield. Col. Stewart wanted to know why a young fella, not yet twenty-one-years old, who had already completed a hazardous tour of 35 missions in heavy bombers, wanted to risk his neck in a Mustang P-51 single-engine warplane. I told him that that the risk in my opinion was less than being reassigned to a B-29 outfit and that I had a score to settle with a certain German flak battery on the west side of the Zuider Zee. Col. Stewart had me transferred that same day.

At the fighter group I was assigned to the 335th squadron commanded by Maj. Pierce W McKinnon. McKinnon was formally of the Royal Air Force he was an early volunteer with the Eagle Squadron, an ex-Spitfire Pilot with more than 20 victories. I flew wing with the Major on a few sorties.

I lived in a two-story brick house with two Officers to a room. We had a Batman who shined our shoes, made our bed and did our laundry, he set the fireplace each day, and did everything for us. We ate in the officer's mess, served by pretty waitresses, with white linen tablecloths and napkins. We are issued seven eggs per week, had white bread, cake and in general lived like kings. The difference between the 453rd Bomb Group and the 4th Fighter Group was like night and day.

They checked me out in the T-6 which I flew for a few hours, and then they showed me how to start a P-51 and gave me a few clues as how to operate 1600 horse power. I found a Mustang easier to fly than the T-6 trainer. I was required to log 50 hours prior to flying combat: this was completed by March 1945. I flew my first combat sortie on April 2, 1945 an escort mission supporting a Wing of three B-17 Groups.

I continued combat operations until the war ended on May 7, 1945. I was over Berlin in a B-24 February 26, 1945; 42 days later I was over Berlin in a P-51, the date was April 11, 1945. I flew one mission to Prague, Czechoslovakia, which lasted almost seven hours. Seven hours is a very long time to be strapped in the tiny cockpit of a Mustang.

On one mission at 27,000 feet over Hamburg Germany my engine started to run rough and completely cut out. I thought this is it, but I got the engine running just before I planned to bail out. It seems that the automatic mixture control had malfunctioned and the engine would only run on high power at low altitude. I flew back to Debden with a buddy escorting me. The British were still far to the south, so we headed out over the North Sea, rather than risk a long low altitude crossing over enemy territory. I explained the problem to my crew chief when I landed; they worked on it, was aircraft WD-G. Somehow the maintenance people adjusted the mixture control the wrong way, and the next day the dammed engine detonated on takeoff and I crashed at the end of the runway. There was lots of flame and smoke, but I got clear of it after a struggle with the canopy, shedding the parachute and life raft on my way out. The flight surgeon gave me a shot of booze and pronounced me fit for service. I flew combat escort the next day and limped back to base on one mag, wishing I had three spare engines.

I flew about 55 combat hours in fighters. A fighter tour at that time, late in the war, was 250 hours, so I flew one fifth of a tour. The 8th Air Force and the 4th Fighter Group continued to train to fight in the PTO. The group was to convert to new P-47-Ms and began exercising against Polish Spitfires who simulated the more maneuverable Japanese planes. Doolittle was in Okinawa setting up 8th Air Force Headquarters when the Atomic bomb was dropped. The second air division, about 15,000 troops, including the entire 4th Fighter Group came home on the Queen Mary in September making port in New York. When we came into the harbor to tie up all of us were on deck, hanging on the rail, when we noticed that there were condoms floating everywhere in the harbor.

## Things Happen

**#1** while on the Tonopah AAB ground gunnery range, the crew got a bit airsick bouncing around at 1100 ft. so Pappy, my thirty-six-year-old engineer removed his lower and upper false teeth and put them in his hip pocket, he got bounced about and sat on his teeth busting both upper and lowers in half.

**#2** Co-pilot Bill Young's heated flight suit was not working (at -40F). We put his feet in the auto-pilot heater cover and his hands in the bomb sight heater. Still we almost lost him; his prayers must have kept him alive over Germany because every minute he would say, "Oh God!"

**#3** Again Bill: He went aft to piss on the bombs; on returning to the cockpit he tripped and fell forward somehow managing to feather the #3 and #4 props! Fortunately we were flying right wing of the high right element in the high right squadron. I did not see this happen, what with my flying tight formation and looking left. I got control of the beast, saw what the trouble was, told the crew to stay on board. Bill got in his seat and we got the air screws rotating and a half an hour later we regained our position in formation.

**#4** Emil, my bombardier, twice released our bomb load early during our first three missions. Emil evidently watched the element lead instead of the squadron lead. Thereafter, I had the nose gunner, Charlie Burnham, drop the load using a remote release switch while Emile helped Bob Busby Navigate.

**#5** Back at Tonopah AAF we were #1 for T/O, watching the Liberator ahead start smoking, roll inverted, and crash BIG TIME. Bill said, "What are we going to do?" I told him we are taking off and that he could grab a first aid kit and toss it out to the wreckage if he thought it might help. Crashes were not uncommon at Tonopah nor were the B-24s at Old Buck. As I recall we lost four lead crews at the end of the runway due to freezing precipitation. High Terrain that climbed faster than a Liberator caused several accidents. However poor leadership—launching overloaded B-24s in freezing rain--.is what killed so many brave men needlessly. After about 12,000 hours flying, I am still amazed that so many survived. The difference between flying the B-24 and the P-51 is not all that different. You go fast to take off and slow down to land-making sure you stay on the runway. The P-51 had a six degree steerable tail wheel that helped correct torque, along with a five degree rudder trim tab and a full stick aft. If you added power too soon and released back stick, there was no way in hell to keep the beast on the runway let alone scaring the ever-loving crap out of the element leader. We always took off in pairs. You had to keep in tight formation because you could not see straight ahead. In a tail low attitude you could see only to the left so if you got behind element lead the added turbulence and poor view ahead could mean real trouble.

With the B-24, the nose wheel and full view ahead, takeoff was a piece of cake. The Mustang is a rudder aircraft. A moderate change of power or airspeed always required a marked change in rudder trim. A power dive at about 400 mph left you standing on the left rudder. My method when on a ground attack was to set power and grab the rudder trim. You got an easy to fly aircraft when you mastered the rudder at various power/airspeed conditions. No such problems in a B-24.

I transferred from the 453rd Bomb Group and was checked out in the AT-6, given six hours, and then shown how to start a Mustang. I studied the "book" but failed to grasp the important element of the extreme aft center of gravity with a full 85 gal fuselage tank. One must burn that tank down to 25 gal or the bird is unstable. With the B-24 the flight Engineer kept the tanks in balance.

My first takeoff was OK; I kept the tail low and added power gradually. Raising the gear was a task because you have to release the shoulder belt to reach the gear lever located low and behind the throttle/mixture levers. I climbed to 400 feet and then took care of retracting the wheels. Takeoff was about 100 mph with coolant in manual at 3000rpm and 61 inches of manifold pressure. (We usually kept the coolant full open to avoid overheating while on the ground then went op auto for both oil and coolant.) I went through the usual drill. I climbed to 10,000 feet, increased speed to 300 mph and practiced some turns, noting very little backpressure was need to make a level turn. I climbed to 16,000 feet to try some aerobatics. I increased speed to 320 mph (max speed for the B-24) and pulled up attempting to loop at 2.5 to 3gs. Somewhere near the top I increased backpressure and the beast tumbled ass over teakettle! I got tossed about a bit, got off rudder and stick, power to idle, and the crate recovered nose low.

I returned to base, called on the downwind leg. 1000 feet, put gear down and half flaps (just like a B-24) turned base and final. Tower then told me to hold altitude and go around. The Group was just now returning from a combat mission in echelon-on the deck peeling up beneath me at 250 mph. the tower told me to stand clear as about thirty to forty P-51s landed within the next half hour with me circling over the runway. The haze plus the low sun angle made it real hard to see the field so they used me as their guide. I finally was allowed to land, after making two missed approaches. (I never had any problems landing a B-24.)

I saw how and why the landing pattern was made out of a constant left shallow turn. The steep climbing left peel-up killed off airspeed plus lowering the gear got the speed down to allow lowering the flaps and got the aircraft down to near landing speed. The ideal pattern allowed you to see the bird ahead and the runway. The objective was to touch down on the left wheel and roll wings level at the same time. No square approaches allowed! (You do <u>not</u> make a B-24 approach out of a climbing turn.)

The Mustang engine tends to foul up quickly at low power so you must not taxi too long. Even at cruise power (240-250 rpm and 40 inches manifold pressure) you must clear the engine at full power-3000 rpm and max throttle. NO such need in the Lib. The P-51 fuel mixture has two positions off and on. No lean, rich or in-between. Fuel mixture is automatic in the on/run position.

**RES.** Mom was raised in Los Angles California and didn't move an inch until she was twenty-eight-years-old, even living at home while she attended The University of Southern California. Mom was the first one of her family to attend college, her brother Sal followed and graduated after the war. Her Dad was comparatively quite well off. Grandpa, Frank Nuno, starting as a mechanic worked his way into owning a shop. He had a small ten acre ranch with a huge wood frame two story house just east of the USC campus, which at that time was on the edge of town. Grandpa raised chickens, sheep, goats and horses on the ranch. He also had a vegetable garden, grew avocados, citrus, nut and fruit trees. I only saw my Mom's Dad twice in my lifetime that I can remember but I was impressed. We boys stayed with Grandpa Frank for a couple of weeks coming and going to Taiwan the summer of 65 and 67. He entertained us on the piano and organ one of those double decked church organs with the foot pedals and hand valves that he taught himself to play.

Grandpa Nuno was an orphan. His father Paco died at the age-of-thirty-three, when he was struck by lightning. Frank was ten-years-old at the time, his mother died when he was born. His father Paco was a famous mural artist in Guadalajara, and his father's father Jamie Nuno was a music professor at Syracuse University, he composed the Mexican National anthem for Santa Anna in 1851. Jamie was a Catalan musical prodigy who was admitted to the Papal Orchestra in Rome at a very young age, where he excelled. Grandpa Frank was very clever too, he was mechanically gifted. Frank went to work for the Mexican railroad as a coaler and made engineer when he was sixteen-years-old. By 1916 Frank eventually worked his way to Los Angles, following the rails, where he took a job in a salvage yard, owned and operated by a Japanese gentleman, Mr. Freddie Fuyoka. Nisi would come to the yard and deal with the owner for parts. Grandpa became fluent in Spanish, Japanese and English, acquired a working understanding of French, Russian, and Italian, so he could easily communicate with the diverse races who came to the yard. Frank's language and mechanical skills made him invaluable to the owner who sort of adopted him. When the owner, who had no family, died he left everything to Frank.

The yard is very busy it's the dawn of the mechanized age. Then came the depression and no one could afford new stuff. Things had to be repaired and recycled or new things made from old parts. This required machining and fit ups to make things work so along with parts Frank would provide mechanical services expanding his business. I don't know for sure but rumor has it that Prohibition was very kind to Frank. Frank and Joyce don't drink. Mom and her two brothers Bill and Sal never take up the habit; but business is business and there is money to be made. Some contraband comes back with the trucks sent to and from Mexico with stuff from the salvage yard. Frank Nuno and Joe Kennedy are in the same line of work. Grandpa later starts an auto repair shop and a sporting goods store that Sal and his boys will eventually run.

After Perl Harbor, when the internment of the Japanese was imminent the Nisi came to Frank to sell their stuff because they believed he was a fair man. Their faith was well placed. In many cases he just agreed to secure their crated up belongings at his yard until everything blew over. One of Mom's best friends was a Japanese girl, the daughter of a prominent business man. Mom's friend was teaching her Japanese. Mom has her Dad's gift for languages. She is fluent in Spanish, Portuguese, English and French. Frank "bought" all this gentleman's business concerns and "sold" them back for a dollar after the conflict.

When we stayed with Grandpa, summer of 65, the ranch was now in the middle of Watts; gone were the chickens, the livestock, the fruit trees were barren, except for the avocados, they were the only things not on his neighbors menu.

Grandma Nuno was an Aguilar. Her Dad was a local patron in Guadalajara. Poncho Villa came to town one day and strung him up. Villa made a bad job of it however. Some of his people cut him down and revived him. Senor Aguilar loaded up all his possessions, his sixteen children, followers and servants, in ox drawn wagons and headed for San Antonio and then on to Los Angles by train where he had family. Its 1916, Grandma Joyce is eight-years-old.

Being the second youngest of sixteen kids, pickings are pretty slim for grandma. Fifteen-year-old Joyce meets and marries eighteen-year-old Frank, who is very handsome and charming, in 1921. Joyce is no longer on the bottom of the totem pole. Mom is the first child of this union. It's November 13, 1922. Frank brings his wife and new baby girl home from the hospital driving an Indian motorcycle with a sidecar. He had plenty of cars and a truck but he took his motorcycle and went roaring around town with his new bride and baby girl rattling around in the hack. His parents and in-laws are horrified and let him know about it when he finally gets home.

Mom and Dad met in high school, Washington High in Los Angles California. Mom is one-year-older than Dad, so she was one year ahead in school, but during the war Dad was graduated, by the school system, a year early for the needs of the country; so Mom and Dad were in the same graduating class. Dad played basketball and tennis in High School, he was drafted before he could sew his letter on his Varsity sweater.

Sal took aptitude tests and was sent to flight school, he was supposed to be a P-38 pilot like the rest of his class. While flying a primary trainer, a Sterman, and attempting a barrel roll, he almost pulls a Raul Luffberry. His seat belt restraint was not fastened properly and he almost fell out. He lost control the plane and the instructor had to take over. Sal washed out. Sal had done everything well up to that point

When Sal almost died he thanked God for saving him and promised the good lord that "If I get down on the ground I'm not going up again". Sal was ashamed of washing out and not wanting to fly. Grandpa was so proud to have a son as a flying cadet and he was sorry to disappoint him. Sal ended up going to electrical specialist school in Chanute field, Chicago to be a B-17 specialist. He later went to B-29 school. Sal didn't go overseas (To Burma) until 1944. Sal and Dad were all drafted around the same time in December 1942. Frank said they lowered the draft age to eighteen and took the whole neighborhood!

Sal did not tell his story to his kids until he was ninety! One member of Sal's class of cadets at Santa Ana came home with Sal on leave several times to Grandma and Grandpa's house on 83rd street. His buddy, a Pole, occasionally wrote Franny during the War. Sal didn't keep in touch with his buddies, cuz he washed out and couldn't face them. Franny received a letter from Sal's buddy who had ended up staying behind as an instructor pilot before deploying. He wrote "I am glad Sal is not with us". When he caught up with the group, they were flying fighter escort for bombers hitting the Oil fields of Eastern Europe and doing ground attack. Three fourths of the original group of pilots had been killed by then. Sal often wondered if he was the only one of his class to survive. If he had not washed out he probably would have been killed. Sal was the youngest in his class and the only Hispanic. Julio Nuno a cousin and a P-51 pilot was shot down by ground fire over Belgium. He left behind a young wife and daughter.

Sal made staff Sergeant before he went overseas. When the regular electrical engineer couldn't figure something out Sal usually could. Sal worked on the C-46, The P-61, B-25, B-26, P-47, and the P-51 as well as B-24s. After Burma Sal went to a rest station in Nepal, then to India to wait to return home. Sal returned March 1946. He and a friend jumped off the train from San Francisco in Sacramento so his friend could see his folks. They hitchhiked the rest of the way catching a ride with a guy in a new 1946 Buick Convertible to Motor Town City where they met the train they were supposed to be on, as it pulled into Union Station. Grandpa drove Sal home to surprise Grandma, she didn't know he was coming home. For Sal this was his great adventure! Grandpa was always worried something would happen to Sal, Franny or Bill but when Uncle Sam called there was nothing to do but answer the call. Frank wanted to shoot Sal in the leg or send him to Mexico so he wouldn't get drafted. Sal had first tried to enlist in the Army Air Corps when he was seventeen but Grandpa and Grandma wouldn't sign for him.

Dad is irked when Sal spins his I was there "war" stories. He won the war fixing planes. Sal is fluent in Japanese, Spanish, French and English he had a couple of Nisi friends in High school who taught him some Japanese. He must not have mentioned his language skills to the Army when he was inducted, skills like that are in high demand at the time. Skills that would surely have sent him in another direction in the Army. Brother Bill was also drafted and trained as infantry. He landed with McArthur in the Philippines. There was a call up in his unit for volunteers who knew how to repair and drive trucks for which there was a great need. So Bill did his duty in the Quartermaster Corps as a truck driver/motor mechanic roaring around the Philippines in a 6 X 6 and later with the occupation forces in Japan. Both boys had worked at Motor Truck, Franks Auto shop and junkyard, they were very familiar with engines and heavy equipment. In a rare stroke of luck the Army had placed the boys where they would do the most good for the general effort.

Los Angles (Shoo Fly Point) is a small City at this time. Mom and Dad were not High School sweethearts; Dad was a hunting buddy of Sal's and a classmate of Bill's. Sal was also on the basketball team with Dad. The classes were pretty small by today's standards so they would bump into each other now and again at Washington High.

Mom has a lot of friends, Ester Williams, the famed movie star and swimmer for one. She also does some Andy Hardy movies; she is barley 5 foot tall making Mickey Rooney look tall. They would all do things together with the gang. The gang went to Lake Arrowhead for the graduating class retreat. They had a picnic and went swimming, everyone swam out to a swim platform offshore, everyone that is except Mom, who can't swim and never learned. Dad swam back, coaxed her on his back somehow and swam her out for a group picture; guess that's where it started. Dad used to tell Mom: "my love for you will ever flow, like water down a tater row ". When grandpa saw that picture he was furious. It was the Fox coaxing the Rabbit onto his back to cross the river as far as he was concerned.

I have looked at Dad's collection of aircraft pictures since I was big enough to open the album. I have seen pictures of his cracked up P-51 WD-G and stared in wonder. There is a before mission photo of him sitting on his fighter's prop spinner looking like a punk kid and another of him sitting on the wing after returning from an 8 hour flight looking like a eighty-year-year old man, haggard hunched over with deep lines on his face from the oxygen mask and the stress. I've seen his gun camera films shooting up seaplanes, strafing trains and trucks, German troops scattering and falling. I have read the books from Dads training class K-42, histories of the 453$^{rd}$ and the 4$^{th}$ where he notes laconically in the margin and on pictures about some of his class who didn't make it and what happened. Some lost in training accidents, bad weather, poor decisions, mechanical failures, or shot down. On one entry, he notes that he saw his roommate going down in his P-51. Dad notes that he must have blacked out at altitude, probably from an oxygen malfunction at 25,000 feet. Dad was calling to him to pull up on the R/T as he spun through an undercast. Another friend, unused to the aft CG trim from the tank behind the seat went into a flat spin and couldn't get out.

Late in the war fighter escorts were taking the fight to the enemy shooting up airfields and trains anything that moved. Strafing attacks were very dangerous, it wasn't the same as air to air combat; you can't outmaneuver the ground and AA doesn't discriminate. Dad destroys two float planes on a lake in a strafing run. Pilots get separated from their lead and get disoriented while beating up the ground. Dad is a good navigator and is always able to find his way home. Many are the transmissions he hears from lost pilots, their calls for help get weaker and weaker as they fly in the wrong direction until they exhaust their fuel.

While flying with the 4$^{th}$ Dad gets to fly Col. Don Blakeslee, the group commander's bird, WD-C noting a custom flap handle extension that is more accessible in a high G maneuver. Dumping flaps can prevent an over-shoot in a dive and can help turning inside an opponent without stalling. He also gets to fly wing on several sorties, for Major Pierce McKennon a 24 victory ace, his bird is named Ridge Runner WD-A.

Dad had lots of quaint sayings: finger four, straighten up and fly right, shape up or ship out, plan of the day, split S, wingover, ground loop, out of commission, head up and locked, milling around, swinehunde The Tiger Never Changes his Stripes, ship you home in a match box, and the ever popular auger in.

Dad came out of the conflict a first Lt. with: the DFC, the Air Medal with six Oak Leaf Clusters, the European Campaign Medal with four Campaign Stars, the WWII Victory Medal, a chunk of flack in his sternum, banged up knees from flack knocking his legs off the rudder pedals and into the cowl, a hatred for Brussel Sprouts, some lifelong friends, a passion for flying and some unacknowledged emotional baggage. He jokes that he is an Ace, he got two of the enemy and three of ours. The war ended May 7, 1945; Dad's birthday is June 17, he turned twenty-one-years-old with 285 hours of combat flying to his credit.

Three hundred and sixty-six airmen of the 453rd went west by the end of the war along with one hundred and twenty-five from the 4th and 80,000 other airmen.

For a contrast to my Dad's rather laconic narrative, written 45 years after the fact, compiled and edited by myself, read Bert Stiles book: *A Serenade to the Big Bird*. Bert was a "mature" twenty-three-year-old Copilot, a graduate of CU, Colorado Springs. Bert received a degree in journalism. Bert wrote articles for several prominent magazines including the Saturday Evening Post during his tour. He wrote a novel about his rather tortured experience in the heavies "slept in the same bed that eight men had slept in before him". Folks with the jitters get transferred to the" Flack

House". The Brits call it LMF, Lack of Moral Fiber. They would break you and transfer you to the ranks to do the most menial of tasks. The stress makes some people behave uncharacteristically in the air in order to deal with it. Not many people realize that flying was strictly volunteer; Bert's Squadron lost half their planes, six aircraft on one mission, that's sixty folks that were there that morning eating breakfast with you, gone. Bert flew Copilot in B-17s for 35 missions and then volunteered for fighters; like my Dad he flew P-51s. Bert was shot down and killed escorting heavy bombers…his young wife published his work, forwarded with his effects posthumously.

## Sky Chief's Official Report

B-24H-25 FO 42-95173C+ 734 E8

SKY CHIEF

NMF

Q- 734 E8

First trace: 19 May 44.

Individual aircraft letter changed from C+ to Q- between 8 & 16 Jul 44.

Force landed 24 Dec 44 at Cambrai-Niergnies, France, - the plane was hit several times near the target, dropped on time, plastered the target and limped out with # 1 engine out, # 4 engine feathered, # 2 turbo out and # 2 & # 3 leaking oil badly. (Mayen) Salvaged 11 Mar 45 on the Continent by 5 SAD5

# 1947 Ferry Command

**JJS.** After I was discharged from Active Duty, 3 November 1945, I immediately enrolled in the University Of Southern California. I completed two years college credit in one-and-a-half years while working at the South Gate General Motors factory, assembling Buicks, Pontiacs and Oldsmobiles. I also flew T-6 aircraft with the Long Beach Army Air Force Reserve on the weekends, which left little time for play; although, I did take time to Bar-Hop with Hub Edwards and Harold Baker.

I was recalled to extended active duty in May 1947, reporting to Tinker AFB, Oklahoma, and reassigned to South Plains AAB, Lubbock, Texas. I was immediately checked out in the T-6, for an area orientation, even prior to unpacking my stuff or being assigned a Billet. This project was the re-deployment of WWII Aircraft, where we flew hundreds of WWII Aircraft, from closed down WWII Army AF Bases to various Depots for Reclamation or Destruction.

## 1947 FERRY COMMAND

I was assigned as Flight Test Maintenance Officer, and cleared to fly any aircraft after a short checkout. I flew as Co-Pilot in a B-25, from Lubbock to Pyote, Texas AAB in 1:15 Hours, and flew as Co-Pilot in a B-17 on the return flight to Lubbock. The next day, I delivered a B-25 to Pyote, with a civilian mechanic as Co-Pilot. My checkout in the C-46 was also Brief – i.e., one hour and four landings. During the period 3 June to 30 September 1947, I flew over 150 trips in these Birds – B-25, C-46, C-47, B-17, AT-6, PT-17, and many others. South Plains AAF Lubbock was a real education. I was loaned to TEMCO Aviation, at Navy Hensley Dallas, I was to deliver C-47s, PT-17s, AT-6s, and B-25s to Kelley AAB, and had to train a few South American Pilots to fly each bird.

I also had to ferry a P-47 and a P-61 from Tinker AAF to Kelly AAF. I had flown P-47's when I was with the 4th, but I had never seen a P-61 Black Widow until I flew one. The majority of WWII Aircraft had Radial air cooled engines. Aircraft all had similar instruments, controls and systems, which made checking out quite easy. The airspeed gauges were all marked to show gear and flap extension speed, and stall speed. All of the Ferry Flights of Aircraft taken out of storage were conducted in VFR – Clear weather conditions. It sounds easy but you must remember that all these WWII Birds had been sitting out in open storage since 1945, making this risky business. Make-ready maintenance for flight was minimum: drain engine oil preservative, fill with new oil, replace sparkplugs, perform an engine run-up, check hydraulics, run wing flaps through a few cycles, then declare it ready for flight. The Pilot, would: make his own: Pre-Flight check, preform an engine run-up, takeoff; climb to safe bailout altitude, make two turns around the flagpole, and if the Beast was running okay, we would head on-course to the destination. I was a real happy twenty-three-year-old aviator, doing what I loved to do. Fly!

In October 1947, right after the establishment of the Air Force as a separate branch of the military by congress, I was assigned to Victorville AAB, California, with the same duty I had at Tinker, Flight Test Maintenance Officer. I immediately began to check out in the B-29.

It was at Victorville AAB that fate jumped in, when I was assigned an Additional Duty as Mess Officer. There were only twenty-three Officers on the base; so, everyone had an Additional Duty. I'll never know why I was picked. The previous Food Service Supervisor, Lt. White, was not one of the Ferry Pilots, he was Discharged *for Cause* (always in trouble); so, unasked I was promoted to Supervisor. I had many, many free meals; I left the operation of the mess in the experienced hands of the Chief Mess Sgt. I didn't think much about it at the time but when the additional rating of Food Service Supervisor was added to my Service Record, that rating would impact my immediate future.

Victorville AAB was a real fun place to live. We had a fine Officers Club, with lots of activity: dances, bingo, cookouts. There was good dove and quail hunting; Lake Arrow Head and Big Bear Lake were nearby to visit easily on a day trip. Both lakes had wonderful water and winter sports. Besides all this, Los Angeles was just a two-to-three-hour drive away. Victorville is located in the High Desert, around 3,000 feet, with warm days and cool nights. We never had to use the Swamp Coolers during the night; when the sun went down, it cooled down quick.

From October '47 to August '48, I flew almost daily in the following aircraft: B-29, AT-11, C-47, AT-6, AT-7 (Beechcraft Twins), plus one ferry trip from San Diego to Great Falls, Montana, in an L-13. My checkout in the B-29, at that time the largest plane in the inventory, consisted of three ferry trips as Co-Pilot, two local training missions shooting landings, some familiarization with emergency procedures, then I was on my own. I didn't even fill out an Aircraft Questionnaire. Of course, we must remember that all Ferry trips were only flown in Daylight VFR Conditions. The Pilot in command was the final authority as to Go/No Go. We had experienced crew chiefs and some real sharp Aerial Engineers, who ran the main Engine/Systems Panels in the B-29.

As Pilot in Command, I was responsible for all navigation. In those early days, we navigated by visual pilotage and Low-Frequency Radio Beacons and Radio Ranges. Some of the places we delivered aircraft to were: Sacramento and San Bernardino, California, Tinker Field Oklahoma, Middletown Pennsylvania, Hill AFB Ogden Utah, San Diego California, Butte Montana, Tacoma Washington, Biggs AFB Texas, Pyote AFB Texas, Kirtland AFB New Mexico, Wichita AFB Kansas, Spokane Washington, Pasco Field Washington, Lowery AFB Colorado, Kelly AFB San Antonio Texas, Wright Field Dayton Ohio, Hammer Field Stockton California, Hamilton AFB California, Inyokern Navy Station California, Lindberg Field San Diego California, Robbins AFB Alabama, Carswell AFB Fort Worth Texas, Dubois Idaho, Great Falls Montana, Lubbock Texas, Helena Montana, Barksdale AFB Louisiana, New Orleans Louisiana, and many more.

I spent many weekends socializing in L.A. My Mom and Pop lived in Gardenia, where I stayed rent-free, kinda like a Bed & Breakfast. I sold my 1937 Chevy Coupe, and bought a new Studebaker 4-Door Champion. I was footloose and fancy free, this was almost like a vacation. I spent several days with Lloyd and Madeline Clark, at their summer vacation home in Laguna Beach, just having a fine, fine time, dating some of my old high school gal friends. Then came the *Berlin Airlift*, which started in June 1948; *orders came down from HQ in Sacramento, transferring me to Germany* . . .

# C-46

Ref Pg. 6-43K Contact Summer of 2003. Can't spell. Can't type. Don't care. I win! It's a C-46 which I remember was a Curtis Sky Train. I flew the C-46 F and D models. One had a Hamilton Standard and the other had a Curtis Electric Prop. Each had different systems. As long as you had plenty of hydraulic fluid along you could usually make it from A to B.

I flew about 200 hours, usually out of South Planes AAFB Lubbock Texas in 1947, a member of a pilot unit comprising twenty three Army Pilots sent there to redeploy WWII parked aircraft to various depots and other places—Most were destined for the storage depot at Pyote, Texas which at one time had about 4000 WWII aircraft parked in temporary storage. We parked a bunch of brand new B-25s there that had only eight to ten hours on them. Some fly by night pick-up maintenance contractor, ex AAF mechanic, would check the: tires, the hydraulic fluid, drain the oil preservative, replace the dummy spark plugs with real ones, crank her up, let it run until the engines stopped smoking, then hand it over to a Pilot. All Pilots were test maintenance types, myself included. We did our own pre-flight and hauled ass. We would always fly two laps around the flag pole and if the beast was still running we went to Peyote or elsewhere. Hard to believe!

One dandy windy day I cranked up a C-46 with a Co-Pilot who had never seen a C-46, Connor Alliegood I think. On taxi out I engaged the onboard rudder lock because the beast was tuff to taxi due to the large vertical stabilizer. On engine run-up I applied max power to exercise the props and mags. A gust of wind caused the rudder lock to pop loose when I was leaning over the center control quad. I was struck smack in the nose by the control column, my glasses were knocked off and I busted my lip; that really pissed me off!

On the way to Peyote at about 8000 feet on top of a real pretty cumulus cloud I took off my seatbelt, slid the seat aft and propped my feet on the control column, turned on the auto pilot and drank a Coke. The air got rough; the control column jerked and the wheel pinched my shinbones nearly in half. I became quite disturbed, grabbed the control column and shook it violently. The altitude control gyro took control and pitched the nose down pitching me with no seatbelt on up into the overhead where I couldn't reach the auto pilot cutoff, located on the floor. We finally pulled out at 4000 feet, auto pilot off and seatbelt buckled. That was the last time I ever mistreated an airplane. The last time I flew the C-46 was in 1967 on a flight from Taipei Taiwan to Tainan on the south end of the island; a nice old bird that had about 10,000 hours on the airframe.

# 1948 Fassberg, Berlin Airlift

**JJS.** I was scheduled to move up to Sacramento, McClelland AFB, with the same job I currently had when I received orders for Germany. We still had several hundred planes to ferry out of Victorville, but, it proved to be that hauling stuff to Berlin, took priority over moving WWII Airplanes here and there. The C.O. at Victorville did his best to cancel my orders to Germany, but to no avail. I had to sell my brand new 1948, four-Door Studebaker Champion to my Dad at a loss; that hurt, it was my first new car. As a parting gift the Air Force required me to ferry a B-29 to Tinker Field, Oklahoma, on my way to New York, my Port of Embarkation.

I fell sick at Tinker, and was put in Hospital for a week. During this time the CO at Victorville was still trying everything he knew to get me reassigned back to California. The Doc at Tinker thought I had Spinal Meningitis, so he tapped my spine in a most unprofessional manner, he had to stick me three times to get a clear fluid. I was never told not to sit up or raise my head for 12 hours. This resulted in severe headaches for two months. The Tinker Doc said I was okay to Travel; so I hopped a flight on a B-29 to Washington, D.C., then traveled by rail to Camp Shanks New Jersey, where I joined two hundred other Air Force types awaiting Air Transport to Germany.

We were flown to Germany via the Azores, in a C-54. Our Occupation Indoctrination was at Marberg, Germany, where we were briefed on the do's and don'ts when occupying Krautland. My orders were TDY (Temporary Duty) to fly the Berlin Airlift as a four-engine Pilot; the Lift started in mid-June 1948, using C-47 Aircraft operating out of Wiesbaden and Frankfurt. Somehow, Personnel saw that I also had an MOS (Military Occupation Specialty) as a Food Service Officer, and assigned me to Oberpfaffenhofen, where a repair depot had been set up to take care of the Airlift Birds. I was assigned as a Flight Test Maintenance Pilot with the additional duty of Assistant Mess Officer.

I settled in at OBI, and got checked out in the C-54 and C-47 on local flights. I was not needed in my additional capacity; the only Mess Hall on base was under a Capt. Dixie, who said he was a P-38 Pilot who saw Service in North Africa. Dixie was just about useless; his claim to fame was his seduction of his housemaid while his wife was in Hospital. 'Nuff on *Dixie.*

After a few days at being an un-needed Mess Officer, I complained to the Base CO that my services would be better utilized as a Pilot. In fact, orders had just arrived from Airlift HQ to send me and another C-47 Pilot TDY to Wiesbaden, to fly the Lift. I flew The Lift in a C-47 September through October 1948, with a few local trips out of Oberpfaffenhofen,

My first trip in a C-47 from Wiesbaden to Templehoff, Berlin, was at night, with a Lift Vet, who had already flown the max of 125 hours for the month. He gave me a complete and thorough checkout into and out of Berlin. Upon return-landing at Wiesbaden, I was given another Gooney Bird, loaded with Coal and a Green Bean Co-Pilot who had never been in a C-47. I was told to get briefed and haul-ass to Berlin right away. My second trip was in daylight. The need for Pilots and Coal was critical. I checked my Co-Pilot out on how to raise and lower the gear and flaps, and how to operate major systems. He and I soon became Airlift Vets, hauling coal – only coal. I settled in at Wiesbaden, flying both day and night. Due to so many changes in assignments, my Form 5 Flight Log was not kept up, which resulted in some trips to Berlin not being logged.

The Food Service (MOS) resurfaced, and I was transferred once again in October, this time to Fassberg, Germany, located halfway between Hamburg and Hanover, in the British Zone of Occupation. Five C-54 Squadrons had recently transferred to Fassburg, to haul Coal and Only Coal to Berlin. I thought for sure I was to fly The Lift in the C-54, and I did fly many coal runs into all three Berlin Airfields, Templehoff, Tegal, and Gatow, in some real cruddy weather.

At Fassberg, I was told to take over the feeding of both the U.S. and the British personnel there. At that time, all that was in operation was a Flight Line Mess, serving Breakfast only, and an Officers Mess. Personnel were fed using both U.S. and British Rations. The Brits also ran a Sergeant's Mess and an Other Ranks Mess; plus a facility to feed the numerous Polish Guards. The poor British had very little to contribute. At the start, there was mass confusion, very little coal was flown to Berlin. The New Air Force Command had transferred five C-54 Squadrons to Fassberg, from all over the Globe: Panama, Japan, and Alaska. The C-54 Squadrons had their own maintenance, personnel and paperwork boys but nothing else. There was no *Supply, no MPs, no Finance, no Base Engineers (Air*

*Installation)*, *just five C-54 SQds* stuck in Fassberg, with very little support services. Food Service was critical. The Personnel Officer was a 2nd Lt, who filed the Personnel Folders in alphabetical order, A-Z. With people being shipped in daily, there was no organization set up for them to report to. No one was paid for the first two months. *A real sorry mess.* Getting organized was a slow process; eventually, we had about 4,200 personnel to tend to.

In due time, I rounded up thirty Food Service types, mostly cast-offs from U.S. Bases in the American Zone of Occupation; some real dull, and some dedicated, professional cooks and bakers. In short order, I got with the Brits and hired about 320 local German Cooks, Bakers and KPs. We took over the British Sgt.'s Mess, and set up the Enlisted Mess Hall in the original German, well-equipped facility. We ended up feeding all U.S. and British personnel on Base, and provided a few bottles of milk for British Dependent Children.

The Main Point to remember is that All U.S. Personnel were assigned TDY for 60 to 90 days; and when their time was up, we all had our time extended to who knows when? This uncertain situation did not help Morale . . . There were C-54s departing and landing every few minutes; due to the lack of an organized Supply, there were many C-54s Grounded for common parts. The parts were on Base, but no Supply set up to issue them. Within a bit over two months, we did get organized with the right people in the right jobs. An old Chief Warrant Officer was head of Finance; and after No pay for two months, I flew him and two Armed MPs to Frankfurt, with a current roster of the Fassberg Personnel, and picked up cash (in U.S. Scrip) for all the troops. My SQd orderly room Sgts typed up a Pay Roll, and with a loaded "45" laid on the table, we lined up the troops and counted out each man's pay; and believe it or not, it came out even.

I finally got some great NCOs, a Supply Officer, a Lt. Adjutant, and a Ration Break-Down Lt. We fed 24-hours-a-day, in five Mess facilities, a Flight Line Mess, an Officers Mess, the SGTs Mess, and two Enlisted Mess Halls. Food was critical for the local German population, so we had to keep a close eye on it. Any German caught stealing was immediately fired. I attempted to get control of the British rations, with little luck. The Brits ran a farm, and had many hogs and chickens; since I had no capability to slaughter and prepare fresh stuff, I was never able to work a deal. Later, it was discovered that the Irish farm Officer was selling the produce on the Black Market. He was hard to convict, he got a ride on a C-54 to Northern Ireland, with the help of our Catholic Chaplain, Big, big Investigation, but Paddy slipped south, to the Free State, where he couldn't be caught . . .

Black Market was the "name of the game" at Fassberg. Everyone had a Ration Card, which was issued each month, allowing one to buy limited items like soap, candy and cigarettes. The weeds were the main means of exchange for almost everything. We paid $1.00 for a carton in the BX, and traded a carton for $15.00 to $40.00 worth of Deutschmarks, depending on the various markets and times. When Ya have an economy that is based on tobacco, you are indeed in poor shape.

I flew the Airlift in the C-54 as often as my job as Food Service Supervisor would allow; many trips to Berlin were not properly logged, because I flew with all five C-54 Squadrons, probably over 650 hours. I also flew the C-47, and was Base Check and Install Pilot in the Gooney Bird. We had a Rest & Relaxation set up in Copenhagen, Denmark, which I flew every month; and I also flew many trips to the American Zone of Occupation. Europe is small; one can fly between the borders, north to south and east to west, in short order. During my tour, I flew to: Paris, Rome, London,

Tripoli, Munich, Glasgow, and all over North Africa.

We had numerous problems, but we never had anyone get sick from eating in my Mess Halls! I had a hard, fast rule: – "Never Feed Leftover Chow to the Troops!" If I found questionable food put in the back of a fridge, I would dump it on the floor and make the one who saved it clean it up.

We had British MPs and Polish Guards, who understandably were feared by the Germans. It was said that one could pay 100 D-Marks to have a Polish Guard shoot a Kraut. It was somewhat dangerous at Fassberg.

Fassberg was a most comfortable base. The stone/mortar buildings were steam-heated, two-story barracks, with modern plumbing and hot water. There was a heated Olympic swimming pool, a three-story stone Officers Club, a tennis court, a rifle range, etc. etc. The main airfield complex were paved 6,000-ft. runways, and extensive paved aircraft taxiways and parking ramps; plus twelve large hangers that a C-54 could fit in with room to spare. There was almost no War damage, only one hangar had slight damage. The field, all areas, and buildings, were well camouflaged; the entire installation was very hard to spot from the air. The Germans had used Fassberg as an experimental base. The village and Base had all the electric power necessary. The British had moved into the best housing with their dependents; later on, when U.S. folks were allowed to bring over their wives and kids, the Brits simply went through the village, picking out the best housing, and told the Krauts to move out, taking only one suitcase and some bedding. Victor's Choice . . . Fassberg was located near a notorious Concentration Death Camp, and the Krauts got no sympathy and did as they were told.

We had quite an unusual number of Commanders. The early ones about went nuts trying to organize troops who were assigned

TDY for sixty to ninety days, then extended time after time. Here were 4,000 American Officers and Enlisted Men, hauling coal to Berlin, to keep alive Germans, in a city they had blown to dust, and were still pissed off at having been shot at and hit over the same city . . . There ain't no damn justice. The Mission was simple: haul coal to Berlin, in a beautiful silver four-engine Aero plane – Are ya nuts???

Gen. Jessie Auton, former CO of the 65$^{th}$ FTR Wing in WWII, was CO for a very short time. I knew him when I was in the 4$^{th}$ Ftr. GP. The last CO was Col. Theron Colter, who brought his wife with him, Connie Bennett, and her dog, a Standard Poodle that ate in the Officers Mess. After Connie B. arrived, the US Officers hauled their dependents to Fassberg too– just one-big-happy-family, and a bunch more mouths for me to feed.

I bought a four-door Italian Fiat in Frankfurt. It enabled me to get about and do a lot of deer hunting, rabbits and birds too. Roe Deer were all over the place; and in some areas, destroying the German's vegetable gardens. The Local Bergermeisters and Jungemeisters begged us to help control the game population, which were eating the fields and gardens. I ate many great Venison meals, prepared by German cooks who were expert at preparing Wild Game.

On the first Thanksgiving of the Lift, each man was authorized 1-1/4 pound of turkey; the Army Quartermaster shipped us 5,000 pounds of frozen turkeys. The smallest birds were nineteen to twenty pounds. Large birds exceeded twenty two pounds. The problem was that we had no ovens in any of our Mess Halls big enough to bake the birds. We rigged up some field ovens, but most of the baking was done in brick ovens, which we built using brick/mortar and German masons which we commandeered from the Brits who were building a Squash Court. I got into some trouble for it, but all was forgiven when the reason was explained. I was pretty much "The Boss," and did not put up with any infractions, backtalk or excuses. I was known as: Snake Scanlon, one of the Berliner Boys; real dangerous if stepped on. We had a most difficult job to feed over 4,000 troops, 24 hours every day; we did it, and We Did It Well!

The Berlin Airlift slowed down and stopped about the end of September and early October 1949. The coal-dusty C-54s went back to their homes. My Supply Boys and I were the last to depart Fassberg. We had a big parade, then folded tents.

When it was all over, there were some odds and ends left that required me flying people and stuff here and there in the only U.S. aircraft left on the field. I made a trip to Berlin, to deliver some Air Controllers to Templehof on one flight. Another Berlin sortie was to pick up a Baseball team. My final trip to Berlin was to pick up some USO entertainers Steve Lawrence and Edie Gorme.

## Neuberg, October 1949

Once again, I was assigned to an active flying outfit, but was sidetracked to a Food Sqdn, with the understanding that I would fly Fighters half the day and attend to Food Service Duties the other half. I drove south in my Fiat with my Dachshund, Pedro, the meanest mutt in Kraut Land, to Munich, Germany. Gasoline along the Autobahn cost $0.15/Gal. I made the trip in one day, from Fassberg to Neueberg, Home of the 86th FTR. GP. When I arrived at "Neuby," there were four Mess Halls under the direction of four Officers, Phil Rulman, Ernie Champion, Gordon Pierce, and a Chief Warrant Officer. The Mess Halls were run by the Most Excellent and Experienced Mess Sgts I have ever been associated with in my 32 years, seven months and 42 days' Military Service! There was absolutely no need for a fifth cook in the kitchen; nonetheless, I was stuck with a Ground Job – which was, as it turned out a good thing. Those Pilots who were in WWII, and were recalled to fly the Berlin Airlift were out of luck. When the Lift was finished there was a surplus of Pilots who were then literally Kicked Out. Those of us who had Ground Jobs were retained.

Neuberg AFB is located at the foot of the Alps near Munich. My first flight was in a T-6, with Capt. Moose Chandler, on 7 October 1949. I had logged T-6 time for many hours, and knew the drill, as well as about the 250 hours of P-51 and P-47 time in both War and Peace. Moose was not aware of this. Nonetheless, this Hot Shot tried to impress me with a Yank & Bank landing pattern. He dove on to initial approach exceeding max-allowed airspeed (200kts), and pitched out with a near vertical bank, idle power at 800 feet, completing the turn to final still at 800 feet, much too high and fast, and had to go around. I guess he thought he was in a P-47. The last encounter with Moose was in 1957, at Johnson AFB in Japan. Moose was dead, the crash caused by a much too tight landing pattern in a T-33, he lost control, too much rudder and side slipping near stall speed. During my first month at Nueueberg, I made twelve flights, four in the T-6, four in the C-47, and four in the P-47, now designated F-47, F for Fighter. On a three-ship F-47 formation, with Moose as lead, he attempted a formation loop, not briefed and stalled out on top; prior to stall, I eased out and recovered without a problem, while Moose and his Wingman almost collided in a spin. I knew that, sooner or later, death would result for such an undisciplined Pilot.

I just about drove the wheels off my Fiat going to each of my Mess Halls every day. I dealt with trouble real quick, and made changes on the spot, usually by making personnel switches; and in two cases, Special Courts Martial! The job of Chief Mess Officer. Was indeed a mighty powerful position, where food was in short supply and a Black Market that was all too easy to get involved in

... I accepted no excuses! I charged one of my Officers, a 1st Lt., with wrongdoing, and had him relieved of duty and immediately shipped out. I finally got a Chief Warrant Officer and two Master Sgts., to take some of the load off me. In the end, we had a Smooth-Running Food Service Outfit!

I acquired a German 8 mm Mauser, had it Sporterized, and mounted with a quick release German Sniper Scope total cost $15.00 U.S. With this weapon, I shot just about every kind of game, including Hirsch (Elk), Wild Pig, Roe Deer, and Mountain Goat. My Dad mailed me a 12-ga Model 12 shotgun, which took care of innumerous rabbits and birds. Small deer were all over the place, destroying the German's crops and vegetable gardens. The Germans could not control any game, since they were not allowed to possess any firearms. The Kraut Officials begged us to help control the game population, literally destroying their fresh food supply. I had many parties where I provided the Game, prepared by my German cooks. . . Airfield Security was again provided by an RAF Police Squadron and Polish Guards, who took no guff from the Krauts. It was still somewhat dangerous in occupied Germany.

During my WWII Combat tours in B-24s and P-51s, in 1944-1945, and later experience, 1948-1951, flying primarily the C-47, I had a pretty good handle regarding European weather, having done all my navigation and flight planning to and from Denmark, England, Germany, France, North Africa, Italy, Austria, Scotland, Greenland, Newfoundland, Labrador, and most of the U.S. I was cautious, and knew when to stay on the ground. The Neueberg Pilots were a very shaky lot, as evidenced by the demise (death) of twenty-one Pilots during the 21 Months that I was stationed there. There were a lot of reasons for so many Fatal Accidents; however, the primary cause was exceedingly poor leadership. Of the 21 Pilots killed in the F-47 and the F-84G; nineteen were 2nd Lts, one 1st Lt, and one Lt/Col, whose wife had just cause to shoot the stupid bastard.

During my time with the 86th Ftr Gp, Clair Chenault, the son of Gen. Chenault Co. of the 14th AF in China and the Flying Tigers was the Commander. Col. Chennault was, in my view, a part-time alcoholic in charge of three Fighter Squadron COs who were absolutely clueless, ineffective leaders. All of the nineteen Lts. came in as partially trained P-51 Pilots, recently graduated from Williams AFB, Arizona, or Nellis AFB, Las Vegas, Nevada. To have come from the summer sunshine desert into the cold, sloppy fall/winter weather of Europe was more than they could handle. The weather, plus the Attitude of every Man is a Tiger and/or . . . What? Do you want to Live . . . Forever? . . . Caused needless death. Flights of four F-47s would take off, join in formation beneath a low cloud ceiling, then climb in tight formation through a 10,000-foot deck of clouds; and maybe all four would make it. Those who lost sight of the leader were "on their own," and would wander about with the all-to-frequent crash, flying into rock filled clouds. I really felt sorry for their young wives, most of whom were recently married. It came to the point where so many "accidents" caused some pilots to turn in their Wings. To get back to my story, I stopped flying the F-47 when the 86th Gp was re-equipped with the F-84 jet fighter. I still continued flying the C-47 all over Europe hauling whatever was required . . . the poorly led 2nd Lts. still crashed by the numbers.

The build-up and collapse of all military U.S. forces of WWII, and the recall in June 1948 for the Berlin problem, Screwed Up the lives of many, many Service Personnel. And then, at the end of the Lift and for the good of the Government, the release from Active Duty of these Loyal Air Crews, when anybody with half a brain could see that the Communist threat was very real. So, what happens six months later? The Korean War starts in June 1950! Sooo, call out the Goddamn Reserves – Again! The Secretary of Defense was a numb-nuts named Brown . . .

Running the Mess Halls was "A Piece of Cake." I sold the Fiat, and bought an English MG, with 19-inch wheels, Fire-Engine Red, with red leather seats. It was a great auto to race about Munich, dodging the Military Cops.

My days of a carefree, MG sports car driving, freewheeling, single aviator came to an end when Francis Nuno arrived on a tour of Europe. We had been corresponding, and planned to meet in Paris. Fran was traveling with a flock of girlfriends, touring Spain, Portugal, and France. I flew to Paris in a shaky Air France Aero plane that was a leftover from WWII. Fran arranged a room for me in a Dorm at the University of Paris. My roommate was from Spain, who spoke French and Spanish; I spoke English and enough German to only confuse the situation. Frances, who did speak French, Spanish and Portuguese, found me, and we did up Paris; walked our socks off, climbed the Eiffel Tower, and had Lunch in the splendid Tower Restaurant. Paris in late July and early August 1950 was beautiful; the weather was perfect; we bought bread, wine and cheese for picnics in the parks. We visited the Paris landmarks, like Notre Dame, climbing hundreds of steps; and strolled the Champs Elysée. We hit several nightclubs, including Madame Pastiche, and The Follies, the Most Famous Nightclubs in Europe. We had fun.

I proposed to Frances on the bank of the Seine River, in the shadow of Notre Dame Cathedral. I gave her a Gold Ring that my Dad made for me as a Good Luck token to get me through WWII. She said, "Yes." Now, what to do!

It seemed quite simple to me. Since Frances said, "Yes . . . We

Should Get Married!" I talked her into coming to Munich, and arranged for Frances to stay with Lt. Champion, my Commanding Officer, and his wife Jo. First, we had to get Fran a German Visa. Then, things got complicated. Frances had to bid her traveling companions goodbye, and rearrange her return passage to California . . . Goodbye, Friends – Hello, Munich. We flew to Munich, and the die was cast . . . call parents, my Mom, Dad, Joyce and Frank Nuno. For a few days, Fran was a bit uncertain, and wanted to go home and get married in Los Angeles, among her friends and family, after I returned from my tour of duty. Complications . . . The Korean War was a factor. I had no idea as to my future as an Air Force Pilot, and where I would be assigned next! The main point was that we were in love. I pressed on, arranging a Wedding date; clearing a big hurdle with the Catholic Church, which required posting Marriage dates some six months in advance, and also cleared it with the Military and German authorities for permission to marry. With a stack of paper documents, we were all set on go. Frances and Jo Champion had a beautiful wedding dress, with all the add-ons like shoes, veil, etc. etc. . . . made to order in a very short time. I had saved my money. I had about $4,000 in the bank, a tidy sum in 1950. So I could take care of all our needs.

First, we had to get married in a German civil ceremony, according to German and U.S. Regulations; this detail was checked off on 11 August 1950. The Ceremony took place in a German Civil Court performed in German. Both Frances and I agreed that the Civil Marriage did not count, and that what counted was our Catholic Marriage in the Catholic Chapel by a Catholic Chaplain, Father William P. Connor. We were married on 12 August 1950, at Neuberg Air Force Base, Munich, Germany. Ernest Champion was Best Man, and his wife Josie was the Maid of Honor. The Chapel was full of guests: Pilots and Ground-types, plus most of the cooks and bakers from my Food Service Sqdn. I was still flying half the time, and covering my Mess Halls the other half. Colonel George Anderson, Deputy Commander, was most kind to fill in for Fran's Father, as he walked Frances to the Altar and gave her to me!

After the service, we had a small reception, with cake and candles, at Lt and Mrs. Gordon Pierce's home, in Ramsdorf, a housing complex near Neuberg Air Base. Master Sgts Romberg and Halowinski, my Food Service Troops, plus all the Fighter Sqdn Pilots, gave us a joyful send-off.

Frances gathered up her stuff, and we headed for Berchestgarten in our 1949 red convertible MG. The honeymoon was a grand occasion, and a fine start that is still in progress as I type this History!

We were married soon after the Korean War started, June 1950. We received the sad news that my brother, Lt Col Robert Emmett Scanlon was killed, 28 September 1950, He was flying a P-51 on a combat mission over Korea. That sad news put a damper on everyone's morale. The last time I saw my brother Bob was when I ferried a C-47 from Germany to Massachusetts. Bob flew a B-25 from San Antonio to pick me up. It's a long flight from San Antonio to Massachusetts returning the same day but we both enjoyed that flight.

When Fran and I returned from our Honeymoon in Berchtesgarten, we stayed in a beautiful hunting lodge, awaiting a permanent home in Munich. We were in no big hurry to move, since the Lodge furnished everything, including all our meals, a private apartment and maid service all at no cost! Since I was senior to those awaiting housing, and was a friend of the Housing Officer, I got priority and my choice of living quarters. We soon moved into an excellent apartment in Ramsdorff, a housing project on the west side of Munich, and only a short drive from Neuberg.

The U.S. was still occupying the country, and could pretty much take over any housing we needed, along with all the furniture, bedding, dishes, etc. We moved into a fully furnished ground-floor apartment that included a Full-Time Maid named Dorothea, who could not speak even two words of English! We had service for twelve in fine china, crystal stemware, silver service and a maid who could cook, clean, and sew. Dorothea and Fran hit it off nicely and she became a lifelong friend.

The Black Market currency of exchange were: tobacco, US dollars, all food items, and motor vehicles of any kind. Germany was severely in need of everything, but it was slowly recovering under The Marshall Plan. The country soon got busy recovering in a hurry.

We traveled all about Southern Germany. We had a carefree situation. I sold the '49 MG, and bought a four-door Ford, that was much more comfortable with winter coming on. We had a pretty active social life, ate at the Officers Club, and at the Haus Der Kuntz in Munich, one of the few large buildings left undamaged. Dorothea, wouldn't let Frances do much in the way of housework or cooking; however, the two got along in every other respect. I tried to come home for lunch with the girls, even though I had four Mess Halls and the Club where I could eat for free.

Fran got pregnant! We foolishly took to the hot spring baths and massages during a stay in Wiesbaden, which unfortunately, we suspected, caused Fran to miscarry. We pressed on, and avoided the many hot spring baths available in the many German towns whose name ended in Baden, which included many of the towns in Bavaria. I was nearing the end of my tour in Germany, March 1951, and was almost certain to be involved in the Korean War. I soon received Orders transferring me to March AFB, California, reporting early April 1951. I had no further Orders, other than to report to Base Headquarters.

Frances was "with Child" a few months prior to our departure from Germany, and had a bad case of seasickness, aggravated by carrying the Wee One. The ship we were on was an old Liberty Ship, from the States returning empty, it pitched and rolled like a cork. The North Sea was not very kind to landlubbers March 1951.

**RES.** Dad is wearing the Bus Driver's uniform now, the blue one, he really liked the Pinks and Greens of the old Army Air Corps but this is the new Air Force established by Congress 18, September 1947. Dad always looks sharp in a uniform, whatever it is. I have a picture of him standing in front of his MG in a uniform with creases so sharp they could cut you. Dad is sporting a Limy fighter pilot mustache now, he smokes filter-less Salem's; he tamps the ends on a large solid gold ring he wears on his middle left finger, it has a 3/4 inch square flat face with his initials engraved in it: JJS, his Dad made it for him for luck. This is the ring he proposed to Mom with. His call sign is Snake. He is rated Senior Pilot 18 December 1950. And promoted Captain January 1951.

I have several pictures of Mom while in Germany wearing out-sized clothes uncomfortably surrounded by a carpet of dead animals and fish.

Brother Bob, who I was named after, was a Major in the Regular Air Force in 1950. He was rated Sr. Pilot. Bob was married with three kids Pat, Dan, and Marianne. He was the squadron finance officer for the FEA stationed in the Philippines. The squadron had just transitioned to P-80 jets; they had turned over their P-51's to the Philippine, and Japanese Air Forces. The squadron had to abandon their shiny new jets that had low endurance and no ordinance hard points. They recovered and flew their old P-51's with 20mm cannon, rocket, bomb and napalm racks to Korea. They flew to Korea to defend the Pusan Parameter within one week of the North invading South Korea, forming the famous Truck Busters, the 18th Fighter Bomber wing of the 5th Airforce and began to fly interdiction missions. The squadron immediately airlifted everyone down to the cooks and typist to a primitive PSP field where they operated out of tents. They named their base, K9, Dogpatch after the Li'l Abner comic strip. A South African squadron fought right alongside the USAF pilots. Chappie James was a flyer with the Truck busters.

Hit by ground fire, Uncle Bob augured in near Mukdang-Ni while providing close air support during his first and last mission on a 100 mission tour, a witness observed that he crashed inverted after a strafing run. Bob's remains were later recovered after the Inchon landing and a push north by the 8th Army Pusan forces. He was promoted to Lt. Col posthumously and is interred in Los Angles.

One story left out in Dad's Airlift narrative is about an incident that happened in Paris. First of all Dad is always packing heat. He always caries a Harrington and Richardson .25 auto. Grandpa Scanlon carried it through World War One and Dad carried it on every mission he flew as a backup gun. While wandering around Paris one night, some toughs carrying knifes and clubs jumped out of an alley and asked for his wallet, Dad reached back in his hip pocket and produced his pistol; they ran away. He later shot out the front tire of a pickup truck that was trying to run him off a dirt road in Montana but that's another story. Dad was a tough customer.

Another story is that in the All Hands Club at Fassberg, he had a sign posted with the rules of the club with the admonition that offenders would be shot. A couple of offenders were actually dragged out of the club by MPs and shot, by him, with a BB gun and banned from the club.

# 1951 Lake Charles, Louisiana

**JJS.** We docked in N.Y. City, and were transported to the Immigration station, then over to the Traffic Office, where we booked Commercial Air to Los Angeles, with a stopover so that I could pick up a Studebaker that I had purchased through the Base Exchange in Germany. Poor Frances was all alone, had not seen her family for over a year, and was not feeling well at all, being pregnant. It was real dumb of me to abandon her, but I had to pick up the Studebaker at the factory in Chicago. Frances held up pretty well, glad to be back with the family; both Nuno and Scanlon Elders met her at the L.A. Airport! Frances was soon feeling pretty good. She was relaxing, back in her old, familiar surroundings, eating her Mom's cooking. The Auto, a Studebaker Champion, was the worst car I have ever owned! Prior to getting the car, I had a short visit in Chicago, with Cousin Joan and her parents, Francis and Anastacia Brady, who I had not seen since about 1930. I took Route 66 from Chicago to L.A., reported to

HQ, March AFB, and was told to stand by, awaiting assignment to who knows where.

Personnel was in a state of flux; it seemed that no one knew who was on first, much less second. Since I was rated a Food Service Supervisor, a Four, Single and Twin-Engine Pilot, with lots of experience in all four ratings, the powers that be took their own sweet time finding me a job, which was okay with me. I spent most of the time in Los Angeles, at the Nuno's and drove out to March Field two or three times a week . . . in Limbo.

I was assigned to the 44th Bomb Gp, which was being formed up at March Field to fight in the Korean Conflict. After Bob died, my mother Viola wrote a letter to General LeMay declaring me her Sole-Surviving-Son, exempting me from combat. I would see no further action. So, I was assigned The Food Service Supervisor of the bomb Wing. A, Wing with no Base, no Aircraft, and about one third the personnel required to man a Wing. Things were in turmoil. I had reported to March AFB in April, and was finally assigned to Lake Charles, Louisiana, an abandoned WWII B-26 training facility, with tarpaper buildings, vintage WWII in June. What a mess. I was in Limbo, in California, from April through June, with one-third of a gang of Cooks and Bakers who had nothing to do. I had a M/Sgt Gotchalk, who was a POW in Japan, and was tough as Wang Leather; he pretty much had the Troops in hand while we all waited three months to make the move to Lake Charles. The Food Sqdn slowly gained personnel, to include Warrant Officer Baxley, CWO Summers, and a real fine 1st/Lt. Most of the Enlisted were raw Recruits, with only a few experienced Cooks and Bakers.

There were enough Supervisor-types on hand, so I spent most of the weekdays and all my weekends in L.A. Frances was Due to deliver our first child in late September 1951; so the delayed move gave me time to go to Lake Charles, and case the housing situation. I fell into a real good deal. The Government backed contractors to build a bunch of small homes close to the back gate of the Air Field. I bought one of them. A brand new home for $8,000. I purchased it with no down payment and no payments due for six months. Just about as close to free as you can get. Lake Charles, in July, was hot and humid, so I bought a Sears three-quarter-ton window air conditioner that kept the Little House Cool and Dry.

The Lake Charles Base was very slow in working up. The 44$^{th}$ and 68$^{th}$ were slowly accumulating Personnel who had nothing to do. The two Bomb Groups had no Aircraft, and did not know if they would get B-29s or B-47s. The runway was being lengthened and repaired. The rest of the Base buildings and Hangar Repair Facilities slowly came along.

The only aero plane was a C-47, and I was the only Qualified Instructor Pilot. Most of the Pilots were WWII Pilots, recalled for the Korean War, who had not flown for the last three to five years. I spent about equal time between running the food service operation and the only flying operation, which consisted of that One C-47 and three other Gooney Bird Pilots.

Lake Charles was a dangerous Town. We had some Service people killed over minor mix-ups. There was a small element of "K-Jons," called "Red Bones" . . . a mix of Black, Indian, and Who Knows?

Since there was a War ongoing, a few officers carried a 38 service pistol. I once threatened to leave a Staff Meeting unless the "dumb ass" Colonel stopped playing with his weapon. I was a Captain. He was a Colonel . . . Much Embarrassed, and forced to put his toy away. From then on, we spoke very little.

Our first child, Kathleen Joyce Scanlon, was born on September 30, 1951 in Los Angeles, Queen of Angels Hospital. All went well. We had a beautiful blue-eyed blonde baby girl. I was able to take a week's Leave from the heat and the mass confusion at Lake Charles in L.A. Of course, even though it would be some time before things were set and ready in the House at Lake Charles, more time would be required before we were able to move a new baby to Louisiana. Kathleen was baptized 11 November 1951. Frances and I simply cannot recall exactly when the new baby got to L.C. I believe that Aunt Nina Aguilar, along with Fran and the Baby, took the train to L.C. sometime in January '52. Nina was a big help. She stayed with us at least a month.

Things were rocking along pretty smooth until Gen. Lemay set aside the Time-Honored-Tradition wherein each Unit/Sqdn would furnish their share of Kitchen Police. Lemay decreed that no Tech Personnel would pull K.P. This occurred at a time when many, many low-mentality people were drafted, and were almost untrainable lamebrains. So when this Policy was put in effect, I was assigned at one fell swoop one hundred and fifty Black Troops who were stuck/rated permanent KPs. Needless to say, all the Sqdns unloaded their Category 4 Untrainable, which caused the Food Service Units to almost collapse. This situation was pure and simple a Catch 22. These men could never get promoted "because, in order to be promoted, one must attend a school". They could not attend school because their IQ was too low. They simply could not pass the tests. We usually had two in Jail, six to eight AWOL, and always a few late to report for duty. I tried to object to this Unfair Policy through the Inspector General. My Reward was a quick set of Orders to attend B-29 Combat Training at Randolph Field, Texas. The Orders assigned me TDY which would not allow funds to move my family, nor my household goods. I informed the Base Commander, in a more or less insulting manner that I was ready to go after He had completed a Combat Tour, informing him of my WWII B-24 and P-51 Combat and Berlin Airlift tours. The CO made the usual excuses; my reply was, "Now is your Chance." I saw this transfer as my escape from six years of Food Service, and a return to flying full-time.

The Lake Charles Tour was a very sad costly mistake for the Service. The Base was rebuilt; the runway lengthened, hangars and repair facilities improved. However, to my knowledge, the whole Base was shut down and abandoned after several floods. Our Louisiana tour was not all bad; we had a new house, a new little baby girl that kept us busy and I was home most every day. Most of the flying was local, or out & back the same day. Most of the local flying was to re-qualify Recalled Reserve pilots, recalled for the Korean War, or Pilots who needed to log at least four hours a month to qualify for Flight Pay. I built a fence in the backyard, and attempted to grow grass. Lake Charles was dripping wet, infested with most known and many unknown bugs and of course every kind of mosquito. It was so wet that Crawdads populated the backyard, they built their little castles at least one or more every square foot.

**RES.** What can I say, Kathleen and I missed being a Cajon by a matter of weeks. Kathleen is the apple of her Daddy's eye a beautiful blue eyed blond. I'm born 13 months later December 26, 1952. Mom was in LA to be with her Mom for the holidays; she thought that she just had had too much Christmas dinner but it was me. I was borne in Queen of Angles Hospital in Los Angles. I was a breach baby. The umbilical was wrapped around my neck and up around my ears; that made my ears stick out for quite a while afterwards. The Doctor was very skilled, he managed to rotate me as he pulled me out; so you could say that I was screwed the day I was born! Perfect, a sweet little boy for Mom and a little princess for Dad. Mom is soon pregnant again with a spoiler.

# 1953 Randolph, San Antonio, Texas

**JJS.** Like I said, L.C. was a big mistake! Here we were, stuck with a houseful of furniture, a little baby girl, a pregnant wife and me holding orders to report to Randolph Air Base, to pursue B-29 Training for the Korean War with no authority to move my family. I had no intention to play this Game. We packed what we could carry in the Studebaker, left the keys to our house with CWO Baxley, our trusted friend, and left the air conditioner on, to keep the mold and bugs at bay. We departed in early August, for Grandma Nuno's in Los Angeles, via the most direct route through West Texas, New Mexico, Arizona, and over the hills to San Diego, to get to some cool ocean air, and were welcomed by Grandma Nuno. I do not know if we could have made it without Family assistance. I stayed a few days, then reported to San Antonio, Randolph Field.

Texas beats the hell out of Louisiana for weather conditions. Randolph AFB was the only B-29 Transition School in the Air Force. I was assigned to the B-29 Transition School in August 1952,

## 1953 RANDOLPH, SAN ANTONIO, TEXAS

most of the students were recalled Pilots. About half the Students had not flown A/C since WWII. Due to my previous B-29 time in 1947-1948, it took me only the minimum number of hours to re-qualify. The B-29 was an easy bird to land, "If ya know how!" The greenhouse cockpit had many windows to look through; the trick was to not gaze through only one, but to look at the runway and let the picture develop as a whole. In short order, I completed the B-29 course with such flying colors that the Transition Course C.O. requested that I be reassigned as a B-29 Instructor Pilot.

We would form up for an AM Briefing at 0500, get aircraft Assignments, pre-flight the big birds, and take off by 0700, for about five hours flying time, 2.5 hours Instrument and Navigation, and 2.5 hours in Landing Patterns for a minimum of ten landings.

The students in this program were extremely varied, some experienced, some Green Beans, recently graduated 2nd/Lts., some eager to fly a four-engine monster, and some scared shitless. There were a few pilots that refused to fly this big bird. Some simply took off their wings and dropped them on the Commander's desk! I had a few who would start shaking as soon as they climbed the ladder to the aircraft cockpit.

We generally flew five hours, 2.5hrs Air Work, 2.5hrs Traffic Patterns, Take-Offs, touch and goes, full stop, and taxi-back; every other landing, was a simulated emergency drill. When you put ten B-29s in the pattern at the same time, it required sharp Control Tower, and real sharp Instructors to accomplish the task, especially, at night! I ended my B-29 tour logging over 1,400 hours, most of which most as Instructor Pilot.

We never had a B-29 accident in the West Field Transition School while I was in the program. We had many, many engine failures, run-away props, flat tires, etc., but no crashes, no one hurt, a major miracle for 1,200 hours in a B-29! My last primary task was Chief of the B-29 Instructors, teaching potential Instructors the ins and outs, the proper handling of such a large, slow-to-respond Bird. I also took over, as Chief Pilot, for Tow Target activities, towing 9x45 foot banners at 20,000 feet and one mile aft, for Fighter Aircraft to fire rockets at. I was kept pretty busy. The reason for the tow-target job was a serious screw-up on the part of the Crew Training School on the East Field. Poor procedures, inadequate preparations, and Crew duties were not clear; this resulted in several deaths when people Bailed out and landed in the Gulf. I towed targets for the first air-to-air rocket Competition at Tyndall AFB, Florida, where F-86, F-94, and F-89s fired live armed rockets; Quite a big deal! Dangerous, too!

My B-29 instructor at Randolph was Pappy Craddock, who was Instrumental in having me transferred to the B-57 Canberra Twin Engine Jet and the T-33 Jet Trainer. Pappy Craddock was a Fine Officer who smoked about three packs a day, and expired while stationed in Japan, from lung cancer. The Cadre of B-29 Pilots in the Transition School were the most professional gang that I had the Honor to be associated with. However, the Jet B-57/T-33 corps at Randolph were pretty weak and accident-prone. There were no Dual-Control B-57s, which probably caused some crashes; however, pilots switching from prop planes to higher-speed jets was, for some, too much to handle.

During 1947 and 1948, I had flown, as pilot in Command of the following Aircraft: L-5, L-13, AT-6, AT-11/C-45/AT-7, B-25, B-26, B-24, B-17, B-29, C-47, C-46, C-54, P-47, P-51, P-61, Spitfire Mk 9, PT-17, AT-6, BT-13 plus many others and felt highly qualified to fly any propeller driven aircraft that I could get started! Having logged 300 hours combat in Bombers and Fighters, The

## 1953 RANDOLPH, SAN ANTONIO, TEXAS

Commander transferred me from the B-29 Transition Sqdn. to the B-57, Canberra/T-33 Jet Aircraft Training Sqd at Randolph AFB San Antonio, Texas. The idea was to include both a Bomb/Attack/Fighter and level Bomber into the training, to use the full capability of the twin jet engine Canberra.

After my assignment as an IP, my status was changed to Permanent Change of Station, and I was then authorized to ship my household goods from Lake Charles to San Antonio, and finally bring my Wife and Kids to Texas. Frances was pregnant with Bobby, and was staying with her Mom, Joyce in Los Angles. It was weeks after my son Bobby was born, 26 December 1952, that we again lived as a family in a home. I bought a house for $10,500 on Breese Blvd., in San Antonio. Frances did not drive. I flew at least three days a week all day long; so I was pretty busy transporting my wife and kids to church, shopping, to doctor appointments, out-to-eat, to swim, wherever they needed to go! The trips between San Antonio to Lake Charles and Los Angeles were a real task, in my 1950 Studebaker. The car had no air conditioner. I was driving back and forth through a landscape similar to the surface of the moon.

We did not do it all by ourselves. Aunt Nina Aguilar came to Lake Charles, and Grandmother Nuno helped with Baby Bob. We also hired part-time housemaids for all the heavy housework. We also had real fine, helpful neighbors everywhere we were stationed. We had lots of help without asking. A pregnant Mom even draws the best out of strangers. We were pretty content with our situation. I enjoyed the flying and Fran was real busy with the babies who needed constant attention. Our neighbor Ruby was usually available to baby sit, so we could attend parties at the Officers Club. Frances certainly had her hands full. Not too long in the swift passage of time, we produced two more little Boys! James Francis arrived on

7 March 1954 and Stephen Gratton Scanlon was delivered on July 11, 1955. Both boys were delivered at Brooks Military Hospital, San Antonio, Texas. The children were a grand delight. They were all healthy, handsome, and about as active as a hatful of grasshoppers. Kathleen and Robert were born in Los Angeles' Queen of Angels Hospital, at about $500.00 each. Jim and Stephen's births only cost the meals Frances ate while in Hospital. The amortized cost being about $5.00 per pound per Baby. A pretty fair price, eh! During my Military career, we were cared for exclusively by Army, Navy, and Air Force facilities with pretty good results in all instances, even considering that we eventually swelled to a family of ten.

I was in the first Cadre of B-57 instructor pilots at Randolph Air Force Base who checked out in the single Canberra available. The Jet was equipped with only one pilot position and control set. There were only six Pilots selected. The "check-out" was a Blindfold Cockpit Check then away we flew! Commercial airline flights took us to Baltimore, and the Glen L. Martin Factory, where we had a party and reviewed aircraft delivery stuff. The following day, all six of us took off for Texas. Only John Giese and I flew non-stop to Randolph Field. The welcoming committees were quite disappointed. The B-57 School Commander, Pappy Craddock, was hard-put to explain the four missing birds. They finally arrived the next day. The start-up of the Twin Jet B-57 Canberra Transition Program was, to say the least, a can of worms compared to all the other training programs I know of and I have a pretty extensive knowledge of training programs! First off, the selection of students was "hit and miss". There was no established criteria regarding prior flying experience for selection. Some Pilots had no multi-engine time, and had to be put through a B-25 twin-engine course. They then had to be trained in Jet Engine operation in the T-33. After that they were required to fly Solo day and night flights in the T-33. Only after that were the students introduced to the B-57; which, in the beginning of the training program, had only single-control B-57s. The instructor pilot that checked and instructed the student in the T-33 was usually not the IP checking out the same guy out in the B-57. There was no continuity or planned pairing of students and instructors. This arraignment caused lots of problems. After finishing B-57 Ground School, we would give the student a Blindfold Cockpit Check, while sitting on the canopy rail; show him how to start the beast, pat him on the helmet; then the IP hopped in another B-57 and called on radio, "Follow me!"

The Canberra was an easy bird to fly, considering the following shortcomings. The J65 engine had excessive delay from idle to 80% power followed by immediate, sudden increase to 100% thrust. You had to wait some ten to twelve seconds to attain equal thrust on both engines. If ya hurried, you ended up with one engine spooling up to 100% while the other was slowly passing 65%. The result was: you end up sideways and off the runway. Controls were a complicated and heavy arrangement of push/pull rods connected to bell/crank torque tubes, and a horizontal stabilizer controlled by a jack-screw with a clutch arrangement. The horizontal stabilizer caused several high-speed crashes. The eventual cause was hard to determine. The aircraft was soon restricted to 250 knots below 10,000 feet until a fix was had. The solution was to install a ten-times-stiffer torque tube to the elevators, so that, if the horizontal stabilizer slipped, you could still control the pitch angle. The 250 knot speed restriction lasted over two years!

In addition to the stabilizer problem the bird had: no nose-wheel steering, no anti-skid wheel brakes, no rudder-boost, a slow-roll rate, a cockpit only pressurized to a maximum ceiling of 18,000 ft., inadequate windshield defrost, and very poor visibility to the rear. The B-57 had good range, and was able to fly non-stop from California to Hawaii, with bomb bay temporary tanks. Airspeed was not fast, in general; cruise was 420kts, with a Mach limit of about .78. The early units were plagued with many accidents. The B-57 school folded in November 1955. The instructors were divided up and assigned to Utah, Arkansas, France, and Japan. I was shipped to Japan.

There were a few notable characteristics worth reviewing from the training at Randolph. Most notable was the jet engine's slow and uneven spool-up. An old pilot, Wreckem Beckem, who was famous for previous crack-ups, jammed on full power instead of waiting for thrust to equalize. His left engine advanced real slowly, and the right engine hit 100% real fast. Wreckem Beckem messed about, pulling and pushing throttles. He ended up 150 yards left of the runway, with the nose parked in the gate to the picnic park at the far southeast end of the runway, in a cloud of dust. I was the first one to get to him. I asked him what happened. He said: "Ya know, this Bird almost got away from me!"

The main reason that the transition Canberra program even progressed at all was due to a sharp Major breaking his leg. Grounded, the Major took over all flight schedules. He installed strict compliance to both the T-33 and B-57 ground and flying schedules and backed it up with Article 15 Military Courts and Fines. His rules were, thereafter, followed to the letter.

To review: student experience factors were not high enough, the horizontal stabilizer and elevator torque tube problem was exceedingly slow to be Aerialized which restricted low altitude flying to 250kts. The primary use for the Bird was low altitude Toss Bombing going in on the deck at 430kts, with a Pitch-Up to toss an Atomic

# 1953 RANDOLPH, SAN ANTONIO, TEXAS

Weapon up to about 10,000 feet and escape via Immelmann or Loop Recovery from Blast/Radiation. Col. Chuck Estes and I held the bombing record, with five practice bombs all within 500 feet well recorded in Korean Records.

I flew the B-57 over 2,000 hours at these Gps: 3$^{rd}$ Bomb Wing Japan, 4677$^{th}$ DSES Hill AFB Utah, ENT AFB Colorado Springs Colorado, Griffiss AFB, Rome New York, and of course, Randolph AFB San Antonio Texas. I also participated in Air Shows and Aerobatic Demonstrations at many locations. The Bird had a sixty-five foot wingspan, and the wing area of a dance hall. I would perform relatively slow-speed loops, Immelmann and Split 'S" maneuvers well below 4,000 feet, at no more than three Gs. I also flew the B-57D with a 110' wingspan up to 72,000', wearing a Partial Pressure Suit.

**RES.** Kathleen wasn't too happy to see Jim arrive; threes a crowd and another boy at that. Dad is TDY a lot, he plays with us when he's home. We have old 9mm films of this time that show Mom and Dad's family plus a lot a folks I don't know spoiling the heck out of us. Some of these films also show a grumpy Kathy already seizing Jim by the throat and patting him just a little too hard on the head. Kathleen seemed to like Stevie, he is a happy good-natured boy. Everything for me is: wading pools, bunnies as pets, and Texas sunshine. Jim is four and I'm five-years-old when we leave Texas for Japan.

# 1957 Johnson AFB, Japan

**JJS.** The move to Japan was not too smooth. In the first place, Glen L. Martin was slow on delivery. There were no B-57s in Japan. We had to fly our Birds over from the factory to Sacramento and have ferry tanks installed. We then had to buck the prevailing westerly winds from California to Hawaii. From Hawaii we flew to Guam, then to Okinawa, and finally to Nippon. Assembling qualified pilots, the necessary Ground Support People, and liquid oxygen supplies at each leg of the ferry route was very slow.

This transfer, being a PCS (Permanent Change of Station), we had our families, household goods and cars, to move across the Pacific Ocean. Johnson was a very small Air Field, with a slick, 6,000 foot runway that had a cliff on each end. Johnson Air Base was occupied, at the time, by three B-26 Squadrons the 8th, 13th and 90th organizations left over from the Korean War. It was presumed that the unit organizations would remain, junk the B-26, and convert to the B57-B. The main problems were: no Maintenance Hangars,

## 1957 JOHNSON AFB, JAPAN

very limited Family Housing, and the base was a long way off from everything! About this time, Air Transport & TAC Air started pre-programming personnel transportation of Pilots and Navigators way, way ahead of time, needlessly separating us from our families. So much so that we sat on our ass in Japan, with nothing to do but wait to re-cross the Pacific in order to pick up and deliver the B-57s from the factory in Baltimore, and fly them halfway around the world, to Japan. In the meantime, our families were stranded and unable to join us until we arranged off base housing because there were no open billets on base.

I had an order to transfer to Japan the week before Christmas 1955. This was real stupid, and I did my best to change my reporting date until after Xmas, without any luck. Several of the B-57 Instructors had to miss Xmas with their families. I had to get busy on arranging the necessary details for the move to Japan because time was short.

The Nuno in-laws provided my family's temporary housing. Franny and the kids stayed with Joyce Nuno. She let us stay in one of her apartments in Inglewood. The apartment was a big gift to Frances and our four youngsters: Kathleen, Bob, Jimmy and Stephen.

We got the moving process started by packing and shipping some necessary household goods and some winter clothing we would need when we got to Japan.

The six of us piled in the Studebaker, and drove to Los Angeles via El Paso, Tucson, Phoenix, and Palm Springs, arriving in LA with our stuff early December 1955. The transfer to station had to be accomplished before 10 March 1956.

I had to drive back to San Antonio, to arrange packing and shipping of Hold Baggage, and all Household Goods, we are shipping a piano! I have to: arrange the rental of our Texas property, give our pen of rabbits to our neighbor Ruby, and do whatever I needed to do to get the place ready to rent.

I drove back to L.A., through the same dull, monotonous countryside, and spent a few restful pre deployment days with my family and our many helpful friends.

It took three days to arrange to ship and deliver our Studebaker to the Port of Oakland, California. I went from Oakland straight to nearby Travis AFB, and got in line for a seat on a Boeing four-engine prop job with canvas seats. It's a real long haul to Japan via Hawaii, Midway, and Guam. I arrived on Base and checked into the BOQ on New Year's Eve, completely zonked, pooped-out, frustrated, and starved. I was assigned a Room next to the Base Lawyer, who staggered in drunk at 2:00 AM. My new neighbor got sick and puked until I thought he would be dead on the floor. Happy, Happy New Year!

I flew the Douglas B-26, to get enough time for Flight Pay, and spent the rest of my time arranging for off base housing so that I could house my family when they got to Japan. I finally gave a Jap builder $1,500 U.S. Dollars to construct a small home with barely enough space for the six of us. The deal was, when Base Housing became available, the builder would return the $1,500, which he did, of course, after I had paid him a reasonable monthly rent. We lived in that tiny bamboo abode for over a year.

I finally received permission to ship our stuff and bring my family overseas, but that plan did not quite meld, permission was not enough. My family beat the pots and pans with Joyce for some time. My presence was required to get things moving. I received permission to return to the States and escort my family to Japan. So, off I flew, once more, across the Pacific, sitting on a canvas chair in an old C-97 four-prop Aero Bird. I arranged rail transport from L.A. to Oakland, and the Military bus took us to the Port facility in San Francisco. The train trip was enjoyable, but the rest of the trip was *the pits*!

Our temporary military housing was completely unsatisfactory, so I moved to a nearby, very nice hotel until the call to board ship with four little kids and a lot of baggage. It was difficult to do baggage-drill so many times since I was the only porter. We were finally escorted to our two cabins, each a four double-deck bed with only a washbasin in suite. The toilets were one deck below. Frances, Kathy and six-month-old Stephen, in a play pen, were in one cabin. I had a cabin next door with Bob and Jim. We transited from San Francisco to Seattle where we picked up 600 Army Troops along with a flu bug. We then headed west. Most of the passengers were sick with all kinds of ailments. Frances was flat on her back for ten days with Flu; Kathleen was quite ill from head to tail, Jimmy had ear problems, and I had a cold. Since most on board were sick, we had no help other than some excellent food service. There were about 800 Military, with dependents, trapped aboard this MSTS (Military Transport) WWII Liberty Ship/Scow for twenty two days. It was a sick ship, crammed full of women and children. The damn boat went past Japan and docked at Inchon, Korea where it off-loaded the Army personnel. It then sailed through the Jap Inland Sea, to Tokyo, the bitter end of the voyage I could write a book about the trip, but it would make ya sick!

The Vessel's arrival in Tokyo Bay was not well planned; there were no docking berths open, so there we sat for another day before we were able to set a foot on land for now over twenty three days. Our sponsors met us, and hauled us to Johnson AFB, located about 30 miles north of Tokyo, where we were well taken care of. Our stuff was delivered to us all in good shape. So after twenty five days traveling, plus other delays, we were home. Home was a 900 square foot, three bedroom house, with a small kitchen/dining room, a front room with a fireplace separated by sliding bamboo and paper doors. Oh, Boy! There was one very small bathroom for the six of us. It snowed the first three days after we moved in. Everyone got well again after getting off the ship.

I was assigned to the 8th Bomb Squadron for about six months, then I transferred permanently to Headquarters 8th Bomb Wing, as Chief Pilot, Flying Safety Pilot, Instructor Pilot and Flight Examiner. I was a busy boy, flying almost every day. I flew primarily as Instructor, Flight Examiner, and did instrument checks. There were very few really outstanding pilots, and even fewer ECM/Navigators. Chuck Estes was our brilliant Chief Navigator. Pilots outstanding were Bill Jessup, Howard Ice, and Bill Bunting, to name a few.

We moved into On-Base Housing after a year in the rice paddies. We had only two maids during the tour in Japan; the first was a loser. Our second maid, Yoko, was a real fine person; she had excellent relations with all in the family, especially our children. Yoko worked long hours, with very few days off, and never failed us in any way. Yoko was with us at Tachikawa Air Field, when we departed Japan with our six youngsters; she was our maid and friend for about three years. Col. Ahern, the 3rd Bomb Group Commander, was also present to see us depart. Col. Ahern was the 3rd C.O. of the 3rd BGP; others were Col. G. Y. Jumper and Col. Jim Tippet.

I was promoted to Major not too long after number four son, Timothy Michael Scanlon, arrived at the Johnson Air Base Hospital May 7 1957. We were all real busy, eating, going to school, eating, playing, and eating. At some point during the day and night, the kids stopped eating and went to sleep. I got to relax and breathe easy at 40,000 feet, in a B-57 Canberra, thing were good.

Life in Japan was very enjoyable, and quite inexpensive. Our Base housing was large enough for all eight of us to have their own space. The Government provided almost all our necessary and desired needs. The Hospital/Dental facilities were first-class. We had a real fine swimming pool, a gym, baseball fields and tennis courts.

The main drawback was the very short and slippery runway. Johnson Air force Base was adequate for the small, slow-speed aircraft of the WWII Jap Air Force, but barely met the USAF Standards. The weather was the same as Washington, DC. It was hot and damp, in summer, cold, cloudy and wet in winter. I was ready to go back to the USA! I was asked to extend my tour to four years. Another little Scanlon happened on 19 June 1958. An allowance was made regarding our departure date to accommodate the newborn baby.

Little William Patrick was born June 19, 1958, just two days after my thirty fourth birthday. Bill's birth kept our record in line with having two babies at each transfer. Kathleen 9/30/1951 and Robert 12/26/1952 in Los Angeles California/Lake Charles Louisiana, then James Francis 3/7/1954, and Stephen Gratton 7/12/1955, were born at Brook Army Hospital San Antonio Texas.

**RES.** Not everything is rosy, Dad is absent a lot on TDY; he is on alert in Korea. The B-57 is the only jet bomber in the inventory at this time, it has a spacious innovative rotary bomb bay adopted by many subsequent airframes. The bomb bay is a perfect fit for the 17 foot long Mk 7 tactical nuclear bomb. Nukes are not allowed in Japan so Quick Strike birds are stationed on alert at Kuansan, Korea.

The 3rd bomb group maintains a regular rotation to Korea. Dad has been a Reserve Officer to this point; he is made Regular Air Force April 1958 and gets his Command Pilot Wings 11 Dec 1958. He is promoted Major January 1959, he is a real hot shot. Dad and Robert Mikesh (later a director for the National Air and Space Museum and author of a book on the B-57) spend a lot of time on the Mito bomb range in Japan practicing their cast bombing technique and escape maneuver to successfully complete what would be a suicide mission otherwise. This is serious stuff. Johnson airbase is now Imura. It is considered the Air Force Academy of Japan.

Memories of Japan are sketchy but several things stand out, like all the shots, getting poked in the thumb with a paper encased lance for a blood test to get my dog tags, now my new prized possession, I gave blood for them, they are just like Dad's, except that they hang down to my crotch, and the horrible trip over. I remember boarding the ship and pushing off, with all the streamers and horns, people waving from the dock, passing under the golden gate. We kids are harnessed up like huskies, leather chest straps and arm loops with buckles and a leash ring in the middle of your back. All the time we were shipboard and not sick in our bunks we were on a leash. I remember a lot of people being sick and being sick myself but not as bad as everyone else. I remember how long it took to get to where we were going, it was the proverbial "slow boat to China" and being very popular with the crew and the ROK Troops on board. It was February 1956 when we left San Francisco passing under the Golden Gate Bridge. Jimmy had his birthday, March 7, on board. The steward made him a huge sheet cake that was shared out. The ship was gray, the sky was gray, the ocean was gray, and our faces were gray. We were in ballast riding high and rolling like crazy. It was a Personnel Transport with us as supercargo. I've since been to

## 1957 JOHNSON AFB, JAPAN

sea and sailed many a small boat and never get sea sick; I attribute this to my early acclimation on this trip. That trip set the bar pretty high. Everyone on board was in uniform except us, there were no other kids to play with for a month. To keep whatever ailment was going around we were isolated. We were quarantined in port when we arrived in Yokohama. A medico checked us out before we disembarked.

I remember the little place in the country, the rice paddies and the farmers. Men and women stooping in the sun gently massaging the mud around the seedlings with their hands. The people working in the fields were old, young people weren't doing this traditional work.

You could see Mount Fuji from our house. Dad's squadron mates give him a signed copper lined honey bucket with a B-57 circling Mount Fuji painted on it for his nuts after he illegally buzzes and loops the top of the sacred mountain inverted in a B-57. Dad flew acrobatics for Armed Forces Day one year. Prior to him taking off Dad helped me climb up into the bomber's cockpit and had me sit down. Dad then told me to flick this toggle switch. When I threw that switch there was a tremendous explosion and a lot of smoke came pouring out of the starboard engine. I was ready to bail. Ha Ha, big joke it's just the explosive charge starter cartridge, a giant shot gun shell that generates gas at a rate sufficient to air start/get the turbine spinning. Good thing Dad has to sit on a parachute pack when he flies, he will need it. I love the B-57 it is an elegant airframe, one of the most beautiful planes ever designed, it is the new Mosquito. It has low wing loading; it flies like a fighter and will out turn a fighter above 40,000 feet due to its low wing loading, the bat wing is the size of a dance floor and the bird carries a greater bomb load than a B-17 with a crew of two instead of ten. The RAF increases the cord length inboard of some more powerful engines to create their high altitude phot recon version while Martin extends the wing spar outboard of some jacked up J-54 engines to give the bird an altitude of 70,000+ feet. Dad always complained of too many trips to 70,000 feet in a constriction type partial pressure suit.

The B-57 is a work of art, there is no USAF equivalent at this time. It is a sleek streamlined beautiful cylindrical airframe a harbinger of future airline designs especially the A model with its faired canopy, like the original English Electric design. The B-57 has a rotary bomb bay, 4 X 20mm cannons that can be angled downward to allow level flight staffing and racks for 5 inch rockets. Prefect for the night interdiction role with its high loiter time of seven hours plus. It is to replace the A-26 plus it has nuclear weapons delivery capability. The plane is painted a sinister matt black with red accents.

Mom has a good time in Japan when she is not in labor.

## Jim loses his front teeth

I remember distinctly the day when Jim lost his front teeth. We were playing cowboy of course. We were in our base housing unit. Dad had an overstuffed arm chair rocker with a lever that would lock and unlock the rocker. We would get on the arms of the rocker and pretend we were pilots in a plane or cowboys on horses. We would rock back and forth wearing dad's old WW II leather flying helmets or cowboy hats, with leather chaps and vests waving our pistols, yelling Yee-Ha and Hi-Yo-Silver, for hours on end. We would also have violent bucking/rocking contests to see who could stay or, or better, buck the other rider off. You had to pull the locking lever to mount and dismount, that was the rule, so the buckaroo who had the lever horse was in charge of stopping and starting, or you could just agree when to stop and go which never happened. We were fighting ove who got the lever horse; Jim kicked me, Jim was big on kicking, he always opened with a kick. I wasn't expecting a kick, so early on, so he got me good. He took off running and had a couple of strides on me, since he kicked me in the gut, but I and hot on his heels. I catch up just as Jim slows to make the corner into the parent's room. Just inside the room I sweep Jim's trailing foot with mine, onto the back of his other leg. Jim trips and falls for ward smashing his face on the edge of Mom's/Grandma Scanlon's cedar chest at the foot of her bed. Jim howls and spits out his four front baby teeth; he is bleeding, his lips and nose are a mess. I feel really bad, that was a cheap shot I pulled. Mom and Yoko are there instantly, Jim can't tell them what happened, he's bawling so much. They look at me. I shrug my shoulders in that annoying manner that Dad hates and say "I don't know, he was running in the house" which is true, running in socks on wood floors can be dangerous. Jim never talks to me about this incident, he doesn't have to; all he has to do is smile at me, if you can call it that, so we get off to a bad start, only fourteen years to go. The good news is , Jim might have fron teeth in five or eight years. We fought a lot. We have 8mm movies of some of the action where Kathleen slaps Jim around and I'm sporting a black eye in a few 35mm stills.

    Mom started me off in Kindergarten right away even though I was only four-years-old so I end up repeating. I was devastated, Kathleen would be two grades ahead of me. I can still see her sneer. I remember breaking in on my Mom sitting on the toilet, throwing myself at her knees bawling. I think I had her pinned to the seat longer than she had intended explaining to me that it was for the best…but I was going to lose my new friends, who would move on, and they would make fun of me and Jim would be in the same grade as me…No appeal, just said I was too small. Great at least I found out that I would still be a year ahead of Jim, he was staying home with Yoko for another year with Stevie

I remember moving on Base to Yuki Village because of the change in scenery, the Orka bomber, the Japanese version of a Nazi V-1 Buzz Bomb only with a suicide Pilot, instead of a clock, compass and altimeter, crazy, baka in Japanese. The Base is surrounded by a high chain-link fence and barbed-wire. The rocks and the tree trunks were painted white; there were trenches around the housing area. I think were whitewashed so and intruder couldn't use the trunks for cover. Another theory is to keep the drunks on the road and to keep them from hitting the trees if they go off the road. Or the airmen just needed something to do, yeah busy work. We would have regular air raid drills where we all run outside with all are neighbors, kids Mom's and maids, then jump in the trenches, duck down and take cover. The trenches were laid out in a zig-zag pattern so a plane couldn't line up on a strafing run and get a bunch of us in one pass; hell of a thing for a six-year-old to know and understand.

Dad bought me a miniature version of a Japanese bike; those suckers are heavy, must have been made of salvaged iron pipe; made in Japan was not a good thing in 1956. The bike was tall. It was a boy's bike of course, with a high bar on the frame. To learn how to ride it, the neighbor kids loaded me up while someone held the bike, my feet couldn't touch the ground; this worked fine as long as they ran alongside and didn't let go of the seat and handle bars. There were no training wheels. I soloed one day, they let go, I was doing fine, staying up, pedaling, went off the end of the sidewalk

## 1957 JOHNSON AFB, JAPAN

and into the air raid trench. I didn't know how to turn. I was not briefed about that on the flight plan or the letting go part,

I don't know of a Christmas where we, meaning Jim and I, didn't get a gun or some sort of weapon/present. I have a chrome full scale Colt .45 1911 cap gun, that I can hardly holdup, in a replica US issue leather flap holster with a two inch leather belt and a USA brass buckle; I wear it all the time. I'm a real bad ass just like Daddy. Something tells me that Mom doesn't buy our Christmas presents.

The trip back was a lot better than the trip to Japan. We flew back on a four engine TWA Constellation, a real sweet plane, one of the most beautiful airframes for reciprocal engines ever produced, that flight took us from Tachikowa to Oahu Hawaii where we had a long layover. We had to hang out at the airport to avoid an immigration hassle, but were met by some of Dad's old military friends. We didn't get to see much of the place beyond the airport, but it was the prettiest airport I had ever seen, with beautiful atriums and Koi ponds, fascinating for little boys; plus we had a great feast at an airport restaurant. We were all pretty well exhausted when we boarded the Connie for the final leg to San Francisco. Ever since then, the hum of a prop plane or a fan in my bedroom is enough to put me to sleep. We arrive at Travis AFB early in the AM. There is a picture of me standing in front of the Connie on the tarmac with my Dog Tags now hanging to my waist. I'm back in the land of the big BX.

Mom's first priority is to get together with her family member, immediate and extended. Someone had the bright idea that all the cousins should go to Disneyland, Sal's, Bill's and Mom's kids, that's about sixteen curtain climbing rug rats with different ideas of what to ride and where to go. I zero in on the submarine ride. The vessel looked like Captain Nemo's sub from 20,000 Leagues Under the

Sea. No one wants to ride with me. One of the subs is out of commission so there is a long line. Mesmerized by the working boat getting underway and submerging I turn around to find everyone gone. I decide that my best move is to stay put and keep watching the submarine. Everyone else has moved off to ride the gondola through the Matterhorn. I am sure that there was a discussion about where to go but I wasn't paying attention. No one notices that I'm missing. Tim is unlucky enough to ride in a gondola with Uncle Bill. In the heart of the Matterhorn there is a boiling volcano, a pool of lava and a devil that the gondolas pass over. Uncle Bill picks up little two-year-old Timmy, hangs him outside the car and makes like he's going to drop him. This so terrifies Tim that he will have a certain fire and brimstone nightmare for twenty five years until he finally figures out where the dream comes from. Thanks Uncle Bill. I am collected on the way out at my last known position. I missed a few rides but I don't care; I got to see the sub. They think that is punishment enough.

# CHAPTER 10

# 1959 Hill AFB, Ogden, Utah

**JJS.** I was issued travel orders from Johnson AFB Japan to Tyndall AFB, Panama City, Florida. Fran went shopping for summer clothing for the entire family of eight. I had sold our Studebaker and bought a new VW bus, which had lots of seats and baggage capacity. I delivered the VW to the port. I got most of our stuff arranged for shipping. I also shipped supplies of immediate need, like diapers, and baby stuff for little Bill. These items were our on hold baggage, destined for Tyndall AFB, Panama City Florida. Not too many days later, a change of assignment came, changing it from Tyndall AFB to MacDill AFB, Tampa Florida. Unscrambling the shipments between our household furniture and hold baggage was not automatic. The plot got thicker. Shortly after we arrived stateside, one more change popped up. I was given a choice between Stuart AFB, New York, or Hill AFB, Ogdon, Utah. Departing Tachikowa Japan June 1959, arriving at Travis AFB, San Francisco in July, we collected the VW bus at Oakland, loaded our luggage, and happily headed south to L.A. and Grandma's house, while our furniture was on a ship expected to dock at Tampa and our hold baggage was headed to Panama City. I selected Hill AFB after arriving stateside. I had some travel and leave time, so quite a bit of my time was spent on the telephone, to get new orders of assignment from Japan changed from Tyndall, then from MacDill to Hill. Many messages were sent, locating our baggage in Panama City, and likewise to locate the household goods somewhere in Tampa Florida and get everything shipped to Hill.

I had to immediately report to Hill AFB for duty as Operations Officer of a B-57 Squadron, leaving the family with Joyce Nuno in L.A. This mess-up caused a family separation of over a month. Oh, well, everything happens for the best.

In Japan, I was more or less Group HQ assistant to the Commander, Col. G. Y. Jumper, and later Col. Ahern. I was the B-57 Pilot Chief Instructor, in charge of Standardization, and wrote many S.O.P.s Standard ways to operate the B-57, and its role in our Atomic capability. The 3rd Bomb Group stood Alert in K8, Kunsan, Korea, with Atomic weapons. I monitored the operation, making frequent trips from Japan with the Group Commander. It was all pretty shaky at the start!

I drove the VW Bus to Hill AFB, Ogden, Utah, and reported for Duty, just as the 4677th D.S.E.S. Defense Systems Evaluation Squadron was flying to storage their last three B-29 E.C.M. Electrical Counter Measures Aircraft. I set up as Operations Officer, and prepared B-57 birds to replace them. Meantime, I bought a brand-new brick house, with a full basement; a real fine home. I still had not received our Hold Baggage from Tyndall, nor our Household Furniture that was on a ship to Tampa, Florida. I did receive a flock of Single-engine T-33 and F-86 Pilots from a Base closing at Scot AFB, Illinois. So, I flew every day, and some nights, checking out the newbies in the Canberra B-57.

Our stuff finally arrived, at the same time my best friends, Vaughan and Madeline Clark, were passing through on vacation, they helped me unpack and set up.

I went to L.A. and packed the VW Bus. I hung a small trailer on the bumper, and off we went to Utah, and a new home. In route, we checked in to the then, grand at the time, Grand Hotel El Rancho Vegas. I took one of their deluxe cottages with six children in tow, one still in diapers. Fran and Dad ordered room service, I had a double Martini.

With the help of a few boys from my office, I had almost everything in order for our arrival. The house was located on a very short dead-end street, just west of the Hill AFB Perimeter. We had most helpful neighbors, who welcomed us with open arms. Of course, most were Mormons whose code was based on family and helping others. The only drawback was a quite fast-flowing irrigation canal, a real danger for inquisitive little boys. Clearfield, Utah, was our home from about June 1959, until my transfer to Maxwell AFB, Montgomery Alabama August 1961.

The C.O. of the Hill AFB DSES was *Not an Able Pilot!* He could only fly VFR in clear skies, in daytime; so I had to assign my Department Ops Officer, George Cap to mother the Boss on *All* flights. Our mission was to simulate a Russian Air Invasion from many possible attack areas. We simulated the *Bad Guys* attacking from far north over the pole, Canada, Bermuda, Pacific Ocean, etc., etc. real useless maneuvers in what I thought was a *Total Waste* of people, aircraft and money... Continuing long after Air Defense Command lost importance.

I ran the entire air operation for well over a year. When the B-57 Sqdn at McDill AFB, Tampa, Florida, folded; their people, and aircraft, came to Hill AFB. We ended up with around forty B-57 B&C models, and eleven B-57 D Models. The B/C bids had 65-foot wingspan, and a max altitude of 47,000 feet; whereas, the D mod had 105-foot wingspan and a ceiling of 71,000 feet. Pilots in the D models were required to wear Partial Pressure Suits and breathe 100% Oxygen. Shortly after the Florida mob arrived, there was some friction. The inept CO and the Florida Boys did not meld. The Florida Boys, under a guy called "Itch," whose first and last name initials were I. H., thought they would take over; and in several ways, misjudged the Boss. The Florida boys were put down a few notches. Even though I. H. outranked me in date of rank, I retained the Sqd. Operations slot. I knew all the Florida Pilots from previous Tours in Japan, and Randolph. I. H. made a fool of himself, and irritated the Boss, to the degree where his next stupid act would result in a Courts Martial.

Military Justice was fortunately avoided when the *Old CO* was transferred, and a new CO, Col. Velfort J. DeArmond, came on board. Col. DeArmond did not know any of us, and wisely delayed making changes, so I remained as Chief of all air operations, even though out-ranked by the newcomer. I had been flying both the short and long-wing Canberras; I knew the mission in and out. No change was made until, after a few months passed, I. H. pointed out our difference in Grade; we were both Majors. Col. DeArmond told me that he was not in favor of a change; however, he felt obligated to go with Military custom in seniority in dates of rank.

Col. DeArmond suggested I be the Assistant Operations Officer, and help I. H. I suggested that I be designated B-57D (high altitude) Flight Comander, and that Major George be assigned to B-57B/C Flight Commander, so that way, Maj Cap and I could run the entire high altitude mission, without the assistance of Itch and his staff, whose primary function was to coordinate the ECM Missions as directed. I ended up with twelve pilots and thirty three B/Cs for familiarization training and support in addition to the eleven D models.

The high altitude pilots required special attention and support! Pilots had to pass a Pressure Chamber Test, be fitted in the Partial Pressure Suit, a tight fitting constrictive suit that was laced on, it had tubes that inflated on the outside of the arms and legs, they would make the suit even tighter applying mechanical pressure in case of cabin pressure loss. Prior to each mission, the pilot was given a short Physical Exam, fed a high-protein meal of steak and eggs, dressed in his partial pressure suit, and required to breathe 100% Oxygen for one hour prior to take-off. We climbed to 55,000 feet in only 15 minutes, and cruised/climbed to 70,000 feet altitude for the mission. If we were deployed from home base and/or landed away from Utah, there had to be, in position: two pressure suit technicians, one Crew Chief to service the Aircraft, a Flight Surgeon (doctor), and one support Aircraft to haul all the support people and Oxygen equipment to location and back. The same assemblage was required to be on standby in the event of an emergency.

Time passed, and a *Baby Girl,* Maureen Theresa Scanlon, was born in the Hill AFB Hospital, on 16 January 1960 on a very cold winter day. Grandma Viola Scanlon came up from Gardena, California, to help us tend to the *Now* seven children. The Mormon neighbors were most helpful; they provided many, many complete ready-to-eat meals, cleaned the kitchen, and vacuumed the place. The hunting and trout fishing was pretty good in the area. I took Bob, Jim and Steve on many rabbit hunting trips. I shot three Mule Deer, and two Elk. I took the Colonel and his dumb son Elk hunting; dumb Dave missed an easy shot. On a late season hunt, on the spur of the moment, Col. DeArmond and I took off from his party at midnight, drove until dawn, to Rifle, Colorado; Shot three deer, and came back home in less than 24 hours. Barbara and "Velfort" DeArmond became good, long lasting friends.

Time speeds on, and number eight arrived on 10 July 1961; Frances named *Him* Gary Anthony Scanlon, initials (G.A.S.). Gary is forever thankful that we did not stop producing babies at Number seven.

**RES.** Utah was cool. Dad had a house built to his specification that he's bragged about ever since, a plain old red brick, loaf of bread house with a basement and a pitched roof but he had sketched it out on grid paper and presented it to his builder; he was paying cash from the sale of the house in San Antonio. He personally addressed every feature and its attendant cost, logging it in a spiral notebook which unfortunately led Dad to decide to finish out the basement himself. The house was located in Clearfield a small bedroom community on the opposite side of the runway from the hangers, base facilities and tower. One thousand yards of prairie tumble weeds, rabbits, mice and wet weather ditches with ducks and seagulls separated us from the runway. You could sit on the back fence and watch: F100's doing touch and goes'; C-119, flying box cars practicing airdrops and of course B-57 B's and D's tearing off to who knows where all day long; it was great. I got buzzed more than once just sitting on the back fence on a go around. You could always tell when a B-57 D was taking off, it was a jet powered glider. The plane would take off at 70% power but would climb at 60 degrees. Dad said coming down was another matter, with a long glide path no flaps, the speed brakes were also deleted, no drag chute. "if ever an airplane needed an anchor and long string to get it down on the runway, the B-57

D was that airplane." Putting the nose down would just increase airspeed and lift, the pilot would have to hold it close to stall speed with only the drag of the gear and the spoilers working for him.

# Hunting

Utah is where we really learned to hunt; and by we, I mean Jim and myself. Dad loved to get out and hunt, just about anything, rabbits and birds mostly and we loved to tag along. It must have reminded him of another Bob and Jim tagging after their Daddy Jim. We had graduated from our Japanese cap guns to BB guns, Daisy model 1938 Red Ryders with a saddle ring on the receiver. We had received all manner of Range Safety Instruction from Dad and weren't allowed any ammunition until we attended a Hunter Safety Course on Base and were deemed capable of handling the weapon in the offhand, prone and kneeling positions. Next we shot to qualify in the back yard with Dad looking on. We were deadly to paper plates at ten yards.

Ok now it's time to find game. Huge Locust are plentiful and an easy kill but a waste of ammo. I see a bird on the fence top; I line up and pull down on him. It's a hit, the sparrow flips to the ground, gets up and flies off, must have just winged him, like Roy Rodgers. There are jack rabbits in the field but they are too fast for us, they sit out of range and mock us. We don't know about range and blaze away, another waste of ammo. Those BBs are dear, we only have one tube between the two of us and we didn't count how many rounds we loaded in each gun or how many were shot. Why Dad always sets up these future disputes between Jim and I, with the words: "you have to share" I don't know; it's like he wants us to learn something, without stating or controlling what that something will be. All for the price of another sleeve of BB's. I shake my rifle and then I shake Jim's. I think he has more BBs than me so now we have to unload, and split what's left or there is going to be a fight. On the way back we discover some mice babies and do great execution. I'm out of ammo and Jim still has plenty, I'm ready to jump him for being a pig and I let him know it. We need to go home and divvy up the remaining ammo right now, then come back, and put more mice in the bag.

We have to split the BBs on the concrete pad in the backyard or we will lose some in the weeds. Tim, looking through the glass sliding door sees us and comes out, we brag and tell him about our hunt, he is impressed with our bag. We reload with everything we have and saddle up to head back to the hunt. Tim wants to tag along, Jim is mean to him and tells him to bugger off, he's too small, and this is big boy stuff. That approach isn't deterring Tim and he follows us to the fence. Jim tries to kick Tim in the ass but misses him before we climb over, using the proper fence climbing technique when carrying firearms that we learned at the Rod and Gun Club Safety Brief. Tim can't climb the fence so we're off. We head toward the runway for a while with no luck when Jim says "there's a rabbit behind us". I tell him "that's not a rabbit that's Tim hiding in the brush". I knew that Tim had been following us, he had just backtracked, gone through the front gate and the neighbor's yard, he had been tailing us for a while. I don't care if Timmy follows us, I like Timmy he's funny, he talks a lot and is always interested in what I am doing; Dad calls him a 30 year old midget. Tim is my test pilot, being the perfect size, he rides my contraptions down the backyard hill with no complaint. Jim says "no, that's a rabbit". Jim is standing 20 feet from me he draws a bead, and before I can get to him, he fires. Timmy yelps then starts howling and crying, I want to punch Jim in the nose but I run to see if Tim got the legendary BB in the eye that we had herd so much about. I see that Tim has a BB lodged just under the skin of his right bicep after I catch him in his screaming retreat back to the house. Tim holds his hand over the wound sobbing, he won't let me look at it closely or touch it. I try to calm Tim down telling him bunch of crap about how Roy Rodgers, is always getting shot in the shoulder and he's ok, it's just a flesh wound. If Tim tells, Jim and I will be a smoking ruin, not even enough to burry in a match box. What Jim did was in direct contravention to the rules set out in the Rod and Gun Club Hunter Safety Course handbook. Jim says something stupid like: "I didn't know, he should have been wearing red" he is lying of course, one of those half-truths Dad hates. Well we need to make a deal here, Tim agrees not to tell for some undetermined future favors. Now we need to get Tim in the house, get that BB out, and some monkey blood on that wound without him crying, and cover it up with a long sleeve shirt. This is a problem; our usual summer outfit was striped engineer overalls and boots, no socks underwear or shirt. We would get so dirty and full of sand Dad would just get us to kick off our boots, drop our overalls and then he would hose us down in the backyard before we were allowed in the house. We were in that condition now. We could see Mom and Kathy in the kitchen through the sliding glass door messing with the babies, so we couldn't get in through the back door, there would be questions. The only thing to do is: leave our rifles, the dead mice, strip, run around front, come in the front door and head straight down the stairs to our room and the downstairs bathroom before anyone sees us that shouldn't. We make it. We go into the bathroom with no door and studs for walls. If Tim screams during the surgery we're toast. Jim gives Tim one of his Lone Ranger plastic nickel plated faux silver bullets to bite down on while I pop out the BB. Jim pins Timmy, who is standing on a stool over the sink with his arms back like a jet plane, he is only too happy to throw a hand over Tim's mouth without asking. I pop it out Tim's eyes are watering, the dam is going to burst. I show Tim the bloody BB, the muffled squeals stop and he relaxes. Tim holds the BB and bites the bullet while we clean him up. Jim insists that we should cauterize the wound but we don't have any gun powder or a camp fire or cigarettes so Iodine and Band-Aids will have to do.

I later tell Tim that I'm sorry that I didn't stop Jim in time and I swear to always have his six, I will always protect him from Jim. We shake hands on it, using the Boy Scout handshake, locked pinkies so you will not lose your grip with your brother when hanging over a precipice; the handshake Dad taught us.

There was a big deep irrigation canal across the street from our house with seagulls and ducks, big game to little boys. The canal was strictly off limits to all personnel, thus its attraction, it had a chain-link fence running along the bank to keep folks out but all it managed to keep out was tumble weeds and it didn't do a very good job of that. When the wind would blow, which was all the time at the foot of the Wasatch Mountains, the tumbleweeds would stack up against the fence making a ramp for following tumble weeds to jump the fence and foul the canal. In the winter when the canal was dry, the city would burn off the tumble weeds that had collected there. The fence wasn't much of an impediment to us boys either. Once over you could sneak down in the reeds and hide. I quickly found out that BB guns were ineffective against ducks and everything else in the canal short of frogs. I just had to get that white duck that was always hanging out at the same spot, that duck had cost me five + BBs. A change in ammo to finishing nails just cut the range, they were also wildly inaccurate. What to do?

I'm in the garage looking around for potential duck weapons when I land on it: a license plate. Last year's Utah license plates were hanging on a nail on the wall. I get a stool, climb up and get one. I take the plate outside and after a few tries, I'm able to throw it and make it stick in the ground. I determine that a better edge would help so I spend the rest of the afternoon sharpening the edges of the license plate on the sidewalk and practicing throwing it in the ground. Jim is not in on my plan, I'm going solo on this one, it's just that Jim always manages to mess things up for me, I believe that it's often on purpose, just to make me look bad.

Early the next day, after losing Jim, I am up and over the fence with my weapon and in the reeds lining the canal. From hunting I have learned: patience, and concentration, how to be quiet and still, something I cannot possibly manage to do indoors or on request or with Jim around. I hear Big Whitey quacking but I can't see him yet. I stay put resisting the urge to climb back up the bank for a looksee. The canal banks are 15 feet high and the canal is not more than 20 feet wide at its widest. I see ripples on the water, he's coming but he is against the opposite bank which is steeper and has no reeds at the base.

I stand slowly and fling my edged weapon like a Frisbee. The plate takes an unlikely diving arc flattening out and hitting the duck in the head continuing on to stick in the opposite bank. Got him!

Ok now, I really didn't think this one out all the way, what with the blood lust and all. My duck destroyer is stuck in the opposite bank of a very deep fast running canal and my kill, is slowly drifting down the canal, leaving a trail in the water, after an impressive gush of gore and lots of splashing.

I climb the bank and try to follow the drifting duck along the bank through the brush and over fences, but it's no use, I can't keep up. I feel bad about killing the duck then not getting it, I don't tell anyone, we could have had duck for dinner, I would have been the hero and eaten the biggest piece. The ducks are safe for now, clearly I need a boat! Better start working on that tomorrow.

Its late summer, Jimmy, Dad, a teenage neighbor kid and I are hunting the general area around a turkey farm. It's all high desert no trees, sagebrush, tumbleweeds and arroyos, some of them pretty deep. The turkey farmer, a friend of Dads, has asked him to take a look around and see what he thinks; the farmer was losing turkeys on a regular basis to an unknown predator.

Dad forms us up in a skirmish line and we sweep the fields. Pretty soon Dad nails a cotton tail or two. He hits a third, it's mortally wounded; a wounded rabbit cries like a little baby. This is my first hunt where I have seen something shot, besides bugs and mice; to hear it cry like that is a shock. I think that Dad is going to shoot it with his pistol like the cowboys putting a horse out of its misery, just like they do in the movies, so I ask him about it. Dad tells me that he doesn't want to waste a round on the rabbit. He tells me to put my boot on its head, grab his back feet, jerk up and break his neck. I do a bad job of it and it takes several jerks with two hands to make it stop crying. I drop the rabbit and take my foot off its head. Just as I step off I see a big black beetle crawl out of the rabbit's ear, the one that is closest to the ground. I think, oh my god, it's the spirit and soul of the rabbit leaving the body and it has come out to haunt me! Dad snatches up the rabbit and puts it in the game vest he's wearing; I lose sight of the beetle. We move out either side of an arroyo that gets deeper and wider as we go, the neighbor kid on one side and we are on the other. Jim and I are close to the edge and looking down into the arroyo. We see a Lynx eating a turkey twenty feet away, we open fire with our BB guns. The cat screams when we shoot him in the face, the kid across the draw can't shoot his 12 gage without hitting us. We cock our BB guns and fire again; that cat is not leaving his lunch and he is really pissed. Dad runs up and lets him have it with both barrels of his 12gage double barreled Sevens shotgun. The cat screams and rolls down a few feet, not done in, Dad pulls out his High Standard 6 inch Bull barreled .22 automatic target pistol and hits him with about three hollow point rounds in the chest in quick succession. My first hunt, how can you ever top that? Jim and I get to carry the cat back to the car slung from a pole by his paws, just like a Kipling ending. We are sold. This cat is a state record of fifty pounds, probably from eating all those turkeys. Dad has it turned into a rug that will hang on the wall wherever we go.

Dad got a chance to go deer hunting on the Nevada Utah border, we are all in for that. Dad has an old Air Force buddy, who has a ranch out there. It's an overnight hunt so we all go. We load up the Ford, hitch up the trailer and head out. There is snow on the ground but the road is clear, it's November, Thanksgiving break.

The ranch is cool, there's a barn and animals all over the place: dogs, cats, horses, cows, chickens, goats and sheep. All these animals are a treat, we have no animals, our house is a pet free zone. Dad's friend has a big family too seven kids, mostly girls with a couple of little boys. The Mom drives the school bus and the Dad drives the local snow plow. Our trailer and all these machines were on location for all us kids to play in and on, hard for fifteen kids not to have fun with all that.

We have dinner after dark; umm fried chicken, a big plate of it, with mashed potatoes and gravy. Everyone was digging in except for one of the little girls who was crying. The little girl later showed us the remains, the heads wings and feathers of her delicious pet roosters. We all felt bad, Kathy thought we should consider throwing up and burying it with a little ceremony to follow, we demurred. It was late. It was cold outside, time to crawl in a sleeping bag.

We awoke way before sunrise. It had snowed a couple feet during the night and we could hear the snow plow firing up. Dad's friend had to clear the road before hunting; they managed to get away before sunrise despite the extra work. We're snowed in so we boys spend most of the day in the farm house basement going out getting cold and wet and running back in to get warm while Mom, the baby and the girls are upstairs doing what girls do.

Dad and his friend returned in separate trucks that afternoon, Dad was driving someone else's truck, and he had a thirteen-year-old kid with him. They got out and entered the farm house. We came pouring out of the basement door anxious to see what they got. We were milling around the trucks jumping up and down trying to look in eyeing some old crates to use. Jim locks his fingers together and boosts Steve into the bed of one truck. There is a dead buck in it that we stare at. We have never seen a dead deer before. Steve sees that there is something in the other truck under a tarp.

We four boys: Jim, Steve, Tim and I drop and crawl over the tailgate of the other truck, the one Dad was driving. Steve pulls the tarp back and sees a dead man, the back of the man's head is blown off, his brain is exposed, lots of frozen blood. Steve is extremely upset, we all are. All the kids want to know what we see. We tell them. They start jabbering and talking all at once. It's cold outside, we don't have our coats on. We all file back into the basement.

Evidently the poor man had shot a deer. The man and his son had walked up on the kill, lying on its side. The man was kneeling between the deer's legs uphill, while his son walked up on the opposite side of the deer. The man, preparing to gut the deer was in the process of handing his rifle to his son across the deer when his son, leaning over to take the rifle, hooked the trigger of his .22 rifle on a tree branch, it was not on safe. The boy's rifle discharged, the hollow point bullet entered his Dad's mouth and exited the back of his head.

The kids were bouncing off the basement walls, everyone was upset, Steve was still gaging after throwing up. I was trying to bring some order and loudly voiced my disgust that they had brought the dead man to the house and left the body where we could see it, causing all the rioting. Well the dead man's son hears me, and goes to pieces. Dad came down the stairs and took me outside for a lecture. The boy and the body would wait at the farmhouse for the Sherriff to arrive, and since I was such an insensitive asshole, Dad loaded me in our station wagon, took me to see the Mom, and made me tell her what had happened; you don't relay that kind of thing over the phone. She hugged me and cried for a long time. I felt very sorry for the Mom. Steve never went hunting, at least that I can remember.

## Gunsmoke

We didn't get to see my Dad's parents very much as we were growing up what with all the moving around and the distances involved. I think that I only saw them twice in my lifetime that I can remember; one of those occasions was when Maureen was born. Utah was close enough to LA for them to drive, so they showed up one afternoon in their new 1960 white Ford sedan, the one with the wings in the back and cat's eye tail lights. Dad bought the station wagon version of the same model car.

Grandpa is a hoot; he gives us all candy cigarettes and bubble gum cigars making him instantly popular. Grandpa was an old school driver, he was not shy about passing cars on the shoulder; he'd mutter something that I didn't quite understand about moving it or milking it before dropping a gear on the column shifter, stepping on it, and making that big V8 roar as we ripped past the slow poke. This was before seatbelts; we would roll around in the back seat during these violent maneuvers squealing like we were on a roller coaster. Grandpa would look back at us with his pipe clenched in his teeth giving us the biggest grin. We didn't know if that was the way he always drove or if he was putting on a show for us, it was the exact opposite of the way Dad drove; over 55 was suicide.

I always remember grandpa drinking beer and listening to a ball game on the radio while he smoked his pipe, and Grandma and Grandpa having a boiler maker an hour or so before they retired. The Grandparents stayed at our place for what seems a month or so before and after the baby was born necessitating a good stock of beer. Several cases of the stuff were stocked under the stairs going down to the unfinished basement where Jim and I had a room.

This was back when four out of five TV shows was a black and white western shoot'em up which we adored, the best was Gunsmoke. We knew all the characters and could get in character down to the way they walked and talked: Marshall Dillon, Miss Kitty, Chester, Sam the bartender, Doc. Cowboy shows were prolific back then. We had all the appropriate costumes and hardware to have our own show: hats, boots, spurs, chaps, vests, toy pistols with holsters, a Winchester 98 and a Springfield rolling block replica rifle, with plastic ammo. From the shoot'em ups we learned all kinds of things mainly that most of life's problems could be solved by judicious use of a Colt 45 Peacemaker and or a fist fight.

## 1959 HILL AFB, OGDEN, UTAH

I was running around outside with the neighbor kids most of the day and didn't even think of why Jim was not out with us; actually glad that he wasn't tailing me around. Dad usually didn't whistle up one of us without the other. He would whistle and we were expected to come running on the double. I joked that my name was Jim-n-Bob, but it bugged me that he always said Jim first and he knew that it did. Find a weakness and exploit it, by habit I suppose. Well Dad whistled recall, Dad could whistle real loud, I heard him way out in the field and came running. We cleaned up and mustered as usual for dinner at 1800. After dinner and policing the kitchen grandpa and Dad took up watching a ball game on the "boob-tube", while Mom and grandma fussed with the babies. The other kids headed to their rooms for the night.

When I came trotting down the stairs that night to go to bed I found out what Jim had been doing all afternoon. He had scooted our beds together and off to one side of the room. In the space he freed up, he had placed a couple of chairs a foot or so away from the wall and laid some boards across the backs of them. He then threw a thin baby blanket over the boards to create the bar of the Longbranch saloon!

Jim was busy getting on his Sam the Bartender outfit: his black Sunday pants, a red silk vest from his Bat Masterson costume, a white dress shirt and my black embroidered clip on bow tie with tails I had had in Japan, he had on black engineer boots and was putting red rubber bands on his arms to blouse his sleeves, just like Sam.

We could pass on Miss Kitty/Kathleen, being there; she usually took over the narrative making it into some sort of play, working up a running script which she insisted that we never got right and must do over. Miss Kitty was also a great fan of dancing and I was her favorite dance partner.

I opted for Marshall Dillon. I put on my grey leather chaps and matching vest, with a red shirt, a light grey hat and my tan rough side out cowboy boots, plus a sweet quick draw tan saddle leather holster with a full scale metal cap gun model of an army colt.

Sam asked me "what'll you have marshal?" I said sarsaparilla. He said that he only had beer. I played along, and said ok. Jim pulled out a small 2oz glass pimento cheese jar from under the bar and started polishing it with a rag that he had on the bar.

Now there is a lot of history behind the pimento cheese glasses. We are not trusted with Mom's crockery. We are restricted to using only: jelly, pimento cheese or dried corned beef glass jars, small jars that our grubby little hands can hang on to and lend themselves to small portions. We break a lot of glasses, fortunately we eat a lot of: peanut butter and jelly sandwiches, pimento cheese crackers with chicken soup, and Dad is a big fan of SOS. We have a constant supply of little glasses that we have to peel and wash the labels off of. When we get a rare treat of soda-pop mom can split a 16oz can eight ways.

Jim finished polishing with a flourish, snapping his rag, and then he pulled out a bottle of grandpa's beer from under the bar. He expertly popped the cap with an opener he had taken from the kitchen earlier, then poured me a half glass with a foam head, so expertly that he must have been practicing. I couldn't believe it. Jim pulled out another pimento cheese glass and started polishing it, breathing on it and looking at it, placing it between his eye and the bare bulb light fixture. Sam then poured a drink for Sam. I thought it a great treat; grandpa liked his beer, Dad liked beer, Grandpa was always giving us sips. More is better right?

Jim is just hilarious, he has never been funnier. Our skeleton wall room isn't so scary now. The shadows of the studs on the wall and the studs themselves make our room look like a prison. Laughter is ringing off the bare concrete walls instead of us arguing and fighting for hours about anything and everything, a nightly ritual. The last ritual fight of the night is: who has to get out of their warm bed, walk across the cold bare concrete floor, climb a stool and pull on the string, hanging from the bare bulb ceiling fixture, in the next room by the stairs, then find ones way back across the dark basement and through the proper opening in the stud wall, without being bit by a black widow spider like the builders son who was bitten in this very basement, or so rumor has it. Instead we are squealing and laughing, jumping up and down on the beds, shaking and spraying beer at each other, sticking our thumbs in the bottle necks to make it spray better. We don't even hear the stomping on the ceiling above us where the parent's bedroom is located, a warning, the usual signal to pipe down, or the hollering that accompanied the stomping.

The next thing you know Dad, in his skivvies, is standing in the door frame, taking it all in with his cold blue eyes. Beer soaked, torn up beds, wrecked saloon, from the now forgotten and forgiven bar fight between Sam and the Marshall, overturned chairs, beer bottles rolling on the floor, toy guns and knives scattered with the bottles, the drunk wet cowboys standing on the beds with foaming bottles in their hands. Dad just turned around and went back upstairs turning out the light in the next room as he left, well at least neither one of us had to turn out the light that night.

Dad was always conscious of the weakness, he never had more than two martinis and two was rare. Dad had seen too many Relations and Pilots ruin their lives and careers with the drink. I'm sure that he thought that we would learn our lesson and we did. We felt horrible the next morning and threw up, everything was sticky and smelled bad. The looks Mom was giving us... there is nothing more disheartening than the look of disappointment in a mothers eyes. The cleanup was closely supervised; we obviously required constant supervision. Kathleen was perfect, she enjoyed bossing us around and telling us what losers we were. Perfect. Sick as we were, we mopped up everything using a string mop and the toilet as a bucket; that mop took the both of us to wring it out. We didn't pass inspection the first go around and had to do everything over, the floor was sticky; we had not used any soap. It took four loads of wash to get the sheets and blankets done. We were also put on permanent KP. Yeah, partner up the miscreants, that always works, there will be no hard feelings and total cooperation. Right.

## The Weakness and Needy Relations

I never thought about it much but I don't really know any of my relatives aside from my immediate family. I always thought it was because we moved so much, about every two years. I got to see my Mother's Dad once or twice in my life and her Mother about five times. I see my Dad's parents four times that I can remember. I see my cousins on two or three occasions while we are in transit to and from the Far East; most of them live in Los Angles. I thought that was the way it was with everyone. I don't know if it was by plan or happenstance. More like the plan was made easy by happenstance.

You make friends, you lose friends; you get good at making friends and burning bridges. You don't care. Pretty soon you don't need friends or even try to make friends, a waste of time. Except now you like burning bridges. All the friends I have ever had just wanted to use me in some way or got me in trouble. The definition of a friend to me is someone who will eventually let you down; I

## 1959 HILL AFB, OGDEN, UTAH

have plenty of examples. Better to be self-reliant. You meet someone on the other side of the planet, no one has to tell you that you will never see that person again. Like your platoon taking casualties, you don't want to know the replacements, who will be soon killed off, the veterans know this and don't want to mix. Military Brats, Base kids, some of the locals, Japanese, Chinese, Americans don't like us very much. People ask me if I was ever in the military, I tell them "yeah I put in eighteen before I was nineteen".

Mom can't let go, she spends a lot of time on her correspondence, remembering everyone's birthday writing to her friends, people from all the places we have been, even her ex-maid Dorothea in Munich and Yoko in Japan. It's a full time job.

Your best friend by default is your brother, he's the only person in the whole world that you have known since you were able to know anything. We have always shared a room. Jim has a mean streak a mile wide, he is always ready to fight. Jim bullies and makes sport of his little brothers constantly. Jim openly challenges me all the time. Dad has spent many hours with Jim's face in his hand looking him in the eyes trying to save his soul. We don't have many games. There is limited access to the family black and white Motorola; it doesn't matter, it's just the Soaps this time of day. Overseas there was no TV maybe limited local Japanese or Chinese programing. There was Armed Forces Radio of course but Dad wasn't in to what the typical GI wanted to hear.

We are all down in the basement. Jim is in a black mood he is making his little brothers cry, I watch him, I want to step in but I know if I do I will catch some licks and kicks. I also have to live with the guy. Finally I can't stand it, he is swinging Steve around by the hood of his sweatshirt. Steve is drooling and turning blue. I grab Jims arm and make him let go of Steve, he spins away and charges me, as usual, with his head down; I catch him on my right knee and kidney punch him till he stands up then I dot his eye with a left jab, he has had enough. Jim now backs away kicking and yelling at me, this brings Dad down in no mood for excuses. Dad is sick of Jim and me fighting all the time. He tells us if we like to fight so much we need to get it on; he tells us to start swinging or he is going to. I demure; I had just punched Jim out and couldn't see the point but then again Dad can hit harder than Jim. While I'm thinking about it, Dad knocks our heads together to make his point. I square away and box halfheartedly, we're crying because we just got our heads banged together. Mom appears at the head of the stairs and says one word: JIM, and it's over. We are sent to bed, Jim as usual rails about how much he hates Dad and warns me not to go to sleep. This is year seven, James Francis Scanlon is to be my constant companion for nineteen years. I go to sleep, Jim is full of shit; I'm his only friend.

**JJS.** Aunt Anastasia, Grandpa's sister was a Nurse. His mother Catherine Scanlon died during World War One while her son Jim was away in the US Army. I have very little knowledge of Catherine. Her eldest child, Anastasia, was more or less put in charge of the two younger boys, Gratton and Joe, because of a dying request of her Mother. Unfortunately both of the boys had the failing and did not live to a ripe old age. Uncle Joe died in Denver, Colorado around 1952 from alcoholism. Gratton did not fare much better and died at an early age. Both boys were drafted into the Army during World War Two, but were mustered out quite early. Anastasia was the family breadwinner, since her husband had the typical Irish failing ...BOOZE.

The only other child of the Scanlon Clan was a cousin George, son of Gratton. Gratton was a real character, full of fun and booze. He used to roar down the street in a Model A Ford cutting the ignition switch off and on which caused loud explosions and scared the hell out of the neighborhood.

Gratton married a floozy when he was too drunk to realize what was happening. This wife #1 had a daughter who was a real smart gal and lived most of the time with her Dad who divorced her floozy type Mom. Cousin George was the second issue of Gratton and wife #1. By 1960 George had five kids, none of which were too bright, a long suffering wife, and a serious drinking problem. Once again the Irish failing! I have made it a rule to steer clear of Cousins in general.

Uncle Gratton moved from Colorado to California in the 1950's and married a widow lady named Lillian. Gratton seemed to settle down somewhat and even got a steady job as a bread deliveryman. Lillian was fairly well off; she owned a nice home, so the story goes, she also had over $40,000.00 in the bank. Grats' luck ran out when Lillian died of cancer and left him all that loot.

Around this period the American Beauty Spaghetti salesman, Cousin George, had a bad accident near Salida, Colorado when he ran his car off the road. He was banged up real bad, and put in a hospital to heal and sober up; DWI was the charge. His wife asked me to bring George to Colorado Springs; they had no insurance and could not afford an ambulance, so I put a pad in my station wagon and hauled poor George home. Soon after the wreck Gratton called George from California asking George to send him $500.00 to bail him out of jail; he also had been arrested for DWI. This was not the first time he had been charged with DWI, in fact his license had been revoked. Chips off the same BLOCK!!! Here we have a poor busted up American Beauty spaghetti salesman with five kids, none of which were overly bright, bedfast with no insurance, his car a total wreck, etc.,…,etc. So quite naturally George asks his

Dad -- ★★-- "What happened to all that loot? The house, the car, and the $40,000.00 bank account??????" Well Gratton's reply was "Son, ya rent a place in Malibu, cater all those parties, and all those women. That ain't CHEAP!!!!!" Gratton blew the loot, but he had one hell of a time while it lasted. My Dad, as usual bailed Grat out, as he had time and again for our "POOR RELATIONS."

Uncle Joe Scanlon came to stay with us after he was released from prison in Michigan. Joe was a hard luck kid. He was tending bar during Prohibition, a period from about 1918 to 1932 when alcohol was illegal. Federal agents caught him at work so Joe was arrested. He could off gotten off because he was only eighteen-years-old, but the owner talked Joe into taking the rap for him, telling Joe that he had to serve only a short time in the Pokey, and that when he, Joe, got out he would be on 'Easy Street'. Well Joe served his time, got out and went to see the owner to receive his reward. When he contacted the bastard the owner said "Joe Who?" Joe was disillusioned at an early age.

When he came to stay with us in Denver in our small four room house, Uncle Joe took over the living room couch. I was about four-years-old, but sure do remember Uncle Joe hanging around. I guess Mother got tired of the situation. Uncle Joe departed and ended up in Skid row, swamping bars for drinks. He died in Denver in the County Hospital a total alcoholic. In fact parts of Joe's innards are part of an exhibit in the University of Denver Medical School! On one occasion Joe was in the hospital with two broken legs caused by a car hitting him while he was laying down drunk in the street. He was near death and in a coma, when the hospital notified my Dad and Gratton, so they flew to Joe's bedside. Dad told me that he was at Joe's bedside and Gratton was at the foot of the bed when Joe came out of the coma, looked at Gratton and said, "Move over Grat, you're standing right under the hole the bats are coming out of." Joe had been unconscious four or five days, but still had the DTs, delirium tremens. Joe had been in and out of Denver County Hospital so often he was probably considered county property, so parts of him still reside there.

Fathers' family members were not pillars of society, but compared to my Mom's family they were Saints from heaven! The Matriarch of Mother's gang was Mary Charlotte. Mary C. was married first to Mr. Hornback, of English extraction, which produced Melville, Viola May, my Mom, and Bertha. These children retained the last name Hornback and were never adopted by Mary C's second husband Alfred Hinds. Al and Mary produced: George, Richard, Russell, Edward, Bessie, and Jimmie Hinds. Of the nine the final six were born losers, although the first three were far from prizewinners. In brief they were all alcoholics except Melville, of which very little is known. Melville was a World War One Sergeant who disappeared during the war without a trace. Both Mom, Viola and Aunt Bertha wilted away in an alcoholic haze. The elder Scanlons, Hornbacks, and Hinds blamed their ills on the Great World Depression. Times were hard during that period; but, my Dad, James E. and his sister, Anastasia worked hard and succeeded while the rest of the Scanlon, Hinds, and Hornback clans went to Pot!

Aunt Bertha married a pop-eyed weakling named Earl Kofford. Earl was supported by a Jewish benefactor who set Earl up in several businesses: a filling station, an egg/chicken enterprise, a laundry, laundry marking machines, dry cleaning, etc, etc. All were total losses. Earl died in jail in an alcoholic stupor after being arrested for stealing and selling clothes he was supposed to deliver for the dry cleaner he was working for. The poor guy had been stealing to support his and Bertha's drinking habit. Bertha and Earl moved in on us in 1936 while we were living in San Diego. They had two kids Barbara and Kent. Kent was another Born Loser; Barbara was quite bright, but also quite homely and overweight. She had a minor leg fracture as a child that was not properly diagnosed. It later resulted in osteomyelitis, a bone marrow disease causing an open festering wound with a foul order, which she attempted to cover with perfume. She was a very nice person, bright, intelligent, and cheerful despite having to use crutches. Barbara graduated from UCLA, married, and had two kids; however, her childhood disease migrated throughout her body. Her husband was well off and devoted to her. Her progressive deterioration resulted in medical bills costing them their wealth, business failure, and finally the disease affected her mind. Barbara jumped off the San Francisco Bay Bridge but missed the water landing on the rocks. Tragic. Kent was an enlisted sailor during World War Two, and the last I heard in 1948 he was a salesman, I don't know where.

Aunt Bertha was always much too heavy and in poor health during the last few years of her life. She was born in 1903 and died around 1962, according to Mom, of cirrhosis of the liver. Her life was a series of ups and downs; they lived high on the hog for short periods when starting a new business, then crashed to the bottom so many times that they took to drinking. Booze finally did them in.

George Hinds was a red headed devil; he was a miner, and was usually employed in copper mines in Arizona, Nevada, or California. George lived with us when we lived at 223 South Irving in Denver. He didn't stay with us but for a short time when he was put in the hospital with lung problems. The Army was in the process of giving George a medical discharge, when he deserted. He had

## 1959 HILL AFB, OGDEN, UTAH

asthma as a boy and working in the copper mines didn't improve that condition. Stupid George could have had a medical disability discharge with honor; instead, he ended up in prison. His wife divorced him, and our family lost track of George.

Edward Hinds was the only gentleman in the Hinds Clan; however, he too ended his life with liquor and expired all alone in a Los Angeles flop house. Ed was a Navy Veteran in World War Two, and had served with honor. Al, Ed's Dad, never seemed to be around, so Ed ended up as the breadwinner and the only son that shouldered any responsibility.

Ed kept a flock of pigeons, and, as the story was related to me by my mother: George, Dick, and Doc (Russell and Richard) had a disagreement with Ed in which Ed prevailed. Those rotten brats were sore losers so for spite they killed all of Ed's birds; those boys were mean. Mean, mean!

Dick Hinds worked for Uncle Earl Kofford when Earl was boss of the Columbine laundry. Neither one knew what they were doing: Dick was "let go" for stealing, and damn near got jailed for his crooked ways. Soon after Earl was canned for incompetence; that's when the Koffords moved in with us in San Diego. Soon after Dick and Ed showed up looking for a handout. The main reason Dad had moved us to California was to get the hell away from all the poor relations. To get back to Dick, he married an Italian, had nine kids all of which turned out to be true Hinds stock, crooks and alcoholics.

Doc Hinds was a truck driver, a crook, and a thief. Doc finally met his end in an auto accident in Los Angeles around 1972. I caught him stealing Mom's Social Security checks; that was after he had swiped all of Dad's tools and dental equipment when Dad died, November 1971. All the poor relations owed a lot of money to my Dad. Dad took a Mod. 12 16ga Winchester shotgun from

Doc as partial payment for a loan around 1931, then gave the gun to my eight-year-old brother Bob. (note: I now have this shotgun. RES ) At any rate we can dispose of the Doc Hinds tale without shedding a single tear!!

Jimmie Hinds was the last of the litter; he was only about three years older than Brother Bob. We saw a great deal of Jimmie Hinds while we were growing up in Denver. We usually associated with Uncle Earl and Aunt Bertha Kofford and their two kids Billy and Joanne. Jimmie and Grandma Hinds were frequently in the loop along with uncles, aunts, and cousins. In later years Jimmie was twice married, was a World War Two Navy aerial gunner in the South Pacific. He drifted here and there and died of throat cancer in the Los Angeles Vet Hospital around 1960.

My parents were the kind of people who would help poor relatives: Dad's brothers Joe and Gratton, and Uncles Ed, Dick, Doc, George, and Jimmie the Hinds boys who were Mothers half-brothers. Much later they took in Larry, Dad's junior apprentice dental mechanic. The last one we helped was Betty Hanahan, the daughter of an old friend; she stayed with us during her last semester of high school in the spring of 1942.

Dad always said that it was not the booze that ruined lives; he said that it was the person that was entirely responsible! I tend to agree because so many of our poor relations were pretty dumb, rotten lazy skunks, drunk or sober.

Dad made home brew beer. The beer was of questionable quality and was quite explosive. Every so often a bottle would explode in the cellar, which could be easily be heard even outside the house. Whenever Dad would send Bob or me to fetch him a beer from the basement, he would be sure to admonish us not to shake it. Sometimes Dad would open it, always over the sink, the beer and

# 1959 HILL AFB, OGDEN, UTAH

foam would hit the ceiling. Powerful stuff! Of course, no one is perfect! Dad, like all Irishmen, had to a certain degree what is known as THE FAILING. A desire for a bit of spirits, but only on special occasions like every day. I NEVER saw Dad drunk. He was not, in any sense of the word, an alcoholic. He was always up and at work on time, he took good care of us and all of his obligations. In some respect Dad was the only real solid, clear thinking, and honest person of both the Scanlon and the Hinds/Hornback/ Kofford/ Kinney clans.

**RES.** Dad built a fence enclosing the backyard of the Clearfield house. He did the work himself. He dug the holes and cemented the 4 X 4 post in the ground the tops were tied together with 2 X 4's and had another one mid span. The frame was faced with plywood sheeting and the bottom edge had chicken wire. Dad meant to put in a yard and garden. He intended to make it rabbit proof. Dad first planted some trees, saplings. There are no trees in this part of the country. He braced the trees against the wind with Bamboo poles. Dad also planted: summer squash, beans, corn and watermelons. He liked to spend his afternoons watering his crop and trees. He got a lawn going too, even if it was only sweet flowering clover and with clover you get bugs and bees, after all Utah is the beehive state. Every one of us are stung, Tim more than most. He develops a strong allergic reaction to bee stings, we usually step on them because we are walking barefoot in the clover. Jim invents a game. You catch a bee by cupping both of your hands around it. If it is dark he says the bee won't sting. He demonstrates and by gosh he pulls it off. We all have to try it and we all get stung. All the watering and bugs bring in the meadowlarks whose song always reminds me of Utah. Everything is growing and looking great. The squash is almost ripe. The corn is three foot high, the watermelons are as big as cantaloupes and the beans are climbing their stakes. Dad only made one mistake. He should have followed Caesar's example at Alesia and built a double wall. We are playing Cowboys and Indians. We have plenty of Cowboy stuff but no Indian gear; no bows or arrows. We pull up the bamboo poles holding up the saplings for a bow, pull down the string holding up the squash to sting it, and use the bean poles as arrows. Jim as usual takes it a little too far. After we are played out and move on to other stuff, Jim returns to the scene of the crime and starts putting arrows in the watermelons, they make a good target, the arrows stick nicely, very satisfying. Dad is pissed when he goes to water his peace garden. Everything is salvageable but the ants have found the watermelons. We hose the ants off the damaged melons, Jim missed some of them, and eat them green with a little salt. Dad is sad and very disappointed with us. It's worse than a spanking. I have never seen him sad.

## The Playhouse

Every pack of kids needs a fort or playhouse to fight over. We started out at the Mayor's house on the end of the block. He had horses, and were there are horses there is a barn and hay. We were welcome at first, we would bring apples and pet the horses. After a while we hollowed out a cave in the stack of hay bales for a secret clubhouse. The mayor saw it and thought it dangerous, so we were banned from the barn.

After the barn incident we started to dig little caves and coveys in the dry arroyo banks between the house and the runway. These weed choked dry creeks behind the house lead to the irrigation canal. After a few cave-ins and the cleanup of the dusty diggers and being banned from the barn, Dad decided to help us out.

There was some lumber and plywood left over from building the house and the fence. Enough material to knock together a pretty cool playhouse on one side of the house. 4 x 4 fence posts were sunk to make a 10' x 10 'pier and beam, plywood sided and decked shed roof playhouse. It even had a little porch and windows on all sides. We loved that house. We moved in some baby furniture from Japan: miniaturized rattan chairs with cushions, TV trays, a bird cage we found in the field and of course Kathleen's pink sheet metal Kitchen set. We boys immediately set about converting it to a fort for Cowboy and Indian games. Cutting firing loops in the walls. We tried to make a secret trap door in the floor but it was going slow with only my official Cub Scout folding utility knife to work with.

The playhouse was close enough to the fence that you could climb the fence and jump on the roof if you were brave enough or if there were some kids around to dare you to jump. There was a mean dog down the block, a huge German Shepard. I had to pass by his yard walking to and from second grade every day, moving a little quicker and cringing as he barked at me and tried to chew his way through the fence. Well, as luck would have it, the beast got loose and cornered about five of us on the roof of the playhouse. None of us thought to shut the backyard gate, on the opposite side of the house from the playhouse, in the race to not be last. We ran to the back door. The sliding glass door was locked, because earlier, Steve was playing the ever popular lock your brothers out of the house and taunt them through the glass game; both the front and back doors were locked. We sprinted to the playhouse it didn't have a door and we hadn't fabricated any of the often discussed window treatments, the present situation making a good case for my wood shutter idea. We didn't need any goading to scramble up the fence and make the jump to the playhouse roof and safety. Being a pack of little boys we proceeded to be jerks, now that we were safe on our perch, teasing the raging animal and whipping him into a snarling frenzy.

Steve hears the ruckus and cracks the sliding glass door, the dog hears it and charges him. Steve slams the door shut before the dog can get him, and this gives me an idea. I had watched way too many hero movies and I decided to be the one to save the day. We had left a couple of bikes in the front yard, one of them I could see from where we were. I thought that I could almost jump on it from the fence. The bike was at the turn of the fence where it was leaning against the house. We had no trouble walking the 2x4 fence top, another one of our games. I yelled at Steve to come to the window across from the playhouse where we could talk. I told everyone that I would walk the fence, get on the bike and draw the dog off, Steve would open the door, and my buds would jump off the roof and run inside while I safely outran the dog on my bike. Everything went as planned, except that the dog was a lot faster than I thought or could pedal; he caught me before I got three houses down the road. The animal bit my right foot as he hit the bike, throwing me to the curb, with the bike fortunately on top between me and the dog, who was snapping and mauling me through the bike's frame and wheel spokes.

The neighbors heard the all commotion. The lady and her son across the street came and drove the dog off with a shovel and a broom stick. The neighbors took me home, mopped me up and called Dad, Mom was not around at the time. Dad came straight home; he debriefed me, and checked out my wounds. I don't know what happened after that but the dog was gone next time I walked to school. The rumor between us boys is: Dad got his Ruger .22 single six cowboy pistol and shot the dog as it came charging at him from under a fence.

# 1959 HILL AFB, OGDEN, UTAH

When winter came early, and it got snowy, Jim decided that the play house needed a fireplace, so he started a small newspaper fire in the birdcage, it got pretty smoky in the playhouse so we ran out and let the two sheets or so of newspaper burn out. Not discouraged at all, Jim decided the fireplace needed a chimney so he climbed on the roof and spent the morning knocking a hole in it with Dad's hatchet. The hole was down slope so it didn't work so well when he started another newspaper fire in the birdcage. So Jim climbed back up and knocked another, larger, hole at the peak near one corner.

About this time Steve and Tim come wandering up, drawn by the smoke and the banging, Steve is dragging a sled with a dead jackrabbit on it that he found out in the field, he wants to eat it, we want to help with that. Fortunately for Steve we have just completed the playhouse fireplace. The rabbit looks big so we determine that we need a lot of fuel to cook it. We stuff the birdcage with crumpled up newspaper and some small branches we chopped off of Dads sapling trees that he had been nursing all summer, chopped them as far up as we can reach or lift your brother. There are very few trees in the high desert and we didn't help with that statistic. We throw the rabbit, stiff as a board, still intact, hide and all, on the top of the birdcage and put a match to the whole affair.

The heavy damp, newspaper, green wood, and wet rabbit fur make a lot of smoke, it's a bit overpowering so we step outside while the rabbit cooks. About this time Dear old Dad comes home from work, he sees the smoke grabs a hose and the fun is over. We're called to attention for a real dressing down; lots of yelling and group punishment is handed out. The playhouse is knocked apart that night. The birdcage is flung over the backyard fence and the rabbit goes in the trash can. It's not over, we were brilliant, inspired, we were so close, we loved our playhouse.

Plywood sheet is heavy and bulky for little kids, but we were determined. The next day we tipped a couple of sheets onto my flexible flyer runner sled and dragged them up to the far corner of the backyard fence. It was quite a struggle, we lost our load a few times, going uphill to the fence. We finally worked out that we had to have one kid in back and a kid on each side to balance the sheet and to keep it from sliding off while I pulled the sled.

We tipped the two sheets against the back fence in one corner to make a lean-to. it didn't take us long to find the bird cage, Dad had crunched it up so the door wouldn't open anymore but you could still tear up paper and drop it through the wires and poke sticks in the same way. Steve found the rabbit in the trash so we were back on, our luck was holding.

Kathleen was watching all these goings on from the kitchen window, in her curlers, pink robe and fuzzy slippers, sipping coco, just waiting for the right moment to rat on us; give them enough rope, Dad always said, and Kathleen always listened to Dad, she was his favorite. Dirty Katrina would be looking pretty good this day compared to those icky boys, especially Jim; girls rule boys drool.

Just as we got her going, Mom might not of have been able to see the white smoke very well against the snowy background and overcast sky but she sure saw four little twerps dancing around a structure that wasn't there before. Kathy was happy to fill in the blanks as to what Mom was looking at.

Mom was out in a shot, wearing just her robe, to the top of the backyard hill where she deployed her secret weapon, the evil stink eye. We froze in place, she had us cold, she was not distracted by the babies per plan, and she was actually paying attention to us, something we usually liked but not just right now. Mom knows leverage when she sees it. She terrorizes us with "wait till your Dad gets home". She makes us kick over the birdcage throw snow on it to put out the fire and bury the lepine floambaeu in the field, where we have a small ceremony. Everyone takes off his hat in respect for the lately departed and interred partner in our tragedy to come, we envy his hole; he is at rest and hidden from sight.

We are very good, we are quiet as mice and stay in our rooms which are immaculate or our version of that. We wait for the hammer to fall for a couple of days, until it dawns on us, she didn't tell, it's a Mexican standoff; we get to keep the lean-to and she's going to keep it to herself. She will have to convince Kathleen not to tell. We have to be careful not to make her mad, but as long as we do what Mamma says, when she says it, and answer yes Mam every time, we are safe, for the moment. Power politics vs. community organizing 101. Power has been taken from the people but there is wiggle room.

## Hanging out on the Flight Line

Dad would take me and Jim out to the tower and the ready room once in a while. They had ping pong tables there and these cool chairs that massaged, reclined, and vibrated; Jim and I would jump in them and play with the buttons when they weren't occupied. The Pilots sat in these chairs while they were getting into their high altitude suits, a two hour ordeal where they were literally laced into their skin tight constrictive partial pressure suits while breathing straight oxygen. The chancer mechanic/flight surgeon in attendance at all times. This was 1961 the height of the Bay of Pigs. Jim and I play ping pong with each other and the pilots. I'm not sure, but I think I heard talk among the pilots about flying at 75,000 feet and seeing SAM's looking like smoking telephone poles coming up and just falling away at 65 to 70,000 feet. Dad always said that he took too many trips to 75,000 feet and speculated on the effects of that on his body. The B-57-D had a higher ceiling, greater payload and endurance than the Lockheed U-2 and was available prior to the U-2. Dad had to be flying these a lot when we were in Japan to be able to assume command of the birds at Hill. The Brit's use a modified Canberra for their high altitude photo-recon work. They gave their planes bigger engines too but make the mid wings, already a dance floor, wider. It's only 1600 NM to Havana from Hill. By my math at 500kts a round trip mission from Hill to Cuba is quite feasible, under 8 hours. There were also nearby logistic B-57 D support facilities at Tyndall AFB Florida. Which makes a high probability for overflight and recon missions over Cuba at this time. This is all purely speculative on my part knowing what I know and do for a living today. There was another story I remember where a rocket on Dad's bird got hung up on the rail over the bombing range, it took a hard right turn and speared through a wing tank. It didn't explode because the rocket didn't go far enough to arm. All in all pretty dangerous stuff.

After school starts for me, Tim comes down with pneumonia and has to be hospitalized. We are very worried about him and go visit him in the hospital. He is in an oxygen tent, wheezing badly. He has to stay there several days. Going forward Tim marks time by his hospital stays. I get to ride my bike to school, first grade for me, but I am soon walking. There are a lot of stickers in Utah so I get a lot of flats. On one of these walks home, taking an off road short cut, I really had to take a crap. I ducked around a privacy fence into what I thought was an empty lot. Well empty except for a staked out goat. That goat butted me just after I finished and before I could stand up, knocking me facedown with my pants

1959 HILL AFB, OGDEN, UTAH

down. I need my bike. Dad gets tired of fixing flats so he shows me how to do it. Jim and Steve start to ride my bike so I get even more flats. Soon my inner tubes are all patches. Dad finally takes pity on me and buys me a set of solid rubber tires making my bike twenty pounds heavier. Steve will always carry a piece of Utah in his nose when he hits a rock and goes over the handlebars doing a face plant in the road. Dad takes a picture of his face when he gets home.

## Haircuts

Dad has been cutting our hair since I can remember and as long as I can remember Mom has had something to say about it. It started in Japan. Mom came home one day to find Kathleen with a Dutch Boy and bangs, Jim and I had military buzz cuts. A Japanese haircutter cut Kathleen's hair. Dad had just purchased a clipper and tried it out on us with mixed results. Mom stopped his cutting before he got to Steve. Mom was furious about Kathleen, she used to spend a lot of time doing up her beautiful long blond hair, furthermore Dad was not to touch a strand of little Stevie's golden locks!

Mom has been intervening and as a result our hair is getting long. My hair is a real challenge because it has all the properties of a wire brush, it stands straight up in multiple cow licks. In order to put off the clipper Mom has taken to wetting my hair down, tying a knot in one of her nylon stockings and pulling it over my head before I go to sleep. Instead of bed head I have helmet head in the morning, it's like matting beaver felt into a watch cap. My siblings tease me about the stocking and the Cops would probably have picked me up with that look in later years when it became popular for 7-11 holdups.

It gets to the point that our shagginess can't be hidden any longer. Steve's hair is now over his collar and ears which is particularly annoying to Dad's sense of military decorum. My hair makes me look two inches taller than I am. Mom is away so it's time for Dad to make his move. Steve is first. In his haste Dad slaps a comb on his clipper a couple of sizes smaller than he usually uses. Dad is half way in before he notices that Steve is getting a real skinning. When it's all over Steve starts crying. Timmy, bless him, volunteers to take a skinning too, so Stevie won't feel all alone. This make Stevie stop crying as Tim jumps into the steel high chair, where all us kids ate chow as babies since Lake Charles for his turn. Tim gets his close crop buzz. Tim and Steve rub each other's head and laugh. I'm next. I am promised my normal half or three quarter inch cut. Dad marvels at my two and one half inch long locks twisting it in his fingers. Dad snaps a comb on the clippers, and in a play to the audience of little boys attending the shearing, he starts in the middle of my forehead and heads for the back of my neck. He doesn't make it. The clippers stop working at the back of my head. They are jammed with my hair and maybe some sand and grit from that arroyo dig cave-in earlier that morning. Dad pulls the smoking clippers from my mop taking some hair and scalp with it, which makes me yelp. Dad is under pressure to complete his mission before Mom returns, now his clippers are kaput; he loses his temper and smacks my head with the clippers telling me to stop crying, like that will help. I catch the edge of the blade in the side of the head and start bleeding freely enough for it to run down my neck and arm, dripping to the ground. Dad takes me inside and puts my head in the kitchen sink. I watch my blood go down the drain. Dad washes me and cuts away the hair around the gash with his mustache scissors. He applies a butterfly, tape with a twist, to

## 1959 HILL AFB, OGDEN, UTAH

close my wound and paints it with "monkey blood" iodine. This will be the second, and unfortunately for me not the last, of Daddy's butterflies that I will receive. The first one was from throwing me on my bed. I don't remember if Dad was mad at me or just horsing around. When I make my bed I always put my prized possession on the pillow. It's a cast iron B-29 with a 24 inch wing span that I can barely pick up, its painted Air Force blue; I've had it since San Antonio. It's the only airplane model that I own; it used to be Dad's of course. I catch the tail square in the back of my head. I make a mess of my pillow and sheets.

There is a lot of haircut fallout between Mom and Dad. I don't feel sorry for Steve or Tim, I am self-absorbed. I don't want to go to school with a partial reverse Mohawk and a shaved patch the size of a golf ball on the side of my head. Mom bought the clippers at the PX and Mom feels that there should some sort of warranty, real, implied or imaginary. She is out for justice, this will take a while but she will wear them down till they eventually cave and give her replacement $9.00 clippers. In the meantime I am to go to school.

I walk to school wearing my knit cap. It's May and it's hot. A pack of boys can't help but notice my hat and it doesn't take long for one of them to snatch it off my head to start a game of keep away. The game gets very popular when the kids see what's under the hat. They laugh. I am exposed and furious. I single out the kid who initially snatched my hat and jump on him, all my anger concentrated on that one pinhead. I get him in a choke hold and punch him in the face until he screams give him back his hat. I go to school still wearing my hat but I don't have any more trouble. I get my haircut fixed within three days when Mom gets new replacement clippers, free from the BX but I still have the shaved patch on the side of my head. Looking back I think that it took a lot of backbone for a little kid like me to climb back into that highchair. Orders are orders.

The basement is still unfinished Dad was going to do a lot of the work himself. He was going to paint and seal the concrete. He was going to insulate and sheet rock the walls to my bedroom and the bathroom by the stairs. He needed to float tape texture trim and paint everything then put in proper light fixtures. He never got further than sealing and painting the basement floor. Dad piled everything in and around my room then mopped and cleaned the rest of the basement. Dad put down gallons of sealant and rust red epoxy floor paint then left it to dry. I think Jim and I still occupied our room maybe with the sunken window cracked. Paint fumes didn't seem to be an issue in 1960. Well, it's been three days curing giving us plenty of time to scheme. All that open floor space was just too tempting. Too small for ball. We all agree it is the perfect roller rink; we are fans of Roller Derby having seen exactly one match. Jim, Kathy, Steve and I put on our steel wheeled clamp on skates impatiently sharing the only skate key we have left. And we are off, blockers and jammers wheeling around the track. It doesn't take too many laps for sixteen one inch diameter steel wheels to completely destroy Dad's paint job. Too late we see the paint peeling up in hundreds of roller skate tracks. Dad sees our tracks too when he gets back from TDY. This is when we hear it for the first, but not the last time: "everything you kids touch turns to crap". The basement is never finished.

# 1961 Maxwell, Montgomery, Al

**JJS.** I had traded in the VW Bus for a 1960 Ford Station Wagon while in Utah and purchased a 16-foot Astro house trailer. We made a few trips into the mountains to test the rig and do some camping; we were ready to roll when I got Orders to attend Command and Staff College in Montgomery, Alabama. Ma & Pa Scanlon, with eight kids, hit the road for Alabama when Gary is only six weeks old. When you are young and healthy, long moves are *No Big Deal*.

I was a student in the Air Force Command & Staff College from August 1961, to June 1962. The College was a good learning experience; however, it took some time to adjust to a desk, instead of a cockpit. I did put in plenty of T-33 flying, more than enough for Flight Pay, and a few Cross-Country Sorties, to maintain Proficiency. School was and still is Not My Bag. We all received well over one-and-a-half years of College Credit, which added to my many here and there College Credits which I earned throughout my career leading up to a Degree in Military Science. *Hooo Ray!*

My Mandatory Study Project was Air Crew Survival. I researched many previous studies, added my own ideas, assembled a kit of actual practical survival equipment and came up with 100 typed pages. The study was selected as a guide for Airmen. Believe it or not, Aircrew Survival Kits now contain exactly what I recommended. Carry on your person whatever the conditions likely to be encountered demand. Shed the parachute, and move out unencumbered with a suitcase full of stuff that you will have little use for. My study is now a part of the Air Force History Library at Maxwell AFB, in Montgomery, Alabama. So there you go!

The South was completely segregated, no mixing of a races at all. This was the first time that our family was exposed to such a separation of races. We rented a nice house, the children went to good schools, the neighbors were a big help, and we hired an almost full-time maid, which relieved Frances of some of the tasks that burden a Mother of eight. We attended many social functions; the Kids learned to Dance at the Officers club, and we all stayed healthy.

**RES.** Maxwell field is where the Wright brothers developed and tested their first airplanes. They wanted a place that was dead level with no trees and no snow. They found it south west of Montgomery where there was nothing but cotton fields.

Dad stayed up nights making prototype survival vests till finally sewing up one he was satisfied with. Dad is a good with needle and thread. He maintains his own uniforms and can sew buttons on like nobody else. His mom taught him how to sew. Pop had to go through escape and evasion school, fielding the escape and evasion kit he had proposed. They dumped him in the gulf; he had to make it to an isolated island in a standard emergency raft with only the kit he proposed. Dad jumped a couple Marines during training and took them prisoner. All he had was a K-bar knife two prisoners and a dead chicken for three days. One pilot was killed while we were stationed at Maxwell, he had to punch out of a disabled T-33. He didn't make it. There was only one Black Officer in the Class, but he was there to represent.

Day one, we pull into Montgomery. It's a beautiful city, with wide tree shaded parkways and boulevards, a sleepy town in the yellow hazy afternoon heat. We go through downtown where we stop at a light. A fellow comes up to the car and asks if we want a Wallace bumper sticker. Dad says no to the sticker, but while we are stopped at the light talking to the guy, another fellow puts one on our trailer bumper. We discover this when we get to the Maxwell BOQ. Welcome to Alabama!

Before we get too settled in we take a trip to Walton Beach to visit Col De Armond. Col De Armond was a P-38 pilot out of North Africa. He was shot down and spent the last year of the war as a POW. Its summer and school has yet to start. He has a nice place on the Inter Coastal Waterway. The Col has purchased a Large Chris Craft; he takes us for a ride out to the gulf, my first ride in a motor boat. I'm hooked. We also get to hit the beach and swim in the Gulf for the first time. Waves, sand, saltwater stinging your eyes and going up your nose. This is Lower Alabama and I like it.

Early on I realized that Dad was quick to temper, a hard ass, who did not suffer fools, and that it was unwise to engage in familiarities with him; much safer to play the loyal lieutenant than the son. Jim preferred to play the fool. One of my childhood friends once described Dad as a lean John Wayne, only the real thing. Dad is a Major with 8 kids. Things are tight, military pay is not very generous.

The house is a rental with a converted garage. Jim and I take the garage, it puts us as far away from everyone else as possible.

Our goings on will be on the other side of the kitchen and living room, hopefully muffled out by the living room swamp cooler. The room only has two old garage door windows on one end, it's a dark cave. To make it darker the trailer is backed right up to the windows and under the eaves killing any chance for light to enter. This dark un-air-conditioned room will be a place of misery for me. I suffer through the mumps and Scarlet fever in this room. The fever puts grooves in my teeth. I also lie in this room recovering from thousands of Red Fire Ant bites. The local bully boy took a badminton racket, stirred up an ant hill with it and shoved it up the back of my shirt when I wasn't looking. I thrashed around on the ground, ran home stripped and hosed myself down, real funny. I got sick, real sick, my back, neck, belly, and waist were all swollen and oozing pus from the bites, some infected from scratching and I was running a fever. The kid's Dad made him apologize; he gave me a rusty little Japanese pocket knife as a peace offering. After he left, that piece of crap knife came apart while I was whittling and it nearly chopped off my index finger when it folded. Thanks pal.

We can always get up a ball game, we have a ready line up for street ball. I am playing catcher. The batter, a big neighbor girl, fouls a softball straight down and its spinning on the plate chalked in the street. I reach for the ball. The batter slams me full in the mouth with her bat on a one handed back swing. I spit out pieces of lip teeth and blood. Mom is screaming more that I am, it's a permanent front tooth. I get a temporary cap, plastic with plaster, then ride my bike back the next day for a follow up, Mom still doesn't drive and Dad is at work. The dentist takes a phone call when I climb in the chair. He forgets to deaden me. He grinds the stump down and puts another temp cap on it. That really hurt. I'm griping the armrests trying to pop them while he works away. I had no pain baseline since that was the first work I ever had done, I didn't know about Novocain. I was nauseous and barely able to ride my bike home. I came back a week later and got a low end oversized stainless steel cap glued on the stump. A cap like that won't make the kid self-conscious or anything, or make him stop smiling. Kids won't tease him, he won't have to fight a bunch of jerks or have to grow a premature thick skin. The boy can get punched in the face a lot and that baby will hang in there. Those porcelain caps are delicate and expensive, the boy is always scrapping. Maybe he can replace it with a natural looking one out of his own pocket when he's twenty or so. It also has the added benefit of making him keep his mouth shut. This conversation is going on right over my head.

I go to school, starting the Fall Semester sporting my new look. I have never seen anything like this place. Stretched out for what it seems miles is a rolling building complex facing the road servicing all grades K through 12. There are people suited up for a full pad scrimmage. I am fascinated. I've never seen a football game before. I loiter a while watching them through the chain link fence. They are wearing all this equipment and running into each other; it looks like fun. It's hot outside, even this early in the morning. The big white boys are sweating like crazy, but they look happy to be on the field and running into each other.

The school is segregated. The only black folks are: the ladies who bring by the milk/juice cart in the afternoon, the cafeteria ladies, and the janitors, they are all really nice. They have good steady jobs with the school system. There is a nice Black lady that comes to our house every day to help out Mom. She's very funny and a great cook; she has a big family of her own but she is here taking care of us. I don't think that I have ever taken notice that people were different colors until it was pointed out to me in Alabama.

I only discriminate between Military and Civilians. I'm having trouble getting a grip on this segregation thing, having come from Japan where I was the only round eye in the room on occasion. We learned Japanese and sang Japanese songs and wore Japanese clothes after dropping bombs on them only ten years earlier. The school in Utah wasn't like this either, the Mormons were real friendly and everybody helped out. They let us Catholic military brats and the Utes go to the same schools as they did. My Mom is a green eyed, raven haired, French Hispanic and my Dad is German Irish he's one of those fair men who turn scarlet red when they exercise. Where do I fall in the scheme of things, All American Mutt? No one make me drink from the colored fountain so I guess I'm Ok in the eyes of the locals. There are no signs saying Irish need not apply. Or soldiers and dogs keep off the grass.

I was flunking out of third grade. I was reprimanded and sent home with a note for looking at the fish tank, and again for looking out the window. I couldn't see the blackboard so I would get out of my seat and walk to where I could see. I was reprimanded for being disruptive and told to stay in my seat. I got real bored. I was falling asleep in class, once even crawling under my desk to sleep. The only thing I liked about school was the juice cart in the afternoon that the black ladies brought. It went like this on the ride home after a parent teacher talk with the principal: Bob can you read that sign? No, Sir. How about the sign on that barn? What Barn, Sir? Bad grades and bad eyes will never make a pilot. I was rated out of commission that day, Dad now had only seven of eight effectives. Tough blow for an eight-year-old to know he's already been written off. Might as well hang a Form 1 around my neck. Out of commission. Don't get me wrong, I worshiped my Dad he's a genuine hero. I wanted to be a pilot too. That was the end of my dream. Now I felt that no one would take me if the military wouldn't. A priest maybe? They wore glasses didn't they? I got my glasses: GI issue with a gray upper and clear lower plastic frame; the lenses looked like coke bottle bottoms, 20/200+. Dad puts them on my face for the first time; I look at him and say "gee Dad now I can see all your wrinkles", further endearing me. The glasses complimented my stainless steel tooth and my buzz haircut nicely, completing my look for the foreseeable future. Never was a fan of mirrors. I don't' think that I even really look at myself when I wash up in the morning anymore. Kathleen has never met a mirror that she didn't like.

I became a great disappointment to my Dad that day through no fault of my own, and he let me know it in subtle ways, by giving me new nick names like: four eyes, idiot, stupid and cousin weak eyes to touch on a few. That was two strikes; I was close to striking out, always worrying about the next pitch and continuously hitting foul balls. Most of my problems were caused by close association with my younger brother, who was unfortunately named after both Mom and Dad.

Mom, who never gives up on anyone, or anything, had me tested. My IQ was determined to be 136+. How they figured that out, I have no idea from the battery of tests I took. Thanks to the tests, the list of my personal failings increase. I'm also a lazy bum who won't apply himself. I am seven-years-old. I think I was just hyper. It's hard to sit still or pay attention. My mind is jumping all over the place, and my legs shake, much to the annoyance of anyone nearby. I can't spell. I'm impatient. I butt in and complete other people's sentences. I run everywhere I go. I can't stand board games or cards. I want to be outside hunting or playing ball, senses balloon then and I like it. They medicate kids like me nowadays. I suck it up. I force myself to sit still, concentrate, to do better. I ignore Jim,

I won't allow him to distract me. I'm behind but I manage to pull out of my tailspin; my grades start to slowly recover.

Montgomery is where Dad gets serious about regimenting us. We were getting out of hand. We were not in Utah anymore. We can't be running around the neighborhood terrorizing the locals. No display of BB guns. No shooting in town. No hole digging. No batting hard balls or rocks willy-nilly here and there. No dragging home of stray dogs or critters. No trapping birds and small game. No fires. No forts. No making spears no knife throwing. No fighting in public.

Kathleen is put in charge of and rooms with Maureen because Mom has baby Gary to take care of. Steve rooms with Tim and Bill he is kind to them and neat. Jim and I are quarantined. You address your elders as Sir or Mam.

We were taught to form up in echelon and at attention when Dad wants to address us. When Dad. addressed you in formation you were allowed only one of three responses: yes Sir, no Sir and I have No excuse Sir. Absolutely no back talk. You are not to answer yeah to anything, if you do, you get a flick in the head and you are asked if you are a Kraut. You get a slap in ranks if you deserve it and you do not flinch or torque your jaw or display insolence or you get another one. We mostly get kicks in the ass because Dads hands are to be preserved for flicking switches in a cockpit. The Air Force has spent millions training him to flick switches. He recommends that Mom save her nails and do the same. We are taught to muster for dinner at 1730, no excuses. We are taught proper table manners. You are not to speak at table unless addressed. No one starts eating until everyone is present and grace is said. God it was like West Point, sit on the edge of your chair and absolutely no elbows on the table. You would have to say: "by your leave Sir" to leave the table, pass by or leave his presence. Rules are established and enforced with a smack upside the head, examples and impressions are made. The upside was, one day we were deemed fit to take out in public.

We go to church and attend Catechism classes on Base at the all faiths Chapple. Kathleen gets First Communion from the Bishop and we all go out for ice cream. There are lots of Official functions plus Hail and Farewell parties at the Base. We get to go to the Officers Club for brunch after church. Saturday night is family night at the O club, dinner and dancing to a jazz band, steak night on Thursdays. We like going to the club and we behave; the training kicks in. We wear coats and ties. We have hats like Frank Sinatra, and the swagger to go with the hats. We polish our shoes, we are clean and our hair is combed. We sit at a table with linen, crystal and silver. There is a kids menu. The food is great with real butter, individual pats. We order Shirley Temples from the bar. I dance the Fox Trot, Polka and the Waltz with Kathleen; I'm making points as her favorite dance partner. There is usually a conga line doing the Bunny Hop. We are acting like grownups. The staff smile and nod at us. Is this what they mean when Mom and Dad yell at us and tell us to "grow up"? If so I'm there.

A beautiful two-year-old boxer "follows" me home from school one day, on the end of a kite string I had in my pocket. I name him Brownie, I have a talent for naming pets. Brownie didn't have a collar so no tags either. We all pet him and feed him treats like bologna. The kite string is no longer required. Dad is not pleased when he comes home; he won't let Brownie in the house and he won't let us tie him to the big climbing tree out front. Dad is hoping that he will just wander off. We check on him often and make sure that he is watered and has plenty of food. In the morning we see that Brownie is laying on the front porch guarding the house and his new found friends, big surprise and with no strings attached. Dad now has a dog problem. Brownie is very well behaved; someone has taken the trouble of training and disciplining him, something Dad should appreciate. We are outwitted and torpedoed. One of the instructors has a little farm in the country; he agrees to take Brownie. The pitch is Brownie is a big dog and needs a big yard, he can run around and play with the other animals on the farm. We all pack up and take Brownie out there, we're all interested in seeing where he is to live thinking that it will require our approval. When we get there Brownie forgets all about us, there are two other dogs to play with; we forget all about Brownie, they have ponies, five of them. We line the corral fence. Before a mare's tail can twitch Jim says "that's my pony" pointing to a pretty white pony with a long silky mane and tail; all it needs is a horn to be a perfect unicorn. We duck under the rail fence and walk up to the five of them. They are used to people and don't shy away when we approach them in mass. They like to be stroked. Jim walks up to "his" pony and has Steve knit his hands and give him a boost up on the pony's back. The pony stands there a bit then throws Jim over his withers and to the ground. We howl and point. Turns out these ponies are not riding stock they all pull a light two wheeled surrey. Dad and his friend pull two carts from the barn and hitch up a couple of ponies. One of them is a Mare who needs exercise, we're told: Her fresh little colt runs alongside as we trot down a red dirt road. Brownie is going to love it here. I want to live here.

Late February 1961 Dad comes home and loads us in the station wagon. We swing by McDonalds, a rare treat. Dad buys a sack of hamburgers and fries. Dad drives out to the levee north of the base where we sit on the car to see what he saw flying a T-33 that morning. The Alabama River is in full flood the highest ever recorded, it is 23 feet above flood stage making the river 58 feet deep. The river forms a loop north and west of the city. We don't see much of Buzzards Island, just a rooftop here and there, quiet, still and sad.

Dad gets orders for Colorado Springs. Colorado is just the other side of Utah. There will be mountains, hunting and snow again. We all know that Dad was born in Denver and that he really liked living in Colorado, we think we will too.

# 1962 Ent AFB, Colorado Springs Co.

**JJS.** There were around 5,000 Students in my class at Maxwell, mostly Majors. I had been promoted to Major on 4 March 1957, while in Japan. I was promoted Lt Col 15 April 1962, when stationed in Colorado. Promotions were slow in the peace time service. Some Officers when faced with up or out, chose to stay on with an enlisted rating so they could retire at their highest held rank. Our good friend and next door neighbor form Utah Emory Lepine was a Major with eighteen years in when faced with this choice, he enlisted as a Master Sgt. to get his Twenty.

To press on, there were many different assignments worldwide; I lucked out and transferred to Air Defense HQ in Colorado Springs, Colorado. The assignments came in time for me to fly a T-33 to Colorado, and do some house hunting. I almost put a bid in on a brand-new split-level ranch, but hesitated to close on it until Frances had a look-see at what was on the market. Soooo, we gathered our stuff, engaged a moving Van, and off we trekked with our sixteen

## 1962 ENT AFB, COLORADO SPRINGS CO.

foot house trailer, which the ten of us lived in for over a month.

We camped in a park, next to a creek and a playground, all in all, a neat get-up while we hunted housing. Fran got enthralled with an older place, which had an add-on second level with two separate gas-heating systems, four bedrooms, and a *dead level flat roof* over the add-on, and very old plumbing. However, the property was close to my job. Sometimes, I took my bike to work. The home had a clear view of Pike's Peak; a view that changed hour by hour, as the sun and cloud cover passed. However, the level roof over the add-on, covering the large family room and two bedrooms, plus bath, was prone to leaks. Once coming back from a family fishing/camping trip, we arrived home to find the entire family room flooded from a heavy rain and hail storm.

I was assigned to Air Defense Command (ADC), a Command that was supposed to replace the old-type Day Fighter Squadrons of WWII P-47s and P-51s, the F-84s and F-86s of Korean War Vintage with the later Super Sonic Interceptor 100 series Birds, Armed only with Air-To-Air Rockets and *No* guns. This was a rather drastic change. There was a big emphasis on Electronic Countermeasures. So, I ended up as Ops Officer of a Canberra, B-57 outfit, primarily equipped to Emit Electric Jamming signals to destroy the intercept capability of the new F-101s, F-86Ds, F-89s, F-102 and F-106 Interceptors.

As the Chief of the Targets Branch in ADC HQ, I was in charge of scheduling three B-57 Squadrons equipped with various ECM capabilities, plus WWII Chaff dispensers, to *fool* the radar guided Interceptor Rocket-Armed Birds. In addition, ADC had about 300 T33s, used for plane targets. In my estimation, the ADC Command milled about for many years, and never did a damn bit of good to protect us from an Enemy using Intercontinental

Ballistic Missiles. ADC had interceptor Sqdns in Tucson, Arizona; in Oxnard, California; in Oregon; Rome, New York; two Sqdns on the East Coast and In Florida. Protecting the United States from What? The entire *Lash-Up of a Ring of Interceptors around the U.S. was a Wasted Effort.*

I did my share of T-33 tours to the B-57 Units located in Arizona, Utah, and New York, to inspect and administer check flights on many, many pilots, with the HQ Authority to stick my nose into all the capabilities of all the outfits. I had been acquainted with Fighter folks of past years who were involved in all units, and the job was a piece of cake, kind of a Flying Club, with no great responsibilities other than Fly Safely. ADC evolved slowly into *Space Command.*

The Springs had changed; everything was growing. However, the Town was still small and uncrowded with a population less than 100,000; no big housing developments; the downtown area had changed very little since the 1920s.

**RES.** We are all excited to go to Colorado. Its summer time and school is out. Alabama is hot. We are dying to see mountains again and we are on the road. We weave our way through Alabama and Mississippi for two days, camping in the trailer and visiting Civil War Battle Grounds, then we cross the Mississippi river into Louisiana over the longest bridge I've ever seen. The next day we are in Texas. They don't have many mobile drill rigs yet. I am impressed with the forest of derricks we see around Houston and in West Texas; they have been erecting them since the 20's, they go on for more than a hundred miles. There's a derrick over every well that's ever been drilled, they leave them up to service the wells. We go north through the Texas pan handle because we are pulling a

# 1962 ENT AFB, COLORADO SPRINGS CO.

loaded down trailer with the old loaded down Ford station wagon, that has never really been up to towing a trailer; it tends to overheat, especially when running the aftermarket air conditioner Dad had installed, and going uphill. We play games, usually laying claim to any shiny thing or pony we see before the other guy. Sounds simple but it can get out of hand with eight kids claiming to have seen it first. Mostly we argue with each other, just to be clever and witty, we think. When we get too noisy in the back Dad cranks the radio up full blast. It's a contest. We all try to scream louder than the radio until he turns it down and we turn it down in proportion.

Jim wants Steve to trade him the middle for the window seat, a bad deal. Steve resists. Jim pinches him and makes him cry. Dad had heard every word because stupid Jim is seated right behind him. Dad throws an arm over Maureen, who is next to him and the back of the seat. Dad tells Jim to give him his hand, he is pissed. Jim puts the hand of sleepy brother Bill in Dads hand and Dad gives Bill a gorilla handshake. Bill's response is immediate and quite loud. It's chaos for a while until Mom, who has baby Gary on her lap, takes control. Mom comforts poor Billy and kisses his little crushed hand while burning holes in the two Jims with her eyes. It's pretty awkward for about 50 miles. Shaming is not a spectator sport. Glad I'm in the jump seat in the back of the station wagon with Timmy and Kathy looking backward. All that business is literally behind the three of us.

It takes us three days to cross Texas stopping at airbases and BXs along the way. We visit folks Mom knows during a couple of stops. Imagine ten people, eight hungry kids ranging from eleven to one-year old, stir crazy from being cooped up for eight hours together, hungry and cranky, pulling up in front of your house and announcing "we're here". I'm embarrassed and uncomfortable when Mom does this; it's super imposing and presumptive. Mom is oblivious, she's just glad to see some friends and they in return must be glad to see her. It's all about Mom. I try to make myself scarce and hang out in the car or in the trailer.

We go through Oklahoma and head east into Colorado. As soon as we cross the state line we are glued to the windows vying to be the first to see the mountains; like Columbus's sailors looking for the new world. The mountains are a long time coming but we see Cheyanne Mountain first, and then Pikes Peak, our new home!

We are real road warriors now. We have breakfast by cutting a cereal box with a knife on the dotted line. We pour milk right in the box and eat it with a plastic spoon standing up. We have sandwiches for lunch when we pull over in a rest area. The trailer is so packed that a kid is passed over the stuff blocking the door to fish out the sandwich makings from the fridge and ice chests.

If we want dinner we help Dad level the trailer with screw jacks and blocks of lumber. We kids unload all the stuff that is on the floor, and put it under the trailer before anyone can get in to even start dinner. There are two double upper and lower bunks in the 16 foot Astro travel trailer. Steve Tim and Bill get the front upper bunk Kathleen and the babies get the front lower which is also the dinette. Mom and Dad get the back lower and Jim and I get the upper rear. I usually get permission to sleep in the car. I want to be alone. I don't want to be: ordered around, lectured, listen to, argue or fight with anyone; it's a rare opportunity and I take advantage of it. I have been right next to Jim for 2000 miles and he has been a constant irritant. He grins when he gets under your skin in the car and you can't do anything about it.

Ent Airforce base is not a Base in the strict sense; there's no runway and there are no hangers. If you want to fly you have to go

# 1962 ENT AFB, COLORADO SPRINGS CO.

to Peterson Field East of town. Ent is an administrative compound named after general Ent; he led a large formation over Ploesti and was the Commander of the 9th Air Force. General Ent was killed in a B-25 crack up. I have always wondered why Air Bases are named after dead Pilots, not a confidence builder. There are: barracks, a swimming pool, a PX, a Commissary, an Officers Club with Ent's portrait on the wall, a Movie Theater, a Dispensary, a Convenience Store, and office buildings. Ent is a Command and Control Center, a link to the nuclear strike hardened NORAD installation tunneled into the heart of Cheyanne Mountain, without having to actually go there. Ent is now the US Olympic Training Center.

Our new house 1134 N Meade Ave. is 4 blocks north of the base. On the corner of Meade and Uinta, it's a brick and stucco house with blue spruce trees that dwarf the house. The house is on a hill, it's a split level. The upper level is Officers country. There are two bedrooms and a bath upstairs, a dining and living room plus the kitchen. Downstairs is Indian country; there is a large family room with a sliding glass door to the backyard, with windows continuing across the back of the house. There is a laundry cove in the family room by the stairs which becomes the center of a lot of activity. There are two bedrooms and a bath off a common hallway at one end of the room. Steve Tim Gary and Bill get one big room. Jim and I get a small corner room. The boys have two sets of trundle beds. Jim and I have bunkbeds. There is a partially finished basement under the main house where the two furnaces are. Jim likes to throw his brothers in there; he locks the door on them and turns out the light to scare them and make them howl. There is also a brick ledge around the house where the brick lower meets the stucco upper, we shimmy around the house on the brick ledge like goats and climb on the flat roof of the lower addition. Dad thinks that's it's the hail storms that make the flat tar paper and pea gravel roof leak but it's probably us kids running around on the roof. It becomes clear to Dad when he pulls up one afternoon and finds all of us on the roof with our Japanese schoolboy backpacks and our baby blankets rigged with clothesline like parachutes; we are lined up along the edge and fixing to jump off the 10 foot roof like real paratroopers. The CO scrubbed the jump.

Timmy had some bad luck right off when we moved to Colorado. We were visiting our Air Force sponsor Col. Shear and his family; they happened to have a litter of pups, Chihuahua Spaniel mix. Timmy went to pet one of the pups, which were all real spun up from all the kids chasing after them, when it jumped up and bit his upper lip, giving him a hare lip; lots of blood and screaming. We go to the dispensary with Timmy. We get a free puppy out of the deal and hamburgers at the Mess hall while we wait for Tim to get sewn up. We got a burger and fries for Tim but he can't eat it so we split it. Bonus. The pup looks like a miniature, female, black, short haired Spaniel. We name her Midget. Midget will be subjected to all sorts of ignominies in the future.

Dad is not enthusiastic, another maintenance item, another mouth to feed. Midget must know how he feels about her, she pees when he talks to her, hell sometimes I can barely hold my water when he talks to me. This of course endears her to him. Dad likes to say that she has a nice hide. He promises to skin her and tack her hide on the wall when she dies. This makes everyone squeal and Dad smile.

One little dog, the sole pet of eight little kids, is not a good thing for the poor pup. Demands for her time exceed hours in a day. Midget is a nervous wreck. Kathleen is continually rescuing the poor thing from her tormentors. Dogs do not like to be dressed

## 1962 ENT AFB, COLORADO SPRINGS CO.

up to play baby in a baby carriage or be stuck up in trees to play mountain lion. Midget doesn't like fire crackers or lighting it sends her into a fit of barking and howling so no hunting for her. The answer is bunnies, Dutch Miniatures, cute little things that look like they are wearing a short white dinner jacket. Kathleen, Jim and I each buy one. Kathy names hers Honey, Honey Bunny. Mine is named Emory fittingly after Maj. Lepine. Jim names his Jack, Jack Rabbit. We have lots of fun building cages feeding and playing with the rabbits which we let out in the yard chase around and tempt with carrots and lettuce. Unfortunately Maureen picks up Honey to pet but Honey squirms and kicks causing Maureen to drop her on the concrete breaking her back, something Maureen will forever regret. Kathleen is livid. Maureen is terrified; she shares a room and a bed with Kathleen. I want to shoot the poor thing. Kathleen says no and lets Honey linger dragging her hind quarters behind her on one side. Honey dies later that day. We had a burial ceremony the next day; the first grave in the pet cemetery. She will be shortly joined by Buddy the parakeet, who tried to fly when wet and broke his neck and my turtle Henry which Midget bites and kills punching a hole through his shell.

We get to see our first rodeo when we get settled in, the Pikes Peak or Bust Rodeo. We go downtown to see the parade too. A mounted Calvary troop in old blue and gold uniforms from Fort Carson and a column of soldiers march along with the floats, Rodeo stock and Cowboys on their mounts. The Rodeo is held at the Will Rodgers/Penrose stadium across from the Broadmoor Hotel. There is quite an impressive crowd. We really like the Rodeo and attend every year that we are in the Springs. Some of our favorite actors from Gunsmoke, Dock and Festus officiate and put on some skits. We are instant fans.

# When it's smoking its cooking; when it's burnt it's done.

Dad often musses out loud that he is "the chief cook and bottle washer" but I have never seen him wash a bottle or cook anything besides greasy eggs. Mom is a great cook. She is always the last one to be seated after making sure that everyone had everything they required. We have to practically beg her to sit down. Mom has mastered her mother's recipes and worked alongside her German housekeeper in the kitchen. Mom took Japanese cooking classes when we were stationed in Japan and we still have the book. She also has a Good Housekeeping cook book jammed with recipes and notes in the margin, all in her beautiful cursive, so unlike Dads hand which looks like an EKG. Trouble is she always had help and this is the first time that she has had to manage by herself. Mom is easily distracted and there are plenty of distractions. She would put something on the range and walk away burning dinner. We would still have to eat it. Nothing is thrown away. There are children starving in India that would be glad to have the burnt offerings. The offerings are never burnt they are just "crispy", even toast. Many a morning I am awakened to the smell and sound of burned toast being scraped into the sink; "It's ok it's just crispy". Everyone moans when she says that about two dozen cookies. Dad kids mom saying "when it's smoking it's cooking, when it's burnt it's done". This infuriates Mom. She always blames it on us. Telling us "I told you to watch it". I like to eat. I like my food unburnt. Everyone on KP hates dealing with burned pans. The only remedy: Brillo Pads and elbow grease, something only found in little kid elbows. If I am to be blamed, it's in my best interest to pay attention. So I am at my mother's elbow at all times when she is in the kitchen. First

## 1962 ENT AFB, COLORADO SPRINGS CO.

rule is, never walk away from the stove. A watched pot never boils but an unwatched pot will surly burn. Preheat, time and check on things in the oven. Timing is everything if you want stuff to be ready at the same time.

Mom has the gift of gab. She loves a pun. We spend a lot of time together in the kitchen with her chatting away about nothing and everything, mostly about people she knows and I do not. So of course she has to tell me their life story and that of their children to bring me up to date. She says "you know" a lot as a segue from one subject to another. Dad calls her you know Nuno. I try to throw her with a pun or a change of subject but she never misses a beat.

After a while I get down all the standards: spaghetti to fried chicken and all the fixings. We all like Chinese food. We have the first and only wok on the block to go with a nice Japanese rice cooker which we are very picky about, no uncle Ben's for we sticky rice fan's. Dad loves beef and peppers. I get good with a knife. There is a lot of chopping in Chinese food. Dad has me carve the turkey, roasts and ham. I perfect my Spanish rice and refried beans with bacon and onions a favorite on Taco Tuesday. There's fish on Fridays of course. Since cooking is not such a chore anymore, now that she has help, Mom shows me how to make some fancy stuff. Deserts are always popular. Sugar. I make: gallons of tapioca and chocolate pudding, pans of bread pudding cheesecakes, German and French pastries, cookies, cakes and pies, all from scratch. Kathleen gets in the act too. She likes to snack on what's cooking; she acquires a taste for steak tartar, when we marinate the beef for Dad's beef and peppers. Another advantage of being the cook is you usually get to cook what you want. Kathleen lands on beef heart, it's cheap and plentiful. She slices it up real thin sautés it with onions and green peppers then braises it in tomato sauce and red wine for a couple of hours then served over white rice. It's her favorite if not everyone else's. I do the same with liver and onions with bacon. Yumm; entrails for entrées all you can eat. Dad brings home live lobsters a couple of times, a great treat. The preparation draws a crowd. When he flies to LA he usually brings back some Mexican candies and treats for Mom, she can have them, and a big wheel of Monterey Jack cheese for us; unlimited quesadillas.

## Divine Redeemer

Our house was one block north and two east from Divine Redeemer Catholic Church and Elementary School. The school was staffed by Jesuit Priests and Dominican Nuns. There was a Rectory and a Convent on site. These were some of the finest people I have ever met. The Nuns reverse my academic decline, they teach me to love books, sentence composition, and mathematics. We get fountain pens and learn penmanship. Mom taught penmanship; her longhand looks just like the example above the blackboard. I want to write like that. I'm still a bad speller, so is Dad and Andrew Jackson so I am in good company. Nothing makes sense, and when I really concentrate, I can spell a word three different ways none of which is correct, the harder I try the worse it gets, but I can spot a misspelled word in context. Go figure. I habitually spell phonetically. Spelling, for me, seems to get in the way of ideas.

    The Nuns make sure that every one of us gets a first class education. The Nuns are wearing the big flying Nun headgear with full robes, but that doesn't stop some of the younger Nuns running around playing kickball and softball with us in the morning and at recess. How Dad swings enrolling five kids in parochial school I have no idea. Dad must have promised a pretty a heavy tithe

## 1962 ENT AFB, COLORADO SPRINGS CO.

for that (the military allowed and subsidized School Choice). We wear uniforms. Boys wear black dress pants and shoes with a white shirt and tie. Girls wear a white shirt and a sky blue school jumper with the DR badge over the heart. Kathleen hates her uniform. I hate wearing dress pants, they don't fit me right; the pockets always stick out. Jeans come in Slim Regular and Husky; I have been in Huskies since Utah. Dress pants do not come in Husky sizes so my dress pants bloom out at the pockets, you can see the whites of the liners. The pant legs creep up with the bloom for that chicken-leg, high-water look. Jim doesn't have this problem; he is slim Jim. We have to change out of our school uniforms and into fatigues when we get home. It's time to do chores. Out of one uniform and into another. We have actual OD fatigues and army caps, flat top Korean War type hats. Fatigues come in Husky sizes. We sew unit patches, rank and MOS badges on our fatigues. I'm a lieutenant with the Combat Engineers, Jim is a Sergeant with ordinance. Jim likes his collar device, it looks like a bomb; he puts one on his hat too.

I go to church every morning before school. We play dodge ball before streaming into the Chapel when the sun is up and Pikes Peak is Rosy Red. I love the Mass, the ceremony, the candles, the beautiful Vestments, the incense and the Priest singing the mass in Latin and the people singing back in reply. I want to be an altar boy. This Church is not like the All Denomination Churches on base that I'm used to. Divine Redeemer has stained glass windows, a beautiful alter and Sepulture where Christ lives. Christ on the cross is very graphic as are the ornate Stations of the Cross. The Nuns help me with my Latin in the hall. I qualify as a torch bearer for high Mass, an altar boy in training. I am going to be a good boy from now on and repent for my sinful ways.

I walk home by myself one day, the air is electric in the house. Jim hasn't changed out of his DR uniform and is sitting on his bed, he won't tell me what's up. Kathleen gives me the skinny. Sister Maria Teresa, Mother Superior has expelled Jim. It seems that Jim was bullying some kid in class and when little 4'10 Sister Agnes Marie swatted him on the back side with a ruler to break it up; Jim spun around and clocked her. Jim is humiliated. He will be the only one of us attending public school. He will ride out the end of the year as a transfer student at Queen Palmer Elementary. He is failing every subject including religious instruction. Jim will be repeating the third grade. But first he must attend summer school or be put back. Jim will be at the same grade level as Steve for the next ten years. Jim never tells me his version of what happened or opens up about it, I couldn't do it if it were me.

About two weeks into Jim's transfer to public school he asks me for a favor. Seems he is having a hard time making friends. Seven kids want to fight him, Jim has agreed to fight all seven kids after school but only one at a time. I am to show up to make sure they don't gang up on him and keep their end of the bargain. I'm not too worried about Jim. I've seen plenty of little kid fights. It's usually a lot of yelling and wrestling until both of the combatants get sweaty and tired. No one really gets hurt except for someone's pride. Jim is an exception to this rule and so am I. Dad has been physically demonstrating some dirty tricks on us usually prefaced with the statement, "let me show you what I learned in Japan", code for hand to hand combat. Jim is already facing them down and talking trash when I roll up on my green Japanese bike. Jim introduces me and tells them what a bad ass his older brother who wants to be a priest is and why I'm there. It's on. Why the seven of them don't rush the two of us is an example of poor leadership. Jim must have had them spell bound with his audacity. Maybe it's my silver tooth. Maybe it's how my thighs seem to bulge in my Divine Redeemer uniform dress pants that makes me intimidating. The first guy steps up. Jim uses the move Dad taught us where you step on the other fellow's foot and then punch him in the face when he looks down; that's one. The next guy is a little more cagy he watches Jim's feet; a mistake. Jim boxes his ear and he is through. The third guy gets wrestled to the ground. Jim gets on his back and bends his head back by the hair. The kid gives up. Nobody wants any more of Jim; he is magnificent, all those sparring sessions with me have paid off. Jim still has a lot of political problems to solve though. You can't fight everyone.

They say "if you can't lick-em, join-em", but what happens if you can lick em? They don't join you; that's what happens. They leave you alone. Friends are not like brothers they don't have to put up with you. Jim is ostracized. He can't get acceptance by fighting them all. Jim has decided to buy his friends. He is very pragmatic.

Jim and little Timmy attend the same school; there are no Kindergarten classes at DR. Tim walks himself to school. Mom showed him where to cross the street, how to use the crosswalk signal and where to go, from then on he was on his own. Jim wouldn't walk Tim to school anyway. He would probably just ditch him if he was made to. He wants to be the big man and doesn't need a little kindergartener hanging on him. Timmy trails/stalks Jim home after school one day and see's what he is up to.

When Mom was living at home her Dad gave her a 1922, the year of her birth, liberty silver dollar coin for every year she had been alive each birthday. Mom had quite a stack of coin in her jewelry box from that custom. We all knew about the coins, had played with them and had heard the story behind the coins. Jim didn't think that a coin or two would be missed.

Jim took a coin and went to school. On the way home Jim cashes in the $20.00 silver coin. He asks the clerk if it its good, the clerk says "sure kid, you want some candy?" Jim takes the candy and gives it all away in true Irish "drinks on me" fashion. Jim becomes immediately popular. This works. More money is required. The stack of coins diminishes. Jim is discovered, he has gone complacent and has not covered his trail. Tim is wise to it and tells. Dad stalks Jim, he follows Jim to the candy store. Dad interrogates the candy store clerk after Jim leaves. Jim is busted.

Dad spanks and lectures Jim he warns him never to do it again. On no account is he to steal again or ever go back to the candy store. Jim can't help it, he has an empire to support. Jim steals money out of Moms purse this time and goes back to the well. "The tiger never changes his stripes". Jim is on Dad's radar scope. Dad is on his tail, he catches him red handed at the candy store stealing and disobeying orders. Jim is pissed and he takes it out on Tim sure that he reported him again; he beats Tim with Dad's Sam Browne Belt. I can't believe it. Jim gets another whipping. Dad asks Tim to select a belt. Tim picks a small thin Wally Cleaver belt thinking that a small thin belt would be easier on Jim than a big wide belt. He wants to reduce the potential for further retribution. Instead the thin belt raises some impressive welts that I put some salve on.

Dad is TDY a lot; this takes the heat off of Jim. Dad flies a medical emergency run. He delivered some inoculations to Fairbanks Alaska direct from the CDC in Atlanta nonstop. The inoculations were to stem an epidemic. Dad flew a world record 6000+ mile nonstop mercy flight on this occasion. Dad brought back a load of Military Surplus Artic Gear from Alaska: sleeping bags, shelter

# 1962 ENT AFB, COLORADO SPRINGS CO.

halves, canteens, back pack frames, garrison belts, bayonets, web gear and entrenching tools, plus a bunch of parachute seat pack pads and a couple of old parachutes, some para cord and a couple of parachute harnesses. Lots of room in the bomb bay.

Jim and I set up a pup tent using two shelter half's in the backyard, we use a couple more shelter half's for a floor. Jim want's out of the house. The tent is set up next to our room behind a blue spruce tree. We throw sleeping bags in the tent and crawl on in with some flashlights and our BB guns. Dad lays off Jim, maybe it's because I am shepherding poor Jimmy in a tent in the backyard in Colorado like his brother used to shepherd him.

Dad has been reading the bible every night but the only quote I hear from him is "they shall be cast out into everlasting darkness where there will be wailing and gnashing of teeth", he says this with a grin. I remember him having me face the wall in the hall to his bedroom kneeling and praying, I also remember assuming the same position with my uplifted chin pressed against the wall and my hands clasped behind my back instead of in front, the what for is lost to me. I used to ask him "what for" and he would answer "I'll give you what for". Me? I'm pouring over a T-33 manual that Dad has been marking up to submit for revisions. I read his comments in the margin and I try to understand what he is talking about. The manual is fascinating especially the drawings. I've hid it, in my room, under my pillow. I read everything, study it and page through it cover to cover several times. So far I am undiscovered. I would give my right arm to fly a jet fighter but I would be all stick and no throttle. I damn my eyes. The fellows that designed and built this beautiful plane, thought everything out on paper, drew it up, and built it must be super clever. I would like to be around people like that. I put it back on a shelf where he can find it.

# The Great Cherry Pie Trial

We had a huge old cherry tree in our back yard, just off the kitchen stoop. The tree was a major attraction for us; not only was it cool to climb and hide in, it bore fruit! Food that grew on trees was something new to us kids. We also had an apple tree and a raspberry patch but they paled in comparison to the cherry tree. In the spring, after it flowered, and the fruit started to set, we were up in the tree tying strips of cloth and foil to the branches, in the forlorn hope that when they fluttered in the wind it would scare off the Robins, which at that time of year were declared our mortal enemies. To assure that the Robins would not get our cherries, we posted a guard up in the tree often accompanied by Midget our tiny yappy Chihuahua Spaniel mix; hauled up by her leash and harness. She looked like a black short hair Spaniel but only 10 inches high at the shoulder. This was serious business, and the guard rotation was strictly enforced. Midget was tied to the tree trunk when we were at school; although a Chihuahua's attention to duty is questionable her loyalty and shared love of ripe cherries was not. This was a big tree, the guard and the guard dog could eat his or her fill and there always were plenty left; I don't know who the largest consumer was the birds, the dog, or the kids.

When harvest time came, we were ready. We had cleaned and saved all ten half gallon wax paper milk cartons we consumed each week that spring, carefully opening the top by unfolding the ends. The cherries were picked, pitted and placed in the cartons which were then placed in our huge top opening freezer in the basement with the bulk frozen bread and hamburger staples from the commissary. There was room for little else. We got most of the cherries out of the tree because a little kid can shimmy pretty far

## 1962 ENT AFB, COLORADO SPRINGS CO.

out on a limb and we were fearless. It's a wonder that none of us fell, but then again we were experienced tree climbers by harvest time. Even the wind fall was rinsed and collected. A half-gallon of cherries will make about three or four pies. Flour salt sugar and Crisco are cheap. The cherries were free and so was the labor. I got pretty good at making crusts for a sixth grader; even my mistakes were consumed with relish. Splitting three or four pies between ten people was a feast and every one could have their fill. As the summer wore on I tried to stretch the cherries, adding tapioca, apples, or rhubarb from the neighbor's garden. But as with all good things the cherries ran out.

Months later that fall Mom did an extraordinary extravagant thing, she actually bought an eight inch cherry pie in a tin pie plate from the commissary. There was a basic incongruity presented here, our appetite was up and an eight inch pie makes for small shares among ten people. Said pie was in a flip top box with a cellophane window and in the fridge. It had a full top crust except for a single cherry sized hole in the center. I know that every one of us checked it out and was tempted, who succumbed first remains a mystery to this day, but succumb he or she did. Well once a crust is broken there is no putting it back and no hiding the void made by the missing cherry. This did not go unnoticed to the constant fridge traffic and so the hole and the void grew, each visitor making it only a little bit bigger until it was a third hollowed out. Not sated by a cherry and a bit of crust, one of our number had taken a package of bologna and impulsively bit into it thru the shrink wrap packaging leaving teeth marks. Evidently, the perp, having second thoughts returned it to the meat keeper drawer. Mom first found the pie and then the bologna announcing those dreaded words, "wait till your Dad comes home" something no one in their right mind really wants to wait for. As the afternoon drags, by we all scrambled to do chores and execute the plan of the day, in order to better our position, which pleased Mom, but made no difference at all in the end.

The time had arrived, the Colonel drove up to the back gate and honked his horn. A gate party of two stopped folding laundry and ran to open the gate to the carport and Dad pulled the Carryall in.

We were all hoping that Mom would wait until after cocktail hour but this was an offence that wouldn't wait; she spilled the beans before the Colonel was out of uniform. We were mustered in the kitchen in echelon at attention and dressed down, the violated foodstuff on the table before us. The offender or offenders were bidden to step forward, but no one moved an inch. None of us were willing to take the fall for the entirety of it all.

Instead of the usual group punishment where the malefactor would be singled out by his resentful peers, Dad ordered us into the family room and directed us to deliver up the guilty party. This was new; not knowing how to proceed but having watched plenty of Perry Mason trials, we elected that format. Now a trial needs structure, starting with a judge. That job fell to Kathleen, who appointed herself, as the oldest, most responsible and by her own opinion beyond reproach due to her position as our constant babysitter and supervisor. Next needed was a jury. Kathleen appointed Maureen and Gary, the babies, the innocents, besides Dad would never accept either of those two as the guilty party, as jurist. There were no dissentions. The judge required a prosecutor to make the case, Steve was appointed to this post due to his previous affiliation, alliance and support of the sitting judge and his unique ability to never be tarnished by any of the general goings on. The quorum was now split 50:50 between the court or demi grand jury and the suspect

## 1962 ENT AFB, COLORADO SPRINGS CO.

pool. The split was along the usual lines. These alignments had long been established, the good and the bad, the sainted and the tainted.

Not wanting to address the problem at hand we all turned to setting up the proper court furniture. After all, you can't have a proper trial without a court room. So we all turned to that. We pushed the ironing machine out and covered it with a blanket for the judge's bench. We made the witness stand out of a stool. The jury sat on the couch and the rest of us sat on the floor behind the coffee table placed before the Bench.

The prosecutor made a motion that the pie, and bologna assault charges, be combined into a single charge, in the interest of a speedy trial and an end to habeas corpus of the judge, jury, and suspects; this the judge assented to. Biting into that pack of bologna was a bone head move. Jim was the only one in the pack that had four front teeth missing I knew he did it. He should have swiped the whole pack to cover his tracks, but he bit into it and put it back in the meat drawer, some sort of primeval marking your territory stuff. Back then we would lick something, maybe a plate we wanted to use, in front of a witness, publicly announce the fact and from then on no one would touch what you licked. Maybe he was thinking something along that line, if you can call biting a pack of bologna thinking.

Jim protests of course. He lies right off the bat and says he didn't do anything. No one believes him. Jim want's to dissolve the court due to the Judge's extreme predigest. I feel sorry for Jim he's really in for it. I move that we collectively use the "my dog ate my homework defense". We have the evidence on the coffee table, we should just destroy it and blame it on the dog. Yeah Midget ate everything while we weren't looking. Both motions are denied.

Steve and Kathleen make their easy case that Jim bit the

bologna by taking molding clay and saran wrap bite impressions of everyone. I must admit that was pretty slick, Jim fights them the whole way, he refuses to bite and threatens everyone in the courtroom. Jim loses by default because none of the impressions taken by the court match the bite in the twelve slice, shrink wrapped, plastic rind, bologna flat pack. We back it up with a secret ballot. There is only one dissenting vote. Jim doesn't want to take the fall for the bologna or the pie. Jim has abused just about everyone in the courtroom at one time or another and this is their chance to give him his comeuppance by sticking him with both offences. If he bit the bologna he bit the pie. It can be shown irrefutably that he bit the bologna and lied about it, so why should the court believe him about the pie? Poison fruit from the poison tree. It's looking pretty grim for Brother Jim. This is the smoothest running railroad job in judicial history.

This is just not right, we should stick together; we are all here together because we stuck together in the kitchen. I say, "If we protect Jim he'll be good". No one believes me. I just don't want him to be singled out like that. Dad is really mad, Jim will certainly get a whipping if Kathleen delivers him up. Jim has had quite a few whippings and it never helps. He just gets meaner and has to take it out on someone. The court doesn't understand this and is in no mood for mercy.

Kathleen presents the courts findings. She shows Dad the clever bite impressions. She tells Dad about Jim lying. She has Dad at lying; he hates liars. They never even get to the pie. I put some ice and ointment on his welts while Jim tells me how much he hates Dad and everyone else and how he is going to get even with them. He's not too keen on me either. He tells me you'll all be sorry when I'm dead and gone.

1962 ENT AFB, COLORADO SPRINGS CO.

# Trust and the Refrigerator

Dad used to play this game with us ever since Randolph: He used to tease and chase us around the kitchen, make us squeal and laugh then grab one of us up and put his catch on top of the refrigerator. That was pretty high up for us little kids, we were actually afraid but we played along, issuing anxious uncomfortable giggles. Dad would threaten to leave his victim up there, turn his back, ignore him and act like he was not there and he couldn't hear the pleas, sometimes leaving the room and the kid on the ledge pondering their fate while he had a smoke. We would continue to scream and holler to be let down, careful not to whine too much; whining was unacceptable.

Eventually Dad would approach the refrigerator with his arms extended and tell the kid to jump arms akimbo. The first prompting was always way too far to jump. There was a lot of hesitation and squirming on the part of the perched. Dad would move a little closer but never within your reach and repeat his prompt leading to more squirming and nervous laughter. Eventually the kid would find the courage to leap and Dad would catch you then swing you around in the air. Lots of screaming. Great sport. We all went through it. There was always a loud raucous audience to witness the victim's discomfort, eventual rescue and the victory loop.

One day I was snatched up and the games began. I could have jumped down on the stove and escaped, I was big enough now to do that but I didn't. After a lot of torture and Dad not getting any closer, I went for it. Dad dropped his arms and let me hit the deck. My glasses went flying across the room when I smacked the side of my head on the floor. I rolled over on my back and looked up at him totally stunned. Dad crouched over me with his hands on his knees and said "let that be a lesson to you, never trust anyone".

That was the end of the refrigerator game for everyone. I couldn't read what was on his face or in his eyes, all that was fuzzy without my GI glasses. I don't know if he was in his cups or if he had a bad day, maybe I was too big or too heavy to play or he flat didn't like me. Maybe he just didn't want to play that game anymore, whatever. He did teach me a lesson that day that I have never forgotten, I was eight-years-old.

**Chores.** When we hit The Springs there wasn't a local levee. No extra help, no maids, no more babies. At the same time we kids were generating a lot more laundry, dishes, and trash, kiddie litter. Gary was still in diapers, cloth ones, dealing with a number two in one of those babies requires some technique; the old single hand flush and swirl. The other hand was for your nose.

We try to share the load but it turns out that it's better if we share some chores and specialize in other tasks. We all rotate KP on a weekly basis between the four oldest. You had to do the pots and pans plus the dishes for ten people every night for a week, once a month, and you had better use elbow grease on those pots and pans, all the dishes were washed by hand by one kid. Paper plates were very popular. We did that for a year or so until Mom purchased a mobile dishwasher at an auction. It got better with the dishwasher but it was still the same duty roster. You had to wash the dishes before you put them in the dishwasher; it was a roll around thing that didn't do a very good job if you didn't.

It's after school, late in the afternoon, I have let the dishes slide. I have last night's pots and pans and the dishes I also have breakfast and lunch stuff to take care of. It's stacked up in the sink and on the counter. It's going to be a long afternoon. I pull the step stool/ baby high chair we have all sat in, up to the sink and get to work.

# 1962 ENT AFB, COLORADO SPRINGS CO.

Somewhere in there I put my elbow on the counter and my face in my hand and doze off. The kitchen is warm; the sun is on my face. I am by myself, it's quiet. I'm tired; there are a lot of dishes to do. Next thing I know I am being jerked out of the high chair. Dad has grabbed me by my hand in a judo hold he often uses but he twists it as I fall out of the high chair and hit the deck. I scream. My hand is doubled back on top of my forearm. My fingers are pointing at my elbow. That's not right. Dad pulls my hand back in place. I scream again and get real woozy. Dad runs cold water on my head and wrist then takes me to the Dispensary, four blocks down the street. My forearm, wrist and hand look like they belong to Popeye by now. There is no Doctor on duty after hours, but there is a very senior Corpsman there to take a look at me. The Col. is in full uniform when he checks me in, he had just come home when he jumped me. He brought me straight to the dispensary without changing. The Corpsman says:" that's a real nasty injury Sir. How did it happen?" Without missing a beat Dad says "I did it". The corpsman doesn't feel qualified to read the X-rays but he can tell that there are no broken bones. He makes a plaster splint, a half cast, because it will have to be replaced with a full cast after the swelling goes down. My arm still hasn't finished puffing up when he puts it on with an Ace bandage. I am sent home. When people ask me what happened I tell them I sprained it. Dad has a pretty bad temper; I just figure it's because he's been shot at a lot. It's my fault for falling asleep on duty. They shoot people for that in the Army. I can understand. Dad usually has a friend who is a Priest. I believe he is sorry for what he did, but he doesn't say so to me. He has been reading the Bible each night lately. Dad has dropped a lot of bombs on people, I'm just one of them.

It's my duty to take care of the trash. I have to sift out the cans and glass then burn the rest of it in an incinerator out back. After the incinerator gets full I have to shovel it out and put the ashes in the ash can. A one armed trash man is about as effective as a one armed dishwasher. I am relived from dishwashing. I can't get my cast wet. Jim is to help with the trash. Jim likes fires. Jim is above all lazy and he won't take any direction from me. I can't fight him with one arm. He would probably make sure to hit my injured arm. Jim packs the incinerator without shoveling it out first. There isn't enough room for all the trash, he lights it off anyway. I tell him he should have shoveled it out, now we have to stand around and feed trash into the incinerator as it burns down. We can't just leave all the trash cans out here and let trash get blown all over the place. It's getting dark and cold too. We have to hang around and burn trash for over an hour now, because Jim wouldn't listen to me, and just turn to for five minutes. We have lots of time on our hands, not a good thing.

Jim picks up a hairspray can. He reads what's printed on the label. It says in big, bold red letters, right on the back of the can: "DO NOT INCINERATE KEEP AWAY FROM FLAME AND FIRE EXPLOSIVE CONTENTS". Jim casually tosses the can in the fire and slams the door shut; he wonders out loud what will happen. We run for it and get behind the trailer on the other side of the carport. The can blows and powders the ashes in the incinerator, knocking a dusky plume of ashes and sparks skyward out the chimney. Jim is a genius we can now get the rest of the trash in and be done with it. As we are pitching more trash in the burner Jim comes across a large bloody bandage, it is two inches wide and three or four inches long it is in a paper sleeve. Jim is not aware of anyone hurt enough to require a compress that big or that bloody in our family at least not lately. Jim sets it aside; he treats it like a relic. Jim comes up with this fantastic bandage lineage while we are pitching stuff in

## 1962 ENT AFB, COLORADO SPRINGS CO.

the fire: it's a compress from one of Dad's wounded B-24 crewmen. Yeah, looks like a belly wound. Dad must have been cleaning out his old trunk, came across it in an old flight bag, a first aid kit with Sulphur powder like on Combat. Dad had finally decided to pitch it. I am half listening like always when Jim is talking. I try to filter what he is saying but my filter is clogged right now. I am still trying get over the fact that he just threw a grenade in a roaring fire that I was standing in front of.

We finish up with the trash and bring the cans back into the house. Jims brings his relic in too. After we distribute the cans around the house, Jim takes his relic on tour. Jim is a regular PT Barnham. I expect him to start charging his brothers a nickel a piece just to see it. Every time Jim tells his story of the relic to another kid it gets better: the wound is more grievous, the bandage bloodier, the mission more dangerous, everyone is more heroic. Jim puts it in a First Aid pouch that has a white circle and a red cross on it. He hangs the pouch from a web belt and hangs the web belt on himself for the tour. Kathleen eventually overhears Jim's pitch; she is skeptical. Kathleen is always skeptical of what Jim says and for good reason. Kathleen asks to see what he is talking about. Jim wants her to pay for the privilege. Jim is very capitalistic these days: he often promises to pay his brothers to do his cores, runs up a tab and never pays up, a dime to do the dishes. Kathleen refuses to pay to see and calls him a liar. There is going to be a fight. Kathleen, like me, knows that this is a classic case of how bad do you want to take honey from the honey badger? Kathleen is still bigger than Jim and can take him if she wants to, but is she in it to win it? Kathleen is smart enough to pick her battles and this is one battle that she doesn't care to fight. She calls him an idiot and walks away for the time being. Jim is going on report.

Mom hunts Jim up after a while. She wants to know what Kathleen is talking about. Mom can always get to Jim. Jim shows her his relic. Mom takes possession of it and that's the end of it, or maybe not. I see Kathleen later. I ask what's up, she turns red and doesn't want to talk about it; she knows something that I don't. Mom is no help either. I finally figure it out when I see a box of feminine products in the upstairs bathroom under the sink, the product looks a lot like Jimmy's belly wound compress. Jim is a jerk. I tell him to his face, all the kids know what it was now; Kathleen is embarrassed and hates Jim more than ever.

Jim continues to "help" with the trash, using his hand grenade cleaning methodology until I notice that the sides of the incinerator are fracturing into chunks, it's falling apart and one side soon collapses. Dad can't miss this, he pulls his car alongside and his door opens to it every evening when he backs into the carport. He takes a shovel and pokes the rubble around, he sees all the blown hairspray cans and blows a gasket. Dad needs to buy a new incinerator. We both get dressed down. Jim is pulled off trash duty. Thanks Jim. Dad gets a load of old bricks. We get to break the mortar off them with a hammer and use them to pave the bare dirt patio around and under the cherry tree placing them on edge. Jim does the hammering and I'm the one armed brick layer; what a team. Idle hands are the devils playground. Braking rocks on a road gang at the age of ten. Great. Dad calls us slave labor.

Dad is TDY a lot. Mom does not drive. We need to eat. We eat a lot. I am drafted. I have my old red flyer wagon; it's painted Air Force flight line tractor yellow. Cap. J.J. Scanlon is stenciled on the sides. I love my wagon. The wagon is heavy gage steel with rolled edges and solid rubber wheels. I think it would stop a .30cal round. This was my wagon in Japan for three months, before Dad

# 1962 ENT AFB, COLORADO SPRINGS CO.

hijacked it to carry his B-4 bag and gear out to his plane on the flight line for those one month TDY missions to Korea. Later he used it to cart his A/C unit for his Space Suit out to his plane for those D model missions in Japan and Utah. I only got it back because the D handle on the draw bar was twisted off when a line tractor hauling it jackknifed it at Hill. Anyway I would haul the wagon the four blocks to the Commissary on Base, shop, load the cart, pay for the stuff and load it in my wagon, usually at least eight half gallon cartons of milk. They would make me take a number and go through the pickup line with all the cars to get my groceries; that was their system. The folks at the Commissary are very kind and understanding when I pull my wagon through the line, they all know me. I see them at least once or twice a week.

Once I had the stuff, I would haul it all the way home up a slight grade with no handle on the wagon; just sharp barbs of steel sticking out from the shank. I made any number of mods to the handle to help with hauling the groceries home, finally hitching up to the axel like an Ox on Wagon Train using some rope and just use the handle to steer. No one ever helps me with this particular chore. This wagon is to be our "war wagon" the chassis for all manner of mobile forts tanks, and race cars. Tim is the Test Pilot for many of these contraptions. On a run down our sidewalk on Uinta Blvd Tim went off the end of the sidewalk out of control crossed the ally, crashed into the neighbor's chain-link fence and caught the shank under the eye in the crash, requiring three sutures. Since Jim shot Tim in Utah he marks time by his accumulated injuries well into the future this is just another one.

We boys are kicked out of the house. I am put in charge of a reluctant, surly, unmotivated work detail, Kathy Reenie and Gary are exempt. We are to weed the parkway. The parkway is faced with broken granite, sharp edged stones the size of a little kids thumb; it's full of chick weed and dandelions. The parkway is overgrown because the previous owner failed to lay down plastic sheeting before laying down the rock. I'm given a twenty gallon trash can and ordered to fill it. Our sin: getting up early on Saturday morning to watch cartoons. We were watching Fireball X-19 and laughing because Zooney, one of the puppet characters, a space alien pet, acted a lot like Billy, and Billy was playing along. Billy was a hoot but just a little too loud with his rendition of Zooney's "Welcome home". We were a little loud too, egging him on and laughing at his impersonation. All that's true, we were making a racket, but I don't know of any kids that sleep-in on Saturday. Saturday morning is the only time they air cartoons. We spread out and get to work. I look up and see that everyone has a pile going except Jim. Jim is laying on the grass, in the shade of our fence, the other side of the sidewalk from the parkway. I tell him to get to work, we have to weed the parkway or Dad is going let us have it. Jim calmly replies: "I can take it. Can you?" I want to punch him in the face and tell him to take that. Instead I whistle up Timmy and tell him to report Jim. Jim snarls at Tim and then moves off the grass to the sidewalk and stares at the weeds. What do you do with that? No one want's to be out here doing this. Jim is lazy. He likes giving orders not taking them. Jim loves to order his little brothers around. He treats Bill like his personnel Batman. He even makes him massage and scratch his back before he goes to sleep. I tell Bill he doesn't have to do what Jim says but Jim has convinced his little brother that it's in his best interest to do what he says. When I go to confession before mass I usually don't have anything to say so I confess some of Jim's sins. I talk about being a bully to Bill and it makes me cry. Three Hail Mary's and an Our Father, all is forgiven. We fill the bin by keeping the weeds fluffy. Jim stuffs wadded newspaper in the bottom of the bin, his contribution, instead of pulling weeds, compromise is an ugly and dangerous thing but the detail is happy to get out of the sun and try the TV covertly; no one is going to tell.

Saturday mornings are best spent at the base theater, it opens at 0730. For a dime you can watch cartoons, a cliffhanger and a movie, out of sight out of mind, no curtain climbing brothers and no responsibility, worth every penny. I mow the lawn with a push mower in order to be able to do it. It's not too bad downhill except you have to dag it backwards up the hill to mow downhill. I am too small to mow uphill.

## The Princess Phone

When we move in, new phones are installed. Mom gets a "princess" phone installed in her bedroom. This is the top of the line coolest phone ever, it is small, compact, and oval. It also has a lighted dial. There is another line in the kitchen and down stairs, wall mounts. We soon find out we can talk to each other just by lifting the receivers. Lots of fun. Until the bill comes in. Dad is screaming mad when he gets it. The deposit for the new phones, the install plus the monthly fee and the long distance charges are busting his balls, Mom just cannot have a phone conversation under forty five minutes. This is an argument that he can never win. Dad has made a lot of promises. He has dragged her away from her mother and her support system. She was the daughter of the second youngest of sixteen siblings. When she was a little girl she used to ride her bike to her aunt's houses where she would be pampered and spoiled. She has borne him eight children. And she will talk to her family and friends. He gives it a go anyway. He says that if you can't get your point across in under a minute, put it in a letter. You can almost hear it: "too much chatter on the R/T". Dad likes letters. He likes to get them and send them even if it's just a short note you dash off and stuff in an envelope. Something he cherished when he was overseas. It was good to get a letter from home every few days; Grandpa used to sign off his letters with "happy landings" after Dad's three crash landings, Dad signs off cheers. Letters he sends to me later in life end with "write when you find work" an Irish depression era thing. We are allowed to only answer the phone and memorize our number. We are to say Col. Scanlon's residence, how may I help you. We are not to play on the phone or tie up the line, read actually use the phone.

## The Family Room

The family room is where the action is, it has: all our toys and games, the guns, TV, the washer, the dryer, the ironing board and a portable record player with exactly two 45 records. The records are: Perry Como side 1 Catch a Falling Star, side 2 Papa Loves Mambo and Elvis Presley side 1 Blue Swede Shoes, side 2 Ain't Nothing But A hound Dog. We sing and dance to all the songs because that's all we have. By far our favorite is Hound Dog. It's like a dance party when someone puts it on; everyone drops what they are doing and start jumping around as if on cue, even Dad likes this song, he likes the part where Elvis sings "if you never caught a rabbit then you ain't no friend of mine". All of our dance moves are current because we watch American Band Stand and Soul Train. We all know how to do: the Twist, The Pony, The Swim, the Mash Potato, the Frug, the Watutsi, the Jitterbug and many others. Kathleen shows us how and critiques our moves.

# 1962 ENT AFB, COLORADO SPRINGS CO.

If you are in the family room and the TV is on you had better be doing laundry. That's the rule. There is a lot of laundry to do with ten people generating it, probably ten or twelve loads a week with all our underwear, diapers, towels, work and school clothes. We usually just stand there with a laundry article in our hands, pretending we're folding, staring slack jawed at Wagon Train.

There are piles of clothes everywhere because it never really gets done. The temptation is to run large loads in the washing machine to knock down the piles; that along with inexperienced operators causes many flood events. When the washer overflows it soaks a large brown pile rug and everything on it. We have to move everything off the rug roll it up and drag it outside to dry. The flood events only serve to double the work putting us farther behind. We mop up the floor and start another load.

While the rug is up we have Ben Hur like chariot races around the room but instead of chariots we use baby blankets. You have your brother sit on a blanket, wrap their legs in it and drag them around the floor and swing them around cracking the whip and colliding with each other.

One cool and possibly dangerous thing that we have in the family room is a five foot long industrial roller type bench iron that Mom picked up at auction. Why we have it I don't know probably a "great deal" that Mom couldn't pass up. One could easily suck in a little kids hand or a piece of your clothing when you operate it. We use it to iron sheets and pillowcases and anything else we can think of. It makes some great creases on your jeans. If you don't use the iron you can put expanding hanger like jean stretchers in your pants when they are wet and hang them out to dry.

Everyone does a turn on the sewing machine; you have to mend, patch and hem your own clothes. We boys can't seem to keep any knees in our jeans. The further down the hand-me-down chain you were the shabbier your pants. Mom would make us ring every bit of wear out of our clothes. None of us like the iron on knee patches that never stay on, swish when you move and make you walk like Robbie the Robot; so we cut up old jeans for material and sew on patches.

We watch a lot of TV and get some questionable ideas from the boob tube. Jim sees an old John Wayne western where the hero leads the bad guys up a narrow canyon. He jumps off his horse and hides him behind a rock. Then when the bad guys come galloping down the canyon in pursuit. The hero pulls up a buried rope and clotheslines the four bad guys together with their horses. Jim decides to rig a trip line at the bottom of the stairs. The stairs have a wall on either side all the way to the bottom to hide yourself and your horse if you have one. Perfect. Fate has it that Grandma Nuno is the first one down the chute. Grandma hears Jim snickering and sees a lump in the out of place rug at the bottom of the stairs. Jim is busted. Grandma calls Jim a little black bird, a negro parrahito, all the other kids are little white blanco birds.

Another of Jim's inspirations comes from Combat in the form of mine fields. Jim sticks straight pins up through the bottom of the brown rug at some choke points intending to run his brothers through the mine field. I would rather he stick to playing "why are you hitting yourself?" with his little brothers. Proud of his work he can't help but tell me his plan. I make him remove them before anyone gets hurt by simply asking why? I share a room with Jim. We have bunk beds now, Jim on top and me on the bottom.

We six boys, and whoever else is downstairs or in the family room, share one small bathroom with a single standup shower stall; hygiene among the troops and general cleaning of the facility suffers

## 1962 ENT AFB, COLORADO SPRINGS CO.

from hard constant use. Jim likes to catch his little brothers taking a crap. There's no lock on the door and Dad forbids us to lock any door, in case something happens to the occupant. If the door is locked you can't help someone who falls in the bath, he says. Jim usually barges in saying he has to pee and can't wait. Jim makes his victim scoot back on the pot and spread their legs; then wizzes between them. His aim is not great, intentionally so. The victim screams invectives. No one comes to the rescue; it's too late anyway. We clog the toilet and toilet paper consumption is too high so I am made to train the boys on the proper four sheet max, double wipe, command approved methodology and supervise the implementation thereof. This is highly humiliating for all concerned but not as humiliating as having Jim pee on you.

There is never any hot water when you want it, especially in the morning after Dad and Kathleen are finished with their morning regimen. The boys are restricted to the downstairs bathroom so they have to shower instead of taking a bath. Showering is tough for a little kid because you are always getting sprayed in the face no matter where you turn in a tiny stall, a three foot kid also can't reach the nozzle to redirect the jet. The result of all this is, some of the boys stop showering. Its's Colorado. If you sweat you don't know it. The humidity is so low we have humidifiers in our rooms to stop nose bleeds; that's in addition to the humidifier in the heater system. One day Dad notices that Tim looks like he has a full dark tan, but its November. Dad spits on his thumb and rubs the back of Tim's neck rolling up dirt worms he also smells of urine; a smell usually masked by Gary's diapers. A couple of other boys fail inspection too. Dad looks at me and says "these boys are going to get scabies"; meaning I have to do something about it. I don't know what scabies are but it must be bad cuz of the way Dad curled his lips and wrinkled his nose when he said it. Yea me, a new responsibility. I am told to drag each pinhead into the shower with me and instruct them in the use of the Mk 1 wash cloth. There are new requirements and procedures to follow. You can no longer let the water run till you are warm, hydrated and comfy, we are to take Navy showers to stretch the available hot water. As explained, you have to get wet and turn off the water when you are soaping up and shampooing. You shouldn't run the water longer than 30 seconds at a time. Sounds cold. You also have to squeegee the stall after you shower. Each kid is made responsible for maintaining his own towel and wash rag. The training takes longer than the time allowed. Some of the boys have never scrubbed the bottom of their feet or between their toes not to mention their nether regions. They have green stuff under their finger nails that has to be brushed out. I now have to clean the bathroom once a week, so it's now in my interest that the boys not trash it out and leave their clothes and wet towels laying on the floor. I tell them that they better improve their aim or I'm making them sit like a girl.

## The Basement

The house has an unfished basement the size of the first floor. It can be both a spooky and a happy place. Spooky if someone locks you in there and turns out the light. Access to the lock and the light switch are outside the basement. Happy because there is lots of space to do things. Dad builds and sets up a ping pong table which is fun except Dad varnished instead of painting the table top making it real slick, the ball stays real low and spin doesn't work so well. Many happy hours are spent playing. There is always someone up for it. In the further corner we set up a BB gun range. We get targets that trap and allow you to recover the BBs. Dad buy's and sets up a reloading station under the stairs, he reloads custom rounds in .30-06 and 8mm Mauser. Dad also dissembles and whittles down the stock of his Enfield in an attempt to lighten it some. Dad spends hours filing, sanding, checkering and finishing it. Mom buys into something called Nutro-Bio. She is going to get rich selling it, trouble is you have to buy bulk and sell it at a markup. Mom never sells any of it. It is stored in the basement. We have to eat a lot of shakes, vitamins and supplements. The basement also has our large freezer. Mom tries freezing bulk, we buy in bulk at the Academy Commissary.

## Shooting

Dad made Jim and I break up our BB guns with an ax for using improper ammunition. We were shooting finishing nails at locust, there were hundreds of them around the house eating the flowers, lilies, marigolds, you name it. I read that the French, during the Napoleonic times, used Bar Shot, long round iron bars called longeron in their smooth bore ships cannon, to bring down rigging, it was more effective at this than round shot, since it was unstable and tended to wobble and spin, so we gave it a try. The nails worked well on the hoppers at short range, the nails would just cut them in half, very satisfactory. Accuracy sure dropped off as the range increased, we didn't need range for this job and we got to save our BBs for worthy targets. Dad catches Jim shooting and asks him what he is shooting at so close to the house, actually at the bricks on the house, at the same time he sees his box of nails. Dad is pissed. Jim rats me out to spread the pain. It may not seem like a big deal but understand that Jim and I are way past BB guns. I have shot everything in the inventory except the Mauser and Enfield. We go out to the range at Fort Carson and shoot all the time. At this point in time there seems to be a sentry in front of every building. The range goes from pistols to cannons. I don't know any other sixth graders that can tear down a Model 12 shotgun. We happily clean the weapons for Dad quite often without him even asking.

Jim is not safe. Dad takes us to the range to shoot his new Ruger single six .22 pistol. A cowboy gun. The Range Marshall issues 50 rounds to each shooter with a target. We take a station and set our targets. Jim is up first; he does a regular Hop Along Cassidy. Jim pulls the gun up to his ear pointing it at us over his shoulder while he pulls the hammer back and then brings the weapon down in a chopping motion and lets fly a round. I swear he was making like he was going to blow the smoke from the barrel too. Dad is on him like the bluing on that pistol. He is dressed down in front of everyone on the range and benched. I shoot his 50 rounds.

Jim is not a good shot either. He cannot hit a target at any distance with a pistol or something that is moving with a shotgun, there is usually nothing in his bag. No conception of deflection. This is despite all the rounds he has sent down range. We go rabbit pheasant and dove hunting in the Black Eyed Susan fields east of town. Jim usually carries the old Annie Oakley .22 pump rifle because his targets have to be sitting still and close. Sometimes I think Dad takes us hunting to beat the brush and shag birds since we don't have a dog, the brush can get pretty heavy by dove season. This day Jim is carrying Dads Ithaca 20ga. pump instead of me, it's the one Dad had when he was a kid. I am carrying a model 12 12gage today, Dad has two of these, one he bought at the PX and one his Dad sent him when he was in Germany. We see a big fat dove on a fence wire. Dad tells me to walk him up and shoot him when he flies. Jim pulls up his shotgun, before I can take a step, the business

## 1962 ENT AFB, COLORADO SPRINGS CO.

end of which is six inches from the side of my head and cranks off a round. Poor firearm discipline. Totally un-sportsman like to shoot a bird on a wire or blow your brother's head off from behind. My ear is ringing and I have powder burns on my face. Jim just couldn't help himself and Dad makes him pay for it; he really wanted that easy kill. Dad takes Jim's ammo from him. Jim spends the rest of the hunt lugging an empty gun, that's heavier than the one he usually carries, trailing behind us with no chance of a kill today or in the near future. He missed the bird in his haste to deny it to me.

Dad takes Jim and I deer and rabbit hunting with him in the winter, checking us out of school. I still have dreams of trying to follow Dad up some frozen mountain creek in the snow lugging a rifle or shotgun. All I see is the back of his legs and his heels. I am breathing hard and terrified that I might disappoint and be a laggard. Dad takes us deer hunting for a week at Camp Hale, the home of the 10th Mountain Div. the highest Army camp in the world. We stay in some old barracks with wood burning stoves in them for heat. Dad doesn't take us with him we just run around the camp and the surrounding woods. I am amazed we never got lost. There ae a lot of old army vehicles around to play in. After a while there are a dozen deer hanging from the gantry in front of the Barrack two of them are Dad's. Jim and I range far on the last day. Dad came back to find us naked and rotating in front of the stove. We had fallen in every creek in the area by walking, nay stomping on the ice. Every stitch and boot we have is sopping wet. Dad had lent his Enfield to another officer who was on the hunt with us he had a shiner and big nasty cut on his eye brow. He took a shot at a deer uphill and got his eye too close to the scope. Note taken.

On another trip Dad got a nice eight point buck on the run. The first round knocked off an antler point but he was able to get off another round at over 300 yds. and knock him down with his Mauser. Trouble was we are way up a mountain ridge and would have to drag/carry him a couple of miles or more. Dad is considering settling for a tongue sandwich. After field dressing the deer we drag him downhill to a lesser grade and out of the trees to a creek wash. There is a bald spot on the hide already. Some hunters on horses see us and come up to us. They admire our buck. They have had no luck. They offer to pack our kill out for half of it. While Dad is thinking about it, a good sized buck steps out of the creek 200 yards behind the mounted riders. Dad shoots him in the neck with his Mauser off hand and tells the cowboys "there's your deer" they gladly pack us out.

Dad also gets a huge 1000 pound plus Bull spike Elk. He hangs and skins it in the carport drawing a crowd of little kids who've never seen such a thing. One little wag asks what they all want to know "what are you doing with that donkey mister?" this cracks him up no end. Dad tans this hide snow white. He makes himself a fine vest that he is never without and mounts the spikes; the rest goes into the freezer where it doesn't stay long.

Dad has a double barrel side by side .410 shotgun, a little baby shotgun that weighs next to nothing that he usually takes rabbit hinting. We go up Manitou pass and drive off into the national forest pulling of a dirt road somewhere. We form a skirmish line Dad in the middle me on the right with a Stevens .410/22 over and under and Jim on the left carrying the Annie Oakley. Timmy is bringing up the rear with his BB gun. He loves to go on the hunts, Dad will usually let him shoot a pistol or one of the other guns before we leave. We can usually get a cottontail or two for supper. Cottontails either run or freeze, like people. People either suck it up and do something about their situation or they crawl under a bush and die. On one of these hunts I see Dad make the best shot I have ever seen. We had just locked and loaded. After spreading out and marching about 30 yards into the field we jumped a cottontail. The rabbit jumped right in front of Dad. Dad still had his .410 slung. He spins it off his shoulder and under his arm, shoots from the hip and nails him before it has gone 20 feet.

## I have a Cousin?

Dad backs the station wagon into the drive and honks his horn, he asks me to take the blankets and sheets out of the back of car and wash them. The two back seats of the Ford have been folded down and there is blood on everything. Dad is just back from Saladia. He brought his cousin George back to the Springs after he went off the road and wrecked his car. George just left the hospital. He didn't want to get an ambulance or pay for a night in a motel or a ticket home. He doesn't have insurance, auto or medical. He was drunk and doesn't need another DUI. Dad makes it plain that he didn't like doing it but George is married and has five kids to take care of our age. I ask if I can call them and maybe meet up with George Junior. I get the number and call George II. He lives north of me by the bluffs. We agree to meet at King Soopers, halfway to his house. I ride my bike to meet him that afternoon. G2 is all smiles but seems a little dull, he wants to take me to his favorite place to ride his bike, dirt roads and lot with little dirt mounds. I am up for that. There is a large, five foot diameter, culvert that he is especially enamored with. We dismount and he proceeds to show me how to torture horned toads in the mouth of the culvert. My cousin has a bright future as a bridge troll. I head home and never see him again.

# The President

JFK comes to the Springs for a meeting held at Ent. It's to be at the Headquarter Building, a secure area within a secure area that is on the base. A ten foot chain-link fence topped with razor wire surrounds the building and grounds. I get to see the president as he is leaving for Pete Field. I get a good view maybe twenty feet from the car by climbing up the fence parallel to the perimeter road/parking lot, with my fingers and toes. He is riding in his Lincoln convertible, the one he would later be assonated in; he's waving to us and smiling. I can only whistle and smile back as I cling to the fence.
I can still remember the day. It was announced on the school PA. I was having a rare tray lunch in the cafeteria. Everyone was upset, some visibly, a few were crying. Classes were called off for the afternoon. Most of us went to the chapel. A mass was said for the president, our leaders and the future of the country.

Kennedy's election was my first cognate exposure to politics. It was a subject of conversation between the parents. I also saw the debate between Kennedy and Nixon on the boob tube, Dad was watching it and commenting. Kennedy was, of course, very popular at Divine Redeemer. The assertions that he would take orders from the Pope were ludicrous. LBJ, the Senate Majority Leader, thought that he should have been the party's candidate, instead of this young inexperienced Senator from Massachusetts.

Dad did not like Johnson. He was a Republican since Truman, definitely an Eisenhower fan. Johnson was a double dipper in WWII. He wrangled a commission as a Lt Cmdr. In the Navy, went on one B-26 flight over New Guinea, as a passenger, where his plane may or may not have been shot at, then wrung a Silver Star out of the service for it that he initiated himself. The experienced fliers of

## 1962 ENT AFB, COLORADO SPRINGS CO.

that same crew went on to fly many more missions surviving not a few encounters with the enemy receiving nothing. After Roosevelt heard about Johnson's stunt, he ruled that acting members of congress could not simultaneously hold a commission without resigning their seat.

Presidential elections are held just before we leave the country. Dad is an enthusiastic Goldwater man, after all he's a Major general in the Air National Guard and an old B-24 pilot who is also a certified in the B-57. You have to wonder what would have been the outcome of the Vietnam Air War if someone competent had been in charge instead of some strong arm populist politician with ridiculous micromanaging ROEs who once bragged "they can't even bomb an outhouse without my say so". Compare that to Goldwater's go all in or get the hell out. Then after dragging it out four years, Johnson abandons his command still in the field. Ask anyone who was alive then how they liked Johnson's "Great Society", the riots, the draft, the dead. Way in the future when Nixon gets serious, it's over in a month after mining Haiphong and bombing Hanoi. If you're going to do it, do it quick and do it fast, use everything at your disposal, no holding back, it's not a game. Run them to ground so they know they are defeated and not playing the long game, engaging in politics and building a support base.

# 1965 Taiwan Defense Command

**JJS.** We enjoyed Colorado for three years, from June 1962 to June 1965, then off to Taiwan (Formosa) China, another questionable part in the scheme of things to thwart the Commies. This was at the height of the Viet Nam War. I was assigned J-3 Air Operations as Senior Officer, Taiwan Defense Command TDC. The island defense was run by the Navy 7$^{th}$ fleet in the north headquartered at Taipei where we were while the Air Force was in the south at Tinian. We were stationed at Taipei.

**RES.** I was in Colorado forth through the sixth grade, Dad had his choice of: Germany for four years, Alaska for four, or Taiwan Defense Command, TDC, for two years. He had already done Germany and had no desire to return to Krautland, Alaska would mean commanding a B-57 Squadron at Elmendorf AFB, Anchorage, but they had just experienced the great 1964 earthquake and there was still a lot of damage to the state's infrastructure; Dad had seen the damage himself, or going to Taiwan. Jim and I were for going to Alaska, just think of the hunting! Dad must have been of the same mind, he would be a Squadron CO. He would be a big wheel on base with over twenty four planes. Mom wanted Taiwan, think of the servants, the overseas allowance, and the shopping potential. Mom enjoyed Munich and Japan, Taiwan promised much the same. Lt Col Poehl, who we knew from Japan takes the Alaska job. We eventually catch back up with the Poehls in Plattsburg upstate New York.

During his tenure as J-3 TDC Dad was in possession of the entire defense/war plans for the theater. There were a number of B57-D models that had been operating off the island from 1958 to 1963 when they were transferred to the Nationalist Chinese Air Force. At least one bird was shot down due to a premature letdown after an overflight. The Nationalists on Quemoy and Matsu were shelling the communists on the mainland on odd days and the Communist shelled on the even days. Very face saving, the war goes on. An estimated 25 million Chinese die of starvation on the mainland during Mao's Great Cultural Revolution in 1965. I hear things; overflights showed bodies lying on the ground. The Chinese are in no shape to interfere in Vietnam. Starvation was endemic, the population was eating the bark and leaves off trees We see movies at Taipei American School of people being "Jet planed" the Revolutionary Guard Students pull their victims arms behind their backs, put dunce caps on their heads and hang signs around their necks denouncing them as counter revolutionary, they are not politically correct, they are capitalist roaders, an ambiguous but serious charge. After all the children know. The accused must make a public apology and confess their crimes against the Party, the Cultural Revolution and not espousing the party line. The crowd is whipped into a frenzy and the poor souls are beaten to death by the crowd; now the community is united in their shared inhumanity, paranoid that they may be denounced and targeted next. I wonder what kind of films the Communists are showing. Mao insists that many people should die in these demonstrations the more the better to magnify their effect. After the students have run riot, burned their books and schools, killed their teachers and erased history they are sent into the countryside to grow melons, because there are no schools, books, or intelligentsia to teach them. Mao is eventually responsible for the deaths of 75 million Chinese; by any definition a horrible person. Mao is the man in charge of the mainland when we go to Taiwan. The Chinese mainland had the most homogenous population in the world (Han Chinese) yet they are able to be xenophobic and act against the perceived other. He takes his clue from the Bolsheviks, anyone who owns land or collects rent is targeted. Tens of thousands of ancient art, sculpture and architecture is destroyed during the Cultural Revolution in an attempt to erase the past and rewrite history remake society and redistribute the wealth. Ancestor worship is strong and must be eliminated. Kind of like tearing down statues from our civil war. We should look at those memorials and say to ourselves that there never can be anything so divisive to take us there again 700,000 souls paid for that knowledge. Blood and treasure spilt to accelerate an end to an institution doomed to failure. Misguided youth spent in a lost cause. Agricultural workers will be replaced by machinery. After all, the increased demand for slaves was generated by the invention of the cotton gin, generating a throughput problem. That and the ban on importing slaves would eventually make the institution unprofitable. A slave costs as much as a house with acreage along with a lifelong commitment to care for another person's needs. Only a third of the slave population were productive; the other two thirds were the old, the children and the infirm. A lot of hard feelings could have been preempted. De Tocquevill commented while on a trip down the Ohio River in the 1830's that the north shore was lined with neat productive farms and churches while the south shore had poor crops tended by a sullen unmotivated workforce. When there was a dangerous job to do in the south, Irish day laborers instead of valuable slaves were employed. What's a man worth?

Taiwan is still on a war footing, and has been since the Nationalist Chinese under Chiang Kai-Shek were chased off the mainland; that was in 1949 an event relatively fresh in the mind of the Island's Administration. It is not uncommon to see Nationalist troops engaging in platoon and company sized field maneuvers especially when the paddies are dry and the harvest is in. Double Ten Day a celebration of the Nationalist occupation of the island is an Occasion for huge Military parades and speeches. Taiwan had been occupied by the Japanese for fifty years. The Island was bypassed for invasion during WWII. An assault on the island would have been tough but logistically taking one big island is better than taking an archipelago, read Philippines or scattered islands. Admiral King suggested this very plan. MacArthur had already shown the effectiveness of bypassing and isolating remote Japanese garrisons. It would have been a hell of an air war; our strength at the time. It would have allowed us to concentrate our forces, ring the island with submarines and provided readymade airstrips. Another England. The Japanese Taiwan air assets were actually taken out before the invasion of the Philippines. The indigenous people and natives were not sympathetic to the Japanese. There were 169 thousand Japanese troops on Taiwan, mostly occupation troops, support personnel and Taiwanese conscripts, vs. 500 thousand battle hardened troops in the

Philippines suppressing the guerrillas and the civilian population. The civilians, Manila and Philippine infrastructure would have been spared. There are still concrete pill boxes and fighting positons at most road intersections. There are camouflaged strong points, little hill top forts, still maned, around the city that I stumble on during my hikes. Yeah it would have been a tough nut to crack but we would have had a big hammer, a smaller nut and would have cut the empire in half interdicted their logistics and invaded their rear.

The terrain is mountainous with peaks rising up to 14000 feet. The east coast is sheer cliff, there is a one lane road blasted into a sheer rock face that looks out over the Pacific for a stretch. Traffic on this road alternates direction every other day. The foot hills are very tropical with bamboo forests and heavy undergrowth. Taiwan also boasts the world's greatest concentration of venomous snakes and vipers. The original inhabitants a mix of: Pacific Islanders, Japanese, Pilipino and Chinese, were aboriginal head hunters who believed they were descendant from the Russel's Viper a particularly venomous Pit Viper locally called a fifty pacer because that is all the further you would get if you were bit. The island was united by a pirate king, loyal to the Ming, from the Pescadores, off the west coast of the island, who refused to shave his head, grow a que, and kowtow to the Manchu. The Chinese never had control of the island. It is too big, remote and the Chinese had no navy. Most of the sailors and fishermen in the area are from the Pescadores. The Island was ceded in perpetuity to the Japanese along with Korea after the Sino Japanese war 1895. This led to the eventual fall of the Qing dynasty. The inhabitants resisted, they were never receptive of the Japanese. It required an army of occupation amounting to over 100 thousand troops. The island was never totally subdued.

# 1965 TAIWAN DEFENSE COMMAND

After WWII 1945 the island was returned to the Chinese by treaty. The local Japanese concerns and the Taiwanese, distinct from mainland Chinese, continued their relationships and profited thereby since the infrastructure hadn't been ravaged during the war. When the Nationalists came over they displaced and nationalized the Japanese concerns and administrators then pushed out the native Taiwanese who were just finding a sense of freedom and expecting American support for their ambitions. There were purges and executions for so called communist sympathizers; the Nationalist making the huge mistake of adopting the repressive paranoid tactics of their mainland rivals. They alienated the locals then replaced an efficient working administration and a flourishing commercial enterprise system with cronyism and graft exported from Shanghai along with T V Soong, Chiang's Brother in Law; GDP tanked.

A lot of Nationalist soldiers, sympathizers and administrators were caught surrendered to or went over to the Communist cause when the Nationalist left the mainland. These poor folks were ill equipped and sent into the maw of the Korean war in 1950 as cannon fodder where they died by the millions, solving two problems for Mao, distracting the west and getting rid of people of questionable loyalty. Chiang, who like MacArthur, promised to return, offered troops to MacArthur for the Korean conflict which were sensibly turned down by Truman. The 7$^{th}$ fleet had to maintain assets in the Taiwan straits for the duration as a consequence. This did not keep the mainland Chinese out of the conflict. The end of WWII did not bring peace to the poor Chinese people.

I attend seventh grade at TAS, a diverse school that Rich Chinese, Americans, and embassy kids attend, grades K through first year college. There are 13,000 students. Before I start, I get a present. Dad gives me a Remington electric shaver and tells me to start shaving. Dad is a hairy man and I seem to be on the same trajectory. My beard starts around my eye sockets. My back starts sprouting, very painful. Dad is always immaculate, he shaves twice a day with his electric razor and keeps his British Fighter Pilot moustache regulation off the lips and not extending past the corner of the mouth. His hair is cut once a week. Dad caries: a comb, scissors, nail clipper, and file in a partitioned, crocodile sleeve in his breast pocket. Military personnel on the island must be in uniform at all times wearing the appropriate seasonal uniform, blues in the winter and tans in the summer. Dad always wears an overseas cap, never his garrison cap with the scrambled eggs on the brim. That shaver irritates the crap out of my face the first time I use it, jerking hairs out and giving me a rash. It gives me acne. Mom buys me a cheap safety razor and some shaving cream which works out better. There are a lot of Courts Martial that Dad attends at a later command for guys who refuse to shave; mostly Black troops who say it irritates their skin. I can sympathize. They also argue that the Navy has relaxed its requirement on facial hair, they even allow full beards. The Navy relaxes its requirements due to the scarcity of hot freshwater for shaving on the small vessels patrolling offshore and in the inshore Brown Water Navy.

Taiwan is an R&R destination. You see military personnel, in uniform with their "dates" going to the movies, at other facilities in the compounds and around the city. Those dates sure seem to wear a lot of makeup and have very short tight skirts. The Moms avert their eyes and herd us away. Taiwan is very conservative; the soldiers keep a lid on it.

Another duty was dealing with the problem of servicemen who had shacked up with a native and had a kid, giving admin a problem. The poor kids may not survive in this racist part of the world.

### 1965 TAIWAN DEFENSE COMMAND

Dad makes a few trips to Vietnam. He attends some theater conferences. Landing on a carrier as a passenger is not to his taste; a controlled crash and he is not flying. Dad also goes to Vietnam proper to serve on the investigating board, looking into incidents involving B-57s stationed at Bien Hoa. One is a mortar attack where five B-57s were destroyed and several others damaged. Another is a ramp explosion where a B-57 loaded with fuel and ordnance, last in line to taxi in a four plane strike package, exploded on startup, taking out ten B-57s, killing twenty eight personnel and wounding many others. Both incidents were compounded due to the lack of revetments for the aircraft because of the limited space available on the apron. The planes were parked wing tip to wing tip for security against saboteurs, Pearl Harbor style. Dad is disgusted.

## The Trip 1965

We all go for Passport photos at the old Antlers Hotel in the Springs, we are to have Diplomatic Passports for this tour. Visas to Taiwan are only issued for twenty four months so we anticipate that this will be the length of the tour. Kathy, Jim, Steve, and I get individual Passports Tim, Billy, Gary, and Maureen are on Mom's. Mom buys two sets of nested Samsonite luggage for the nine of us Dad has his B-4 and will be accompanying us. And then there is Midget, Kathleen's darling pup and all our bunnies to deal with. The bunnies, are given to a breeder and later retrieved when we saw how they were kept and treated. The bunnies were given to friends instead. Midget will be kenneled and shipped out of SFO with us but we will not see her for thirty days due to a mandatory quarantine; predictions are that she will be absolutely nuts by then.

We will be departing from Travis AFB San Francisco but need to get there first. We are taking the White 1963 283 V8 4-speed standard transmission Carryall, equipped with an aftermarket air conditioner, with us. Dad likes his truck; when we get to Taiwan we find out the Navy likes them too; they all have factory installed air conditioners. We are giving Uncle Sal back Mom's 1957 Ford Fairlane station wagon. She only got to drive it one year. Sal is coming out to get it. If Mom had it any longer she would have bought and brought more big stuff home from auction. A lot of the stuff Mom bought was crappy stuff she intended to leave overseas, she didn't want to damage any of her good stuff in the move, due to past experience. She buys an old 1940's wringer washer and a piano among other things. Everything must go into three large wood crates which will be trucked and shipped to Taiwan; the crates need to leave way before we do, so the plan is to dawdle on the way to our new posting.

There is a lot of stuff in the house now from all the auctions Mom has been going to, we are certainly overweight. What to take? What to store? No chance of getting rid of anything where Mom is concerned. This seems to be an issue between Mom and Dad, lots of yelling, we retreat to our rooms or go outside. We have a 16,000 pound allowance, anything over that Dad has to pay for. Mom wants to ship a piano. Her Dad gave her a piano that she shipped to Japan and back; that was not good for the piano. Mom wants to put her Dad's piano in storage and ship a huge mahogany upright grand piano that must weigh 1200 pounds that she bought at auction and has tuned for just that purpose. That's almost one tenth of our allowance. The issue? Mom does not play the piano, at least I have never seen her play. She used to play but doesn't have the time anymore. We kids have been taking piano lessons from the Lady next door, Mrs. Hilton for two years. I cannot get past lesson six mainly because I will not practice and have to start over once again every week. I just can't seem to sit there when the sun is shining and I'm not good at route memorization. My hands are meaty and my fingers are short, baby hands. Jim wants nothing to do with what the rest of us are doing and Mom can't make him do it. I see him try to play something when no one's around, he doesn't want to compete or be shown up by anybody. Steve and Kathy are brilliant of course, they are learning to read music, still a mystery to me. Kathy is into lesson plan book three or four and Steve can play the Baby Elephant Walk. It's chopsticks for me, sort of apropos. Is it confidence that we will get better, does she know that we will need something to occupy our time over there, is she wise or is Mom just being stubborn?

It rains for two days just before we leave; a record breaking deluge. Colorado gets twelve to eighteen inches of rain, one year's worth in two days. The South Platte and Arkansas River basins flood. Everything south of the Palmer divide is flooded. Fountain and Turkey creek flood. Fort Carson loses a couple of bridges and 450 feet of rail siding. There are flooded homes with only the roof showing just off the elevated highway as we head south. We drive west of Fort Carson on the foot hills to Canyon City to keep out of the flooded areas. We are on the old Santa Fe Trail. We follow the Arkansas River west then the Rio Grande south. Both rivers are in full flood over the banks in places. Devastation everywhere along the rivers.

We get to see our cousins on both Mom and Dad's side of their families on this trip. We all line up for a once in a lifetime photo with the Scanlon cousins: Jim and Bob's kids in LA; It will never happen again. We see Moms brother, our Uncle Bill, and his kids in

Albuquerque, on the way to the coast. Uncle Bill is a salesman for Winston Salem; he sells tobacco products. Bill is always handing out a three pack of sample cigarettes. He bites on a pipe and grins, but he doesn't smoke or drink. Bill has the gift of gab, very important for a salesman. Bill is also in great shape, he can do feats of strength like ripping phone books in half. Bill and his cousin Rudy were two of the original Muscle Beach Boys. Rudy is the first cousin to get married, he is a LA Cop, infamous for packing heat to his wedding. Uncle Bill is able to do a pushup on fingertips and toes, sounds simple, except his arms are stretched out straight in front of him; I've never seen anyone else do that. He must have won a lot of bets with that one.

It's July and it's hot in Albuquerque so all us cousins except Jim, go the pool where they have a five meter diving board, a first for me. Jim wanders off in the nearby desert with one of his cousins looking for lizards and snakes, he gets into some giant red ants and gets bitten. We all get sunburned of course because nobody used sunblock back then.

All eight sunburned sleepy kids pile into the Carryall and head west the next day; the aftermarket A/C is not keeping up so we chafe and sweat before the sun is very high in the sky. The Carryall is a three door unit with three bench seats, the windows in the back slide back only partway the ones in the rear not at all, so we bake in the back especially the kid in the very back where there is no seat, just some old parachute seat pack cushions stitched together to make a mat, it's a luggage space; the spare is back there too. The tire wells intrude into the back bench and the short second bench only allows three passengers per bench. There isn't room for more than Gary up front with Mom and Dad; so that leaves one kid in the back of the bus when we have a full load. There is always a race and an ensuing fight for prime seating, the looser gets thrown in the back. Sometimes it's not so bad if you want to sleep and it's no too hot; there are pillows pads and blankets back there. Billy usually gets this seat and he usually doesn't care. This time Jim is ordered into the back where he theoretically won't bother anyone. Jim was caught being a little too aggressive getting on board. Dad turns off the aftermarket A/C and lets us open all the windows, its better but not as good as the Ford where you could open all the windows all the way around even the tailgate window.

In the afternoon we run into a sand storm, another first. We have to wind up the windows and flick on the aftermarket A/C; it's not too bad. The sun is blocked by the flying sand; it looks like dusk outside the car. Before long it starts stinking in the car like we hit a skunk or something. We debate the probability of skunks, in New Mexico, in the desert, at high noon, during a sandstorm. The smell doesn't get any better like it should when you get down road of a skunk. Maybe we hit it fresh? Dad declares it non–skunk, it's pretty heavy in the back then it spikes; Jim has taken off his tennis shoes. Jim hasn't taken a bath for several days. He didn't go swimming with us the day before and he's been wearing his nasty black converse high tops with no socks. Dad gets out in the flying sand and ties Jim's shoes to the trailer bumper, about as far away as he can get them without throwing them away. We toss Jim a big box of Toweletts and he's told to get busy on his dogs. We're off again like a "heard of turtles".

We lay over in Flagstaff and wait for the sun to set before crossing the Mojave a very popular plan. We run around Flagstaff in a park enjoying the cool mountain air until it's time to eat dinner, Hot Dogs Beans and Lemonade. After KP we load up and were off again. This is one of the better legs of the trip: its downhill, the aftermarket A/C is working better in the dark, all the Indians are sated and sleepy, and Dad made Jim take a sponge bath in the park after licking his thumb and rolling dirt off the back of his neck with it during inspection. As we are rolling downhill we see a semi go off the road and roll over in a cloud of dust about five miles ahead of us. The driver must have dozed off. The road is straight as an arrow. We pass the accident. Several cars have already stopped. We press on.

We wake up in Bakersfield and make our way into LA. It looks like late afternoon all day because of the smog, the air is brown and the sunlight is orange, the plants must like it. I've never seen so much foliage as there is here, even along the highways. Dad is commenting on the traffic whizzing by he says it is too dangerous to go more than 55mph. that it is suicide to go any faster in this traffic. As he is talking we get pulled over by a CHP Officer who tells us that we are going too slow and causing a stack up and to speed it up

We go to Dad's parent's house and then we are farmed out. The little boys go to Sal's and the girls go to Grandma Joyce; Jim and I go to Grandpa Frank's. This is odd we manage to find out that somewhere in there Mom's parents have split. Rumer has it that Frank took up with some "Apache" who later hit him on the head with a frying pan then left him. This is Grandma's version. The Catholic Church doesn't recognize divorce and Grandma who is quite devout doesn't recognize it either. Grandma now has two luxury apartment complexes in Inglewood that she owns and lives at while Grandpa is still at the old ranch house. We have a gathering of all the Scanlon cousins at Grandma and Grandpa Scanlon's house. Bob and Jim's kids. We take a picture with Pat, Dan, Maryanne and Craig Anderson. Bob's widow Betty, had remarried a real nice gentleman, Bob Anderson.

We spend a week in LA doing round robin visits with relatives then we are off to Travis AFB with our matching luggage; we leave the trailer at Grandpa Nuno's house, the carryall will be shipped out of San Francisco. We fly a commercial Continental Airlines flight, a sleek new Boeing 707, a real treat. The plane is a military charter, it stops in Hawaii, Guam, Clark field in the Philippines, and finally Taipei Taiwan. We have a long layover in Hawaii, one that allowed us to get away from the airport to a friend's house where we were fed and feted and we kids could run around and be kids. We take off at dusk and land in Guam in the middle of the night where we stay on the plane. We deplane in the Philippines while it is serviced. Stepping off the plane and onto the stairway it hits you, the humidity hits you like a soggy baby blanket followed by a powerful stench. The humidity is impressive coming from the high plains where it is 10%. I was always amazed the you could watch a puddle on concrete dry up right before your eyes in the Springs. The stench is a bouquet of rotten vegetation, stagnant water and feces plus a few dead animals. Crossing the tarmac on the way to a hanger where we are to wait I see some of the biggest insects I have ever seen in my young life, like a 1.5 inch wide 8 inch long centipede looking thing and 3 inch long flying water roaches. I wished that I had my BB gun.

We get back aboard and takeoff for the short hop to Taiwan. The A/C on the plane feels great after the hanger. The search for air-conditioning and heat relief will become a personal quest in the near future. I have a window seat and my face is pressed against the glass as the sun is coming up. Taiwan is socked in, solid cloud cover, nothing to see, we let down through the clouds and then were there. The runway is along the Tamsuei river right downtown. CAT airlines (China Air Transport) started up by some of the old Flying Tiger pilots operates out of this airport. CAT is rumored to be a CIA run operation.

Our sponsor Commander, later Captain, Cunningham and his family meets us at the airport and tell us what's what; they prove to be a lot of help since they have up to date knowledge about the posting and the local situation. They give us a TDC brief, a booklet, about the posting. After we clear customs and gather our luggage a couple of Navy carryalls, arranged for by our sponsors, transport us to the Liberty House Hotel right downtown, where they have booked accommodations for the ten of us. What we see on that short trip leaves us slack jawed. We see: carts pulled by crapping oxen and water buffalo driven by little boys, Pedi-cabs, smoking little Toyota Blue Bird taxies, even smokier two stroke motorcycles with ladies sitting side saddle on the back, and bicycles with of all kinds of baggage tied to them, competing for the same road, all riding their horns and bells incessantly. Then when stopped in traffic we see a young lady hike up her skirt straddle and squat over the roadside ditch, a binjo, and let fly.

We check into the hotel, a structure with a two story frontage with rooms and a courtyard in back surrounded by a high concrete wall and some low bungalows. We take one of the bungalows. Some of us, Mom Dad Kathleen and the babies, sleep but some of us are too excited to and want to explore. The first thing we find out the door are giant snails crawling on the walls, big as a hen's egg. Jim of course has to smash a few, they break like a hens egg too, then he chases his brothers around trying to smash one on them; lots of screaming, good first impression for the locals. I move off to the lobby and pester the clerk. He is a nice fellow who introduces me to green tea. He is very patient with me; he is the first Chinese fellow I have talked to and he teaches me some Chinese phrases. Meanwhile Jim has hit the main drag with his brothers in tow, full entourage mode. He proceeds to break two rules. Don't eat street vendor food and don't use US currency you are to use NTs. Jim buys the boys lemonade using a dollar bill, what a deal, drinks all around; he buys it from a guy who probably dipped the water out of a ditch, as a rule all water if not bottled or distilled is to be boiled. Tea is very popular due to the necessary boiling. The water pipes are often run in the same binjo ditches as the sewage. Tap water is not to be trusted and is boiled as well. All clearly stated in the USTDC country brief we all had access to. Needless to say, the boys are very sick the next day. Mom gets to find out where the Dispensary is and learn how to get around in a Chinese taxi cab day two. The clerk has taught me right, left and straight ahead in Chinese, so I go along as interpreter and for moral support. It is the first of many trips to the Dispensary. The taxi is ripping along doing 60 over the bridge. I try to tell the driver to slow down, something I hadn't learned from the clerk, when I notice that the speedo was marked in in kilometers, then I'm distracted trying to remember the conversion to mph and almost miss the turnoff. We usually travel in a convoy of two taxis because they are so small and we are so many. No word of Dad's truck or our stuff yet. It only cost the equivalent of .25 cents in NTs (the exchange rate is 40NT:1Dollar) to get from the hotel to the club where you can get SOS for breakfast for the same amount.

We stay in this hotel for two weeks or more, this gives us plenty of time to explore the city. We are right downtown near one of the old City gates; it looks like the Arch De Triumph, it's huge you can go inside and climb up but it is locked up securely, realized after a determined assault. The Japanese have built most of the permanent structures in the city including the main road through the city, the government buildings and the Fuhsing Bridge over the Tamsui River. Things get interesting as you move off the main streets to the side streets and alleys. There is a myriad of little shops and businesses with the extended family living above the shop. In the evening they all pull out a little charcoal brazier or hibachi the size of a mop bucket that they place out in front of the shop in the alley or street. They squat on their haunches, lite the fire with paper and sticks stoke it with coal or reconstituted and ventilated fly ash then fan it. There doesn't seem to be very many chairs in China everyone "sits" like that. Dinner is cooked right there in the street you can take a stroll then and see what everyone is having for dinner; it's about the only time during the day that the air smells good. Little kids grin at me as they gnaw charred fish heads they have with their seaweed and rice. I'm twelve-years-old, 4000 miles from home, a couple of days in country, wandering around by myself in Chinese alleys, watching people eat monkeys and snakes. I make mental notes of where to go next when I have an opportunity. The best place is the large open air circular market by the railroad tracks. It's exciting when the train comes through honking its horn, everyone scrambles to get out of the way but not too far or too fast in order to save face.

## Our House

We are living out of suitcases on top of each other and eating dinner at the hotel until we get squared away in our new house, which is being built and nearing completion; we are to be the first tenants. Eating Chinese style at the hotel restaurant for dinner is a treat, we sit around a large round table that has a lazy Susan in the middle where all the dishes are set to share out; there is always plenty; they just keep it coming. We have breakfast and lunch at the Officer's club, swim in the attached pool, and lay in the sun marking time. While we are at the hotel with time on our hands Dad takes all the "big kids" to a Chinese optometrist, his office is so small we read the chart behind us from a mirror. It turns out that Jim, and Steve, need glasses too; no pilots there. A year later Tim gets specs after he gets low grades in school like me. I'm no longer singled out for verbal abuse in that respect. Kathleen is 20/20, she has Dad's coloring, eyes, and temperament; too bad the Air Force isn't recruiting Female Fighter Pilots at this time. Bill also breaks his front tooth, getting out of the pool at the O club; he slips and bites the coping. Bill gets a nice porcelain replacement. Mom didn't want to ugly up two of us. We also have time to check out where we are going to live.

The house is located in Tien Mou North West of Taipei on the road to Beitou. It is a new sub division next to an existing BOT compound. The BOT or Bank Of Taiwan properties are owned by Madame Chaing the wife of the Generalissimo and her brother. Madam Chaing owns a lot of property that she rents to the military. There is another compound on Grass Mountain where the rich Chinese go to live in the summer to escape the heat. The compound is fenced all around; there is a small store, a theater, a swimming pool, a tennis court, a basketball court, soccer and baseball fields plus a hundred or so single family dwellings. Our subdivision is outside the fence, it's a private rental. We have a seven foot high concrete wall, with shards of broken glass mortared into the top of exterior walls, surrounding the place. Five hundred yards of rice paddies separate us from the BOT compound behind us.

There is a bomb shelter on the edge of the paddy; it's a five foot diameter concrete pipe sunk fifteen feet in the ground. It has as steel trap door with metal ladder rungs down to the bottom where there are six fold down metal jump seats, there's one for each house. Ours always has water, mosquitos and bugs swimming around in the bottom and probably a snake or two. The water table is pretty shallow where it is planted. Hard to imagine hunkering down in there in the binjo water with the snakes and bugs while being bombed. Others are dry. These shelters are also distributed around the compound where people might gather. We are the second house on the street. There is a Chinese village across from us. A large binjo runs right behind the house; everything in the country eventually goes into the binjo. All grey water from the house, when you take a bath, brush your teeth, do laundry or wash dishes, all that water drains into a gutter that is on the sides and back of the house; from there it drains through the back wall into the binjo. Separate pipes are provided for the toilets but it goes to the same place. This collection drains in to a large partially covered septic pit adjacent to the paddies. The farmers dip up the filth on the bottom of the pit and put it in their "honey buckets" which they use to carry the "night soil" out to their fields one at each end of a split bamboo pole to use as fertilizer. The overflow water from the pit feeds a buffalo wallow next to the main road to town; where it connects to other binjos on the side of the road. There is a potato chip factory on the right that I like to visit and a series of two story brick store fronts on the left where our street tees onto the main road. There is another village and more paddies across the road. This intersection is where you can catch a Navy Shuttle or the School Bus. Except for the store fronts on the road the rest of the countryside is rice paddies and little villages or extended family farm plots. The paddies continue to the hills either side of the river and then continue up the hills in terraces. The terraces are so steep and small and the dikes so thin and delicate that there is no way to get a plow animal or motorized tiller up there to work them, they are all worked by hand. The dykes on the terraces need constant maintenance more so than the level paddies. The dikes can breach from a heavy rainfall when they get too full and over flow. The cascading flow may washout more terraces below. Rice is everything. Filth is everywhere. All arable land on the mainland has been under cultivation for two thousand years. China has a history of famine and mass starvation. One of my favorite books is *The Good Earth by Pearl Buck*, a Novel written by a European woman from a Chinese man's prospective. Time here is marked by the growing of rice. In the winter the paddies are dry, in the spring small plots are flooded and seeded. The paddies are flooded and tilled with manure. Water buffalos drag rakes, boards with wood pegs that the barefoot farmer stands on, thru the watery paddies tilling the watery soil. The farmer gathers the shoots and plants them individually spacing them about a foot apart, backbreaking work. The fields are kept flooded until the rice matures and goes to seed. It takes constant effort to maintain the paddies and keep them flooded. When the rice ripens the patties are drained and the rice turns brown. It is gathered by hand in sheaves and threshed using a foot pedaled thresher. The rice is spread on a hard packed earth and raked to dry. The rice is scooped up in shallow straw baskets and tossed to separate the chaff from the kernel. The women do this work. Everything you eat has to be boiled in water or oil. You can get pinworms if you go barefoot. We get shots for that every six months. Everyone in the country is either barefoot or they wear flip flops. Their food is rolled in filth. After having filth ladled on it to grow. How they stay healthy I have no idea. The mosquitoes are horrendous but malaria doesn't seem to be a problem. We are inoculated for: Smallpox, Cholera, Typhus, Tetanus, Diphtheria and Polio.

When we show up to check the progress on the house workmen are finishing the terrazzo floors and plastering the walls. We are also to get a screened in porch and a grass lawn. I watch some workmen install the hot water heater. The heater is located outside in a covered breezeway next to the kitchen on the side of the house behind the carport. This area has brick bins for wood and coal plus storage for 5 gallon glass drinking water bottles. The breezeway also has two large sinks for laundry and a place for Mom's 1948 washing machine. It has a separate hand cranked roller wringer you can swing over the sink; someone is going to get a work out. The water heater is coal fired. You have to start a fire and wait awhile if you want a bath. Someone will be shoveling coal in the morning for this crew. I usually fire and draw my own bath. I tend the fireplace in the winter. The water heater is a bare metal boiler so later on we have men make up some insulation. The workmen take a 1 inch diameter hemp rope and wrap the boiler with it. After tying off the rope ends they mix up some plaster and cover the rope with it and make it smooth with their hands, trowels and water. Don't know what the R factor on that insulation is but you won't get burned if you touch or bump into it.

The house is a large single story four bedroom two bath concrete and stone house with a tile roof. It has a large living room, a spacious dining room, a big kitchen and servant's quarters. There is no central air in the house, just a fireplace in the living room and a window A/C unit in Dad's room. Thankfully there are celling fans in most of the rooms.

Our furniture shows up and is quickly jammed in the house. The first order of business is to set up the beds. The second is to get the mosquito nets up over the beds because they are driving us mad. Third is to start hiring help.

We hire a head Amah Hakim and her husband Shu as cook and later a fellow named Jack is hired to be our houseboy, after Shu gets a job as a cook at TAS. We also hire a young girl named Mei lei to help with the wash and in the kitchen. Jack kills a fifty pacer in the driveway with a shovel his first day proving his worth, we watch its body squirm and the severed head biting air. I start carrying an old

12inch butcher knife that I found and made a handle for in my belt at the small of my back after that. I practice throwing it until I am fairly proficient. It will also work as a small machete. It is just the thing for harvesting bamboo shoots. There is nothing like a fresh bamboo shoot. That stuff that comes in a can is disgusting.

Mom's piano did not travel well, there is water damage and she raises hell. Next we know the piano is in the carport. Three Chinese have shown up with a mirror. They break the mirror, take the pieces of glass and with their bare hands scrape every bit of finish off the piano. The workman apply a hand rubbed finish to the piano over the next three days till it gleams. The finish is now worth more than the piano. It is moved into the house and someone is sent over to tune it, he came back a month later to see if it has acclimated and is still in tune.

We are settled down for the night when we are treated to a Local Hail and Farewell tradition. Out house is firecrackered. Our sponsors have got together some folks for a party. They snuck into our yard and rolled out firecrackers all around the house then lit them off, filling the gutters with firecracker husks. We boys gather the unexploded ones and save them for later. The visitors bring adult beverages and food. We go to our rooms.

Mom wastes no time in furnishing the house. She buys a French provincial twelve person dining table with two lazy suzans, a matching hutch and sideboard all in teak. She also purchases a teak living room set that is of Mandarin design, the cushions are covered in rich jade green velvet. The coffee and end tables are in laid with green marble slabs. A marble found only in Taiwan. The marble is green with black and white veins. It is green because of the rich copper deposits in the north end of the Island where the marble is quarried. There are also hot Sulphur springs there. The teak comes from the

Philippines, huge logs are floated up by sea in rafted up masses chained together pulled by seagoing tugs, when the weather is good of course. The living room furniture is covered in plastic and is verboten; kids and dogs sprawl on the tatami mats. Nice rugs come later. They are custom order and handmade most having 100 knots per square inch or more. The chairs of the dining table are covered too. The rule is established that everyone must muster for dinner at 1730 or there is hell to pay. I paid one day. Dad sits at the head of the table, mom is to his right and I am to his left. I came home late from baseball practice one day, I'm playing Pony League, missing grace. Everyone has started eating. Dad is somewhat irritated, excuses are not allowed and I make none but offer an apology. I further compound my sin by laying my right arm on the table. I am tired. I just hiked three miles cross country to get home after a two hour practice under the blazing sun. All I am thinking about is a tall glass of iced tea with lots of sugar and lemon. Dad snaps at my insolence and presumed familiarity. He takes the fork in his hand and stabs it into my forearm just below my elbow. Maybe he didn't mean to wound me so bad but wound me he did. It looks like a four fanged snake bit me, blood is running freely down my arm. Of course I yelp and squirt which makes Dad even madder. He tells me to knock it off, he didn't hurt me. I am not allowed to leave the table to take care of myself. Mom brings me a wet towel with ice. Silence, no one says a thing…I have no appetite now, neither does anyone else.

Just after we move in the boys say that there is something crawling around the ceiling above their room. Dad wants me to crawl up there and take a look. I borrow a neighbor's BB gun and climb up thinking it may be rats. Before I go up Dad regales me with a story of him shooting stray cats that got under their house in Denver and stunk up the place spraying everything. He said that he first tried his

.22 but got a ricochet so he crawled back out and borrowed his Dad's .410. He shot several feral cats with it. I didn't see a mature cat or any rats but I did collect several flea infested kittens. Unfortunately, one fell between the walls and died. It was undiscovered until it started to stink. We had to get management to open up the wall, get it out and repair the hole. Out houseboy took the kittens. What he did with them I don't want to know. The boy's room had an additional stench to the general smelly atmosphere for a good while.

## Bored Games

Dad says, "Children require constant supervision". We are basically unsupervised. We are, instead, implicitly made responsible for each other. You had better know the answer when you are asked, "Where's your brother?" what to do? Kids don't need supervision, they need entertaining. Ball games and swimming are good but you can't do that all the time. You cannot park the little ones in front of to the boob tube, there is no TV programming except an occasional Graham Kerr or Skippy the Bush Kangaroo episode, kinda like Flipper but without water, Australian TV. There is Chinese TV with their nonsense and Darky Toothpaste commercials. Once saw a stripper pulled offstage before she could pull her top off; no eight second delay at this time. Amateur hour for two hours a day. There's Armed Forces Radio but that goes only so far entertaining the little ones. We have a ten inch diameter two foot high barrel of Lincoln Logs. We build Fort Apache using every stick in the barrel while Dad watches from the couch. When we finish, Dad pulls off his shoe and "bombs" our effort while making a whistling sound followed by a boom. Our protests only make him smile. By now I have played every game Milton Bradly ever came up with from Mouse Trap to Monopoly a million times, every card game ever invented down to card houses just to entertain. I hate every minute of it, meaningless games of chance, statistics; chance signifying nothing, a waste of time, artificial winners and losers. You cannot influence the outcome by skill, work, effort or experience, frustrating. Leaving things up to chance is a losing game, the house always wins. I play checkers, Chinese checkers, Mom Buys a Ma Jong set, I never learn play, too tedious but I marvel at the inlaid chest and the handmade, multiple Ivory pieces, the materials and craftsmanship are exquisite. I start playing chess. Dad teaches me the rules, the ability and the limits to that ability of each piece. I like it. Dad says it's like flying, you have to think three moves ahead and have at least that many alternatives, all at 500kts and no turning back. Be aggressive to win. It's the game for me. We are the only ones who play. Mom likes what she sees and buys us a sweet two foot square, six inch high, teak inlaid chess table with a drawer for our cheap plastic pieces, it's the game not the pieces for me. I start winning. I take Dad's advice. I think four or five moves ahead at 500kts. I start winning consistently. Dad kicks over the table one day after an impending, inescapable win for me. That is the last time we ever play…I am thirteen-years-old. Luckily Bill and Tim pick up the game I love to play.

## Bugs

There are all kinds of obnoxious insects in Taiwan, Mosquitos of course but for disgust nothing beats the water roach. This cockroach is about two plus inches long and it flies. They are everywhere and you know where they have been. We attend High Mass downtown in the Cathedral then go to the O club for brunch as usual. We are all seated hands in our laps, in our Sunday best, seen and not heard. All that's on the table is water and iced tea. Dad is holding court, looking at each of us in turn, in the eye. He is in the middle of a running commentary on the sermon. He usually has something to say about that; we are in rapt attention, of course. He has his right hand wrapped around a tall glass of iced tea he has been sipping from. Every kid on my side of the table, sees it, a huge flying roach, executing lazy figure eights behind and above the diners on the other side of the table. No one says a thing. Not a blink, not a twitch. Not allowed. The tea colored roach climbs toward the ceiling and then goes kamikaze right in Dad's glass as he is absently lifting it from the table while looking to his left. Dad is taking a long draught when the thing gives him a kiss and crawls on his mustache. Much to my surprise Dad doesn't jump, yell, curse, or spill his tea. Dad pulls the bug off his face with the napkin in his lap and crushes it after placing his glass on the table. No one dares say a thing, we don't have to; he knows that this will be legend in the backrooms.

The house is extremely hot. The Master has the only window unit in the joint; a point of contention. It is pure torture to lay there in the still air sweating under a, mosquito net, when you can hear the A/C, salvation and relief, humming down the hall. No relief for you. The mosquito nets over our beds kill most of the breeze from the ceiling fans. Kathleen's room is the first one down the hall from the Master, so close to heaven she can get an occasional cool taste of it wafting under the door. She also shares a bed with sweaty little kid, Maureen. Cold air sinks and falls and there is so much falling it is squeezing under the door. In a rare instance of cooperation Jim and Kathleen scheme to get even. I am not in on any of this neither are any of the other kids. Dad likes to sleep in the buff. He usually sits around in the living room in his skivvies and flip-flops until his room cools down. Mom usually wanders around until everyone else has gone to bed. Jim has a mayonnaise jar with a giant roach. Kathleen and Jim sneak in and dump it between the sheets. We all hit the hay, school tomorrow. Jim is laying on the top bunk snickering. I don't ask. Dad finally goes to bed, undressing in the adjoining bathroom. He rolls into bed and under the covers without turning on the light. After not a few minutes I hear yelling and stuff hitting the floor down the hall, Jim is laughing now. Kathleen puts me wise the next day. No one would ever believe that those two ever conspired or cooperated on anything.

We get neighbors, the Nelsons, they have four boys our age and a baby girl; Dad says they are cannon cockers (Army). The Nelsons were at the Liberty House Hotel the same time that we were there and move in to their house shortly after we do. Their place is the first one on the block adjoining ours. Mrs. Nelson is a head taller than Major Nelson. It is summer the time for military family rotations, when the kids are out of school. We boys hit if off right away. Big John Nelson becomes one of my best friends, his brothers: Mark, Jeff and Paul are the same ages as Jim, Steve and Bill. John and I are both Boy Scouts. We run a telegraph wire between our houses and rooms and exchange messages by Morris and a secret code we make up. One problem though. The Nelsons are to attend the Dominican School where as we will attend Taipei American School, if you go to Dominican you have to arrange your own transportation. Taipei American is run by the US Government so Navy/Government transport is provided. Mom has only been driving a year and she is sure as hell isn't going to drive Dad's four speed manual transmission truck through the mayhem that is Chinese traffic, so Dominican is out for us. Mrs. Nelson chauffeurs her husband and kids around, she and Mom make trips to the commissary together. Mrs. Nelson is a tall read head and Mom is short brunet they make quite a pair, Mutt and Jeff. After a while I find a way to go cross country down dirt cart/foot paths and through some villages, bamboo forests and rice paddies to get to school on foot, it's no more than three miles, so I'm not tied to the bus. The only things to watch out for are snakes and the village guard geese which honk and hiss; you never run from them or they will bite you on the ass; one just has to stare them down while backing away with a stick in hand. The cross country route comes in handy when I start playing soccer and Pony League Baseball, school busses also run an irregular schedule in the summer. There is no way to hike to Dominican which is located on a hill above Taipei, just too much traffic and too far away.

None of the Nelson boys have back pockets on their jeans. They have a big German Shepard that they picked up locally; they named him Binjo. The boys thought it great sport to put a treat or rag in their back pocket and get Binjo to chase them and nip at the snack or rag; you could put bologna in your brothers hip pocket and get the dog to bite his butt ha ha. They called this, the Christians and the lions. We would sit on the partition wall, it didn't have broken glass, and watch the goings on safely out of the lions reach. This morphed into making the dog play tug of war with your hip pockets. The dog and the kid took turns swinging each other around. One night a thief went over the Nelsons gate, big mistake I don't know how far he got before Binjo was on him. In the morning there was shredded cloth down the side of the house beside a window and blood where he went over the back wall and caught some glass. Binjo didn't even bark. Our Midget is all bark and no bite.

The Nelsons have BB guns, I don't have mine anymore since Dad made me take an ax to it, so I like it when John comes along when I go hiking in the paddies and hills; he can shoot any snakes or water rats we might run in to instead of me trying to kill them by throwing a butcher knife. You are screwed if you are on a paddy dyke and there is a snake on it, nowhere to go. I haven't had my BB gun for a year now and really miss it. I have already stepped on a cobra and that scared the crap out of me. John and I go on a hike. We are going to hike up the road to Beitou till the road ends and keep going into the mountains, a hike we have made before. We have day packs with snacks and web belts with GI canteens of water. The day is hot and the road is getting steep as we pass the church and leave most of the storefronts behind. I tease John about being such a poor shot for an Army brat. I tell John that he couldn't hit the broad side of a bus. We hike another fifty yards past a large traffic circle with a lot of trees in the middle when here comes a bus barreling down the hill. Chinese buses go as fast as they can. They are usually packed, overloaded even, standing room only, they bounce and sway all over the potholed bumpy heavily crowned road, which makes them lean over the binjos. John had his pocket picked on a Chinese bus when we went downtown, after ignoring my advice. I put my money in my underwear with the family jewels. Out of the corner of my eye I see John impulsively crank of a shot from the hip at the speeding bus as it passes. I call John an idiot and ask him why he shot at the bus and ask if he hit it. John says he doesn't know. I tell him that I still think he is a bad shot who can't hit a moving bus with a BB gun while I shoot skeet with my Dad with a double barreled 12ga. shotgun. We hike up footpaths through the terraced hills as far as we can go. The rice paddies give way to mango and papaya trees, at the top of the little mountain are mandarin orange groves. John shoots at some mangos in a tree but can't hit anything. I check his gun and sure enough the sights are off which I remedy using a Swiss Army knife and a wax paper sandwich bag. John is delighted and shoots several mangos just to see them leak, kind of wasteful but John is my friend and there are a lot of mangos. Anyway he can now hit what he is aiming at. We come across a farmer tending the orange groves. We greet him in Chinese and pantomime that we would like to have some oranges. He gives us a dozen oranges and we than him profusely for them.in Chinese, bowing and saying she she ne. Smiles all around, he laughs. We probably look pretty funny to him. Two sweaty pasty faced boys, one quite large, toting firearms, wearing some military gear and knives trying to look fierce but failing. There is a fast running stream called the Red River, it really is red due to the high iron content. We follow the river down taking a different route back home. It's slow going because the terrain is so rough we have to backtrack a couple of times to find a place to cross when the terrain pinches the river and our way down the mountain. We finally break out of the riverbed and go cross country across the paddies and finally make it back to the Beitou, Tien Mou road. As we approach the upper traffic circle we seem to be attracting a lot of attention followed by a general hue and cry. We are soon surrounded by an angry crowd yelling at us. Out of the crowd come three Chinese cops, they have been looking and waiting for us. They seize and disarm us. John and I are marched to the police station just off the traffic circle. In the station house is a fellow sitting in a chair, holding a red spotted cloth to the side of his face. The fifty-year-old man jumps up, points a finger at us, and rattles off a torrent of Chinese; well I guess John didn't hit the bus. The cops make us empty our pockets and take our military IDs then they put us in "jail". The "jail" is a fifteen foot deep hole in the ground with a grate over the top of the hole. A section of the grate is hinged up and a ladder is put in the hole. We are made to climb down the ladder, the ladder is pulled up and the grate is lowered and locked. Crap. There is one other fellow in the hole with us. Language barriers keep us from asking the usual, what are you in for? Our cell mate knows what we are in for since he got to hear everything and is kinda eyeballing us; not to worry John may look like Howdy Doody but he is a head taller and fifty pounds heavier than our cell mate. Seizing our IDs leads to MPs and an interpreter arriving at the police station. The MPs spring us because we have diplomatic passports. The MPs take us home in their car, give us our stuff back and have a word with our parents. We are in deep trouble at home and subject to deportation, Dad could get in trouble too. Everyone concerned lets us sweat, we never learn of any sort of resolution.

## Taipei American School

Just before I start seventh grade at TAS I get a present. Dad gives me a Remington electric shaver and tells me to start shaving. Dad is a hairy man and I seem to be on the same trajectory. Dad is always immaculate, he shaves twice a day with his electric razor and keeps his English fighter pilot moustache regulation off the lips and not extending past the corner of the mouth, his hair is cut once a week. Dad caries: a comb, scissors, nail clipper, and nail file in a partitioned, crocodile sleeve in his breast pocket. Military personnel on the island must be in uniform at all times wearing the appropriate seasonal uniform, blues in the winter and tans in the summer. All Dads uniforms are tailor made in Hong Kong, Tokyo and Shanghai, he is just under six foot tall and his weight is a steady 175 pounds; his motto is: you should push away from the table while still a little hungry. Dad bought a fancy Rolex in Hong Kong when we were in Japan. He regretted spending the money he isn't one of those fliers with fast cars, big watches, and big egos to match; he returns the watch claiming that it was inaccurate. Dad says that he doesn't need anything fancy; Mom is all about fancy. He still wears his Army Air Force issue Bulova watch navig type A17A with the hack function, it must be a good luck talisman. He wears it on his right wrist, his stick hand. He doesn't want to snag it on any of the myriad of small switches in the cockpit he manipulates with his left hand. I copy his habit. Dad always wears an overseas cap, never his garrison cap with the scrambled eggs/darts and farts on the brim. That shaver irritates the crap out of my face the first time I use it, jerking hairs out and giving me a rash. Mom buys me a cheap Gillet safety razor and some shaving cream which works out better. There are a lot of Courts Martial that Dad attends at a later command for guys who refuse to shave; mostly Black troops who say it irritates their skin. I can sympathize with that. They also argue that the Navy has relaxed their requirement on facial hair, they even allow full beards. The Navy relaxes its requirements due to the scarcity of fresh hot water for shaving on the small vessels patrolling offshore and in the inshore Brown Water Navy. There are also a lot of problems with troops hooking up with the locals and having kids. Selfish people with no self-control breaking all the rules and making problems for everyone. These are very racist societies and these kids will have no chance in life. Especially the Black/Chinese/Vietnamese kids.

The teachers at TAS are for the most part real good. They are mostly Officers wives. I especially like my science teacher. She takes us from botany to astrology. One day we will be dissecting pigs eyes the next its Van De Graff generators. From Moles to the speed of light. Lungs tomorrow with real pig's lungs. We learn about: DNA, viruses, vaccines and antibodies. My Spanish teacher is from Honduras she is with the embassy and tells us much about that region and its history. PE is fun. I learn: archery, boxing, wrestling, soccer and baseball. I sign up for Pony League. I win a medal in the 40yard dash. I take shop and learn about wood working, I make a teak rifle stock; it's way too heavy and hard to work even with the sharpest rasp. I also learn:

stick welding, soldering, wire splicing, wood lathe, table saw, router etc. you can ride the bus downtown to the MAGG compound to do ceramics and race slot cars with the GIs.

I am into Boy Scouts, we have out big camping trip at Camp Mc Chauly a military reserve on the north east coast. We march out into the hinterland just off the beach. The main attractions are some 16 inch naval shell impact area craters. The shells generate huge 60 foot craters in the sandy soil that we somersault down. We play king of the ridge. The shell is the mass of a Volkswagen. We pitch our tents on a slight rise; careful not to pitch them under trees. In case it rains you don't want the trees dripping on your tent for hours after it stops raining. We set up our WWII shelter halves put down ground cloths throw in our packs and sleeping bags then trench around our tent with an entrenching tool 8 inches deep in case it rains.

We start up a campfire then heat up some Vienna sausages and beans in our canteen cups for dinner. We sit around the fire feeding it and telling tales. We spook each other warning the new guys not to go wandering around in the dark because Commie frogmen saboteurs have been known to come ashore here slitting throats and blowing up stuff.

We turn in only to be awakened at 0300 by a steady drumming on the tent canvas. Its rain, heavy drops, the tempo of which steadily increases, followed by some stiff gusts. It's soon pouring down. We are up and looking for leaks with our flashlights. John looks at me and says, "I've heard that if you touch the canvas it will leak where you touch it". He then goes on to check out this supposition by poking the tent in several spots. Well it leaks where you touch it alright but it's nothing compared to what's coming in under the tent now that our ditch work has overflowed. It's howling and blowing rain sideways, tents are going downwind. Everyone grabs what they can and we head down the trail into the wind. There are some screened in barracks about a two thousand yards from our campground that we passed on the way in that we now break into. We put some tents over the screens on the windward sides of the building; he wind is strong enough to hold them in place. At least we have a roof over our head and a windbreak. My mind wanders back to that red line at the Officers club, the one near the entrance, the one that is 6 feet off the ground marking the high water line for Typhoon Gloria. That storm came right up the river and inundated Taipei. The river valley funneled the storm and the storm surge. If this isn't a Typhoon I don't know what is. I've already experienced an earthquake, waited out in the bathroom at home; what's next? We are all good Scouts and make the best of it sharing out our food and heat in the way of Colman lanterns, Sterno cans and huddling together. The sun rises and the storm shows no sign of abating in fact there is some local flooding as the roads and ditches overflow becoming raging streams. Late in the afternoon some Navy and Chinese Army 6 X 6 trucks come out to rescue us. We pile into the canvas covered back with what stuff we had with us and head out. Most of the way back the water is up to the axles, it's still raining. Looking out the back of the truck, there's no fly, I see the Chinese folk struggling with the flood waters. It's a wet land but it would be hard to get used to this. They are perched on their roofs or on the upper story of the common two story roadside structures. No one is coming to rescue them.

## Riding the Bus

You get everywhere, north, south, downtown and to school, on a short gray Navy Shuttle bus driven by a Chinese Soldier. Chinese drivers "talk" to each other with their horns. The roads are heavily crowned and potholed. The bus always feels like it is going to roll over on the foot bicycle and motorcycle traffic on the right. Wags stand in the isle pretending that they are surfing while doing a poor rendition of Wipeout. Buses downtown have everyone from kids to GI's. There are over 50,000 foreign, US personnel and their dependents on the island.at this time. The bus to school has ages five to twenty there are some very interesting conversations on the bus, debates, battles of wit, to, yo mama and cut fights. Everyone knows that the first guy to lose his temper or curse loses the argument. The passengers laugh at the loser, when he loses it. The whole bus is judge and jury they know that sticks and stones will break your bones but words will never hurt you. Words are not weapons they are words, weapons are weapons. It is good to know the difference. At the same time everyone is Kumbaya, peace, love, no judging. Children aren't borne racist that's something that is learned. Times have changed. Lots of race baiting and judging these days. Tim's best friends are: Mark Phillips a half Chinese kid, Mark Payne a Cherokee, Steve Buckles an albino, and David White a black kid.

I am on the bus to go home after school, Kathleen climbs aboard too; she's not looking so good. She plops down beside me throwing her books in my lap then she doubles over groaning. She tells me she has some horrible cramps. Every pot hole causes her to yelp and moan. She is nearly passed out and drooling by the time we get to our stop. I help her limp down the street yelling for help. Mrs. Nelson piles Mom me and Kathleen into their red Plymouth Valiant sedan and we are off to the hospital. Mrs. Nelson is a great driver so we make it there in record time. The Corpsman at the desk wants us to sign in and wait our turn, I protest making a scene. Then Dad shows up and she is admitted. Kathleen has an emergency appendectomy, it was ready to burst. Kathleen spends a couple of days in the hospital out of caution against peritonitis. We go and visit her. I try to make a joke to cheer her up. Wrong. You don't try to make someone with stitches in their gut belly laugh. Soon enough she's home.

## Music

Everyone is into the Beach Boys, Beatles and the Stones. It seems like every week someone shows off his pirated copy of the latest hit album on the TAS shuttle bus. You can tell the records are pirated because they are multi colored clear vinyl. The material is so soft that it gets scratchy pretty quick. This gives rise to the reel to reel market. You record the record the first time you play it and so get a third generation recording. Even Dad, in a rare show of extravagance, buys a stereo FM/AM record player reel to reel console from the Naval Exchange (NX). He likes Herb Alpert, Mitch Miller and the New Christi Minstrels, we like the Mamas and the Papas, Kathleen's record. One recording Dad plays that I like is the "Grand Canyon Suite" a beautiful piece. Sukiyaki and Brazil 66 is always playing at the O club. All the kids grow their hair longer now or at least it's hanging in our faces, and we all want to play guitar. Guitars are cheap and plentiful but low quality. I managed to talk Mom into buying one for me, at auction, in the Springs before we left; it was only five dollars but it was a nice steel string Spanish cutaway acoustic. This guitar was too hard to finger right off. The action was real high and it need work. I bought a cheap nylon string classical Philippine guitar, one at the NX. I need it for guitar lessons at school. The music teacher is forty year old Dr. Ma. I like Dr. Ma, he teaches me, Walk Don't Run, a current hit song the Ventures play. That gets me to practice my fingering. Dr. Ma is hard to understand sometimes. Besides his thick Chinese accent, he has episodes. Dr. Ma still has a Communist bullet in his neck; it makes him stutter and twitch his head to one side when he wants to talk sometimes. That doesn't stop him from showing me some sweet licks and some technique. Unfortunately Dr. Ma succumbs to his wound and passes before I could get any further, he was an excellent teacher. There will be no replacement for him.

## Serving Mass

We start off attending church at the Catholic Cathedral located downtown, between the Spanish and Brazilian Embassies. Usually followed by brunch at the O' Club. After we move to Tien Mou we start attending a St Vincent DePaul Church about three miles up the hill from us on the road to Beitou, Maureen and Gary attended daycare and kindergarten at the school across the street from the Church. The little school is staffed by Chinese nuns. Maureen and Gary wear the school uniform: a navy blue smock with a white apron over it just like all the Chinese kids. The rector is an Old Dutch priest who I get to know quite well. Dad has volunteered me and Jim as altar boys. Father John has taken up his offer with a vengeance. We attend two services Sunday and always serve high mass for Fr. John. The Church is not air conditioned or heated. We always serve in full vestments; welcome in the winter but oppressive in the summer. Fr John sings the mass in Latin and I respond like the nuns have schooled me; Jim fakes it, he is OJT. The service is always overly long. Fr John gives the gospel readings and the sermon in both English and Chinese. Jim and I straighten out things and clean the church afterwards. We arrive early to set things up in the vestry, set up and dress the altar, light the incense pot and candles.

The nuns take care of the flowers. We leave late after we chat with the priests in the rectory where I learn that priests like to drink. Then there is the three mile walk home. All in all a long day.

We have Fr. John over for dinner often; Dad likes to chat him up. We have to mind our manners. Fr, John enjoys coming over, like any priest, he likes an audience and here he has ten souls to save. For as long as I have known, Dad has always had a priest as a friend, starting with the Chaplin that married him in Munich; he later went Pheasant hunting with us east of Denver before we shipped out. After a few Brandies, Fr. John notes the rice straw tatami mats we have spread over the terrazzo floors at our place. He declares that he has a fine rug that was donated to the Church but he hasn't a room big enough for it. The rug will be perfect for our living room. We have no heat in the house, save the fireplace. It's winter and there is a chill in the cold damp air. Fr John will not take no for an answer; he will have it delivered. What's delivered is a dirty brown jute rug that bristles like a fuller brush door mat. You can't walk on it unless you have shoes on. Forget about laying on the floor in front of the fire. I want the tatamis back. Along with the rug comes a letter from the good Father putting the value of the rug at $400.00 and would we be so kind as to make a donation for said amount? The donation was forthcoming but invitations to dinner and DeKuper were no longer. I continue to serve mass. I like Fr. John.

## Bringing Home the Babies

The Amahs usually take and bring home the babies from school. School for them is St Vincent DePaul day care with the Nuns. I go up and walk them home once in a while if the Amahs are busy or absent. I don't mind the walk. Lots of stuff going on in the street.

Lots to see. There is new construction. Red brick is the standard building material. The brick is so crappy someone has to keep hosing the pallets to keep the misshapen things from crumbling under their own weight. The walls going up are fronted by rickety bamboo scaffolding lashed with sisal. Coolies go up and down ladder ramps, with no rails, balancing hopers of brick and mortar. Most stores specialize. There are quite a few stores with cheap electronics, brass and wood things, fake native stuff, headhunter masks, knives, bows and arrows, carved chests. Stuff made of camphor wood, what first drew Japanese traders to the island. There are wet markets with pigs sawn in half hung alongside geese and chickens all with flies that a little boy futilely tries to wave away, his job. Shu buys some local produce and meat but most of our stuff comes from the Commissary. There is usually a Chinese opera on one corner of the traffic circle, all gong and reedy pipes, crowds milling and pressing against the stage, wandering in the street. The Chinese do not get in line for anything. Vendors hawk their wares. One has a board of dried squid nailed to it, they look like boot soles. The only snack I ever buy is sugar cane. You can get a two foot cane for next to nothing. Using a machete the vendor will cut it fresh from the stalk. He then takes a small bat with nails in it then smacks the cane twisting it to peel off the tough purple bark. He starts it peeling you finish it. You bite off the cream colored woody pith and chew it to get the sugar water out and spit out the pulp. There is a lot of spitting here. Beetle nuts are disgusting. They stain the mouth and teeth blood red. On holidays and special occasions there are men walking barefoot on hot coals, performers with long needles in their faces. There are men who appear to lick the edge of a long knife then paint bloody prayers on rice paper for a fee. The purchasers' burn the notes at their ancestral family alters as prayers and offerings to heaven along with joss sticks. I like getting the babies, they are real tight. Reenie and Gary ally themselves to make a unit always watching out for each other. Bill is no match for the two of them. They look real cute in their school smocks, like little dolls. They make me smile with their chatter, talking about what they see. Good kids. I usually have to alternate carrying them, their legs are so short; they tire easily. At least it's downhill. The following year Reenie is going to Kindergarten at TAS and rides the bus with us. That must be sad for them...first time that they have been apart. My load is a little lighter but not brighter bringing only one baby home.

## The Sand Pebbles May 66

They shoot the movie *The Sand Pebbles* while we are stationed in Taipei. I love that movie. The gunboat steams up and down the Tamsuei River for their shots; we get a glimpse of it now and then. The movie comes out long after we are back in the States. The river scenes showing the fishing junks, the hills and buildings along the river, the crowds of Chinese (no shortage of extras), brings it all back. The Navy was very supportive of the cast and crew during the filming. The company puts on a musical at the MAGG compound. Major Nelson comes back from Vietnam after a six month tour a Lt. Col. He takes all of us to see the production. His kids go to school with Steve McQueen's kids at Dominican. We never see the principals on stage but the cast is very talented and put on a great show centered on the bar scene, except there isn't a fight just dancing and singing. I wonder what the Navy thought of the film when it finally came out. There are a lot of controversial ideas and issues of the day in conflict and laid bare, brilliantly embodied by the principal actors.

# Bill's Accident June 66

We got acquainted with Chinese burial procedures and customs. Ancestor worship. There were jars of bones by every Chinese farm with little shrines, houses with deities. How they got there was quite a process. When the respected elder died there was a New Orleans style parade, brass bands and everything, with a giant ornate teak coffin with upswept ends on the lid drawn by horses, oxen or water buffalo. The length of the parade, the number of professional mourners, and the number and variation in the bands, was a sign of the wealth of, and respect for, the recently departed. Lots of gongs Chinese stringed instruments and flutes making a tremendous din. The roads were heavily crowned to shed rain, there were no shoulders just a beaten earth bike and foot traffic trail with Binjo's on one or both sides. The buildings were set back from the road in most places. There was a buffalo wallow on our side of the road to Beitou and the parades of course tied up traffic coming and going. Most of my brothers and the Nelson boys were on the other side of the main road watching the parade. I am sitting on our partition wall watching the same thing from the house The Nelsons have Binjo with them. As the tail of the parade passed going down the hill from Beito, a bus stopped for the parade signaled the boys to cross. Little Billy leads off with Binjo nipping at his heels, Timmy and Steve are right behind. At the same time a carryall with some Chinese Buy Sell men pulled around the slowpoke bus to head up the road in order to beat oncoming traffic. The carryall hits seven year old Billy, missing Tim and Steve by inches. Bill catches the hood of the truck on his forehead and is pitched back ten or fifteen feet down the road smacking the back of his head on the pavement. All I see is the truck pulling around the bus, I then hear tires squeal and a bang.

# 1965 TAIWAN DEFENSE COMMAND

Mark Nelson comes running down the street saying "Bill's been hit by a truck". I jump off the wall and run to the street. The crowd of villagers who were watching the funeral parade are now looking at my brother laying in the road. There is a pool of blood, under his head slowly draining toward the binjo. Head wounds bleed a lot. Bill looks horrible, he looks dead his eyes are open and staring at nothing. I have seen this before. There is a three inch gash on Bill's forehead and his face is covered in blood. A Chinese kid steps out of the crowd and dips his finger in Bills blood. He turns to the crowd, displays his finger, says something, then laughs. They don't think it's funny and neither do I. I go for him but the crowd parts and closes to cover his escape while my friends and brothers restrain me. Someone brings towels and blankets from the Nelson's house to cover and clean up poor Billy. Father John beats the ambulance to the scene and gives Bill Last Rites there on the street, the saddest sacrament I have ever attended; the good priest and I are crying. Billy was born in June, overseas, in Japan. This just cannot be the end of him. The ambulance shows up. Bill is taken to the airport packed in ice, put in the back seat of a T-33 and medevacked to Okinawa. Mom and Dad follow on another flight.

It's the height of the Vietnam, War. There is a Head Trauma Center at Kedna. Bill's near scalping looked pretty bad but the real damage was when the back of his head hit the road. Bill has eight burr holes drilled in his scull the size of a quarter to relive the pressure on his brain. He is put in an induced coma and placed under ice blankets to keep his temperature down to thwart hemorrhaging and clotting. Bill will spend nearly eight weeks in an ICU; with Mom and Dad at his side; and another month in recovery.

Its early summer. We are not in school. What now? Dad often says that children need constant supervision. We are basically unsupervised all the time. Captain Cunningham's mother moves in to help with the situation. Mrs. Cunningham starts imposing her brand of discipline, we are not impressed. Mrs. Cunningham thinks that we should be constantly on our knees praying for our brother and our poor parents. Kathleen resents this as an imposition. Kathleen thinks that she can do this gig, the staff is there to help and we can do the necessary shopping, after all she is going to be a Sophomore in high school. Kathleen organizes us. We have secret meetings, we go on strike, make signs and picket the Nanny. We call her Fart Blossom and protest her authority, she is not the boss of us. This well-meaning matron cannot handle us, the staff is amused with our antics which doesn't help. Our kind sponsor's mother is gone that night, she turns over funds Dad left for our support to Kathleen.

When Bill comes out of it, and comes home his motor skills are damaged and he has acquired some nervous ticks. Bill flicks his big toes till he wears holes in his socks. He also flicks his thumbs incessantly and has an eye tick. Bill's speech is also slurred but that may be due to the liquid Seconol, a barbiturate, he is dosed with to keep him calm. Bill is of pale blue white complexion now, almost translucent, you can see the veins in his face. He used to get a deep tan; Dad jokes and calls him spook. Bill wears white rimmed dark glasses when he goes outside so he won't be over stimulated. Bill has to wear a motorcycle or a lighter baseball batters helmet all the time now. You can see his pulse move the hair over his burr holes when he is not wearing his helmet. Dad tells his friends that "you can pick him up like a bowling ball now". Billy is glad to be back with his brothers and we are glad to have him back. Jim had better leave him alone. Bill goes back to school in the fall but he has to repeat the second grade; it's not without a great deal of effort.

Now that the parents are back the investigation into Bill's accident can proceed. It is here that I learn that the fellow who gets his story in first controls the narrative. I am questioned by the MP's none too gently. They question my version of the carryall pulling around the bus after the parade passed. I know that I would have done it, the bus goes slow, makes a lot of stops and it's a two lane road. This was the driver's one chance to pass for miles. The driver's version is that he started to roll after respectfully waiting for the funeral procession to pass; being at the head of the stopped traffic and Bill just ran out and hit him when he began to roll. They say that it would be impossible to do what I said that they did given the traffic and the stretch of road available to pass. Dad, much to my surprise, sides with me. He drags the MP's and the Chinese Cops out to the scene of the accident where they mark off a bus outline and recreate the maneuver I described after stopping traffic, with Dad driving his truck proving that that was what happened: the driver, got impatient, pulled around and hit Bill. Bill couldn't run or get off the road without falling into a buffalo wallow. The bus driver could see what was happening, so he was still stopped after the funeral parade passed; same with the oncoming traffic everyone was stopped to let five little boys and a dog cross the road; everyone except the Buy Sell men who hit little Billy.

We had had a lot of discussions about Chinese justice and what to expect, speculation and rumors mostly. We had all herd the story about killing someone's chicken. You paid for the chicken, all the eggs it would have laid, all the chickens it would have raised and all the eggs that they would have laid. Then there was the story of the military family that hit and killed a baby girl. After being put through the wringer it turned out that the girl was already dead; a small nail had been driven into the base of her skull and she was tossed onto the road in front of their car. Dad is not a litigious person he is happy that Bill survived and the Air Force has taken care of us. He thinks that dragging us through court proceedings without end is not worth whatever outcome that he can envision. In the interest of International Relations and the mental health of alcon we leave this incident in our wake. Billy, although younger, is bigger than Tiny Tim; Billy had excellent vision…

## Jack's Sister's Wedding February 1967

Jack, our houseboy's sister is to be married and we are to be the Honored Guests. The Chinese prefer Boys to Girls; you have to pay someone to take a daughter. The new wife is subservient to everyone in her new family. Better to marry the number one son than number six. There is to be a big feast and we get to take a place of honor at the table. Jack's village/extended family is relatively well off. Their place abuts the mountains. They grow rice of course and button mushrooms in caves carved in the rock face behind the village that they process and can. They also raise pigs, huge monsters that they keep in a restrictive pen where they eventually cannot turn around in. they feed them rotted vegetables, fruit, leftovers, and spoiled fermenting rice with the husk on. Jack gives us the grand tour. He is justifiably proud of his extended family. We have polished our shoes. We put on coat and tie and comb our hair running some Brylcreem through it. We splash on some Aqua Velva. After the tour we all funnel into a long low room. There must be about fifty plus people there. We are twenty percent of those present. Most of the folks at our round table speak English, they are smart and it is seen as a way to get ahead. The food starts coming. It is great and varied served family style on a lazy suzan, you take a little of each dish. Tim manages to pick out a fried rooster head, comb and all with the neck. There are oohs and ah's at the table. Our hosts explain that it is very good luck, like catching the bouquet. Tim is number one bachelor. Tim does us all proud by chowing down on it. Applause follows. We are a hit.

## Timmy the Flying Squirrel March 67

Everyone is very fond of Tiny Tim. Tim has that nebulous quality called, "a good attitude". Tim is in the middle, he's no threat to the older kids and the little guys usually take his lead. If it's ok with Timmy, they'll go along. Steve, Tim, Bill, and baby Gary share a large room across the hall from Jim and I. Our rooms are separated by the bathroom at the end of the hall. It's a zoo next door. The boys have made up a game, it's called Superman. They pin up the mosquito nets over their beds and pull out the lower trundle beds. Next they place the steel haircut/baby high chair in their huge oversized closet and using it along with a boost from a brother, one kid would climb up onto the closet shelf. The object was to see who, with or without cape, could jump from the shelf, hit the first upper trundle bed, and bounce from there across the room to the other upper; if you fell short you just landed on the lower trundle beds, great fun, lots of squealing and laughing. The boys only did this when the parents were gone of course. Tim is particularly good at this game. Tim is also part of our bike circus where he leaps from bike to bike or rides two bikes at the same time a foot on each with a hand on both biker's shoulder. Tim was climbing up and making the jump but he was the only one in the room. He slipped when he was climbing up and got hung on a sixteen penny nail sticking out of the wall. It caught him in the crook of his right arm. Steve and Bill hear Tim screaming, run in and "help" Tim down by pulling on him, ripping a four inch gash in his arm. Another trip to the Dispensary. About a week later, before he has his stiches out a couple of neighbor girls tug on his arm and pop the stitches and its back to the dispensary. Tim is hitching a ride alongside one of the Nelson boys behind the three wheeled motorcycle coal cart. You ride up on the moving cart on your spider bike, grab the cart's side rail with one hand and you are off! What could go wrong with this? All the kids did it, much to the amusement of the cart man collier. Tim loses control, crashes and hit's a neighbor's wall at a fair clip getting all skinned up on the poured concrete road. He is lucky that was all that happened.

## The Fort

Just when you think that everything is on an even keel, another event. Never boring when there are ten people involved. Jim and I in a rare spate of cooperation have built a little fort. We both like forts and this is not our first one. We have some sheet plywood some 2 X 4 lumber and Dad's Skill saw. I am taking wood shop at school so I take advantage of my training. We put up a floor on one corner of the yard bridging the back and the partition walls putting up some braces below it. We frame up some low walls cover them with ply and make a roof out of the remaining plywood. It looks like an Army watch tower/fighting position. We make a ladder to climb up and down. All the little guys love it. You can catch a breeze up there above the seven foot tall walls. Even Midget gets hauled up there.

One afternoon, I hear lots of screaming and run outside, it's coming from the fort. When I arrive at the corner of the house I see Reenie, Gary and Bill up in the fort, they are very agitated and yelling. Jim has a handsaw; he is under the fort sawing the floor in half. Jim has taken away the ladder and has the little guys trapped up there. I ask Jim what in the hell is he doing. He tells me that he is taking his half of the fort. He turns away from me and turns back to his task. The little guys start screaming again so I hit Jim upside the head with a hard right. His head bounces off the wall and he falls in the binjo gutter. Jim doesn't want to fight much after he gets up. I throw a few left jabs to help him make up his mind. It's a good thing the fight is out of him because my right hand hurts like hell. I tell Jim that I won't rat on him for torturing the little guys if he doesn't tell about me hitting him. The babies all agree. A deal with the devil. Jim snarls at me and moves off around to the front of the house his head must be really ringing. He usually doesn't give up that easily. Jim's fights are traditionally followed by a lot of verbal rage and abuse. Doesn't bother me as long as he's not swinging at me, sticks and stones you know. I think that the whole point of this debacle was to piss me off and provoke a fight. I have no idea what I did to him to bring it to this point.

 The lumps on either side of Jimmy's dumb ass head are covered by hair. No such luck for me. After about three days my hand looks like a big red grapefruit and about as useful to me. I keep thumping it but bite back my yelps. I am trying to do my homework left-handed with a fountain pen. This is not working out so well. I show Mom my hand and she takes me to the dispensary. Just shows how much of her attention I draw. I have a hand that looks like a medieval mace and have been taking meals with her for three days eating with my left hand. The doctor says I have broken three bones in my hand. My hand is broken again and reset. My fingers are curled around a firm wad of packing and gauze to make a fist and then the whole thing is plastered up with no fingers poking out. If only I had this sucker on when I hit Jim. No baseball, tennis or swimming for me.

## The Grand Hotel June-July 1967

That last month in country is great. We stay at the Grand Hotel. The Grand Hotel is owned by Madame Chaing. It's beautiful, the best hotel in town, five star. It has, of course, one of the best restaurants in town. The hotel has a lot of art and artifacts I assume came from the mainland. I have been to the museum in Taipei. The National Museum contains some of the most beautiful things ever created by the human hand. Madam Chaing had been crating up national treasures since 1936, when the Nationalist Chinese first started retreating from the Japanese onslaught. It never came out of the crates and so was good to go when the Nationalist left the mainland for Taiwan. You can spend days there and not appreciate it all.

Dad is a member of the Tennis Club here. He has been playing on the Hotel's red clay courts for two years now. Dad usually plays tennis instead of eating a proper lunch. The pro here is the national champion of China, his name is Tzi and he is amazing. The games are pick up. You have to play doubles if there are people waiting. There are only two courts and a covered grandstand. We sit in the stands; compose our matches and comment on the play. There is a locker room and shower. You must wear white and a collared shirt to play. There are ball boys here so play is fast. The ball boys, kids really, sweep the lines and water the court manually with big water sprinkler cans. They swing the heavy cans back and forth between their bowed legs. They water the courts when you switch sides to keep the dust down in the heat. Despite the sprinkling all our white shoes and socks are orange. There is cold water and salt pill dispensers by the stairway to the pool to help the players deal with the heat.

Dad has always played tennis as long as I can remember. He played Varsity in High School. I started playing tennis here in Taiwan. I taught myself on the concrete courts of the BOT compound by first hitting a ball against a wall. Dad would throw around a hardball, play catch with me, but he said he would not play tennis with me until I could beat my Mother. Mom was not taking us out to hit it around either. She would go to the hotel with Dad and hangout on the courts with the swells on the weekends. Mom plays socially at best but only if they need another female player and there is no one else.

Kathleen, Jim and I try our luck at the BOT concrete court with the chain link net and no wind breaks. Bare concrete courts with no surface paint are fast, the ball stays low with little bounce skidding actually. We are very bad. Jim keeps cranking the ball over the fifteen foot backstop fence. You have to walk all around the compound security fence to get it so we get adept at scaling chain link fencing. Jim will not get his balls until we run out; always an argument. Kathleen and I learn control because we don't want to hit the balls over the fence. Kathleen gets up on the curve before Jim and I. We have real crappy Chinese rackets with loose stretchy nylon strings. Despite our crappy gear we steadily get better as we play sets. I like hitting with Kathleen more so than Jim; she's also better than me. Kathleen quits playing after I start beating her consistently; a shame. I still play with Jim, it's something to do.

I start hitting with Mom. She is a good sport. She doesn't get mad or go for the winner or charge the net or slam her serves; she's just happy to keep the ball in play and get it back. I try to hit the ball to Mom so she can get it. Who wants to make your Mom run or crank balls at her? Mom and Dad both get new Dunlop racquets strung with gut in Hong Kong so I get their old ones. They are a couple of Slazenger German made racquets, my game immediately picks up. Mom is not a problem to beat so Dad starts hitting with me. Dad's game is full of spin and chop. His favorite shots are the drop shot, the lob and the overhead, perfect for doubles. Dad has a great spin serve that pulls you wide or spins into you.

I love playing on clay at the hotel. I have to learn everything all over again, the ball hops higher and plays slower than concrete, you also have to slide into your strokes when you run down a ball. You also have to watch your footwork taking off to chase one. I get to see and hit with some good players. You have to play up to get better. I also get some tips from the Pro who is on the court all day. There is a fifteen-year-old girl phenome that everyone wants on their team. She is a lithe beautiful blue eyed tan blond with game. I'm too embarrassed of my game to ask her to hit with me. Jim and I have the place to ourselves during the week at certain times. Jim's game has improved some but he still has a flat sidearm serve; hits everything too hard for his chop strokes and sends the ball sailing. Still going for the winner every stroke. The little guys play ping pong nearby on a table covered by an open veranda. When not on the court we are in the pool.

We spend the days exploring the hotel and never see all of it. We play tennis and swim the days away, ordering mini hamburgers and Chi Sway soda pop at the clubhouse. Teenage cover bands play the latest rock tunes as we dance in the evening. Everyone is in transit; the place is full of military and diplomat dependents during the day. For a change of scene we go to the Officers Club and the MAGG compound downtown. There is also an amusement park with rides by the bridge located unfortunately across the river from a smelly duck farm. There is also a fish hatchery with tanks of all kinds of tropical fish to see.

JJ Scanlon, Aviation Cadet May 1943, San Antonio, Texas

JJ Scanlon, top left with his B-24 Crew, Tonopah, Nevada

Major Robert E Scanlon, Italy 1944

Lt JJ Scanlon 4th Fighter Group

Lt Julio Nuno 1944

JJ Scanlon, Munich, Germany 1949

Wedding Day, Munich, 1950

Japan, 1958

Jim and Bob, Utah, 1958

JJ Scanlon, B57-D Hill AFB

JJ Scanlon and B57-D Hill AFB

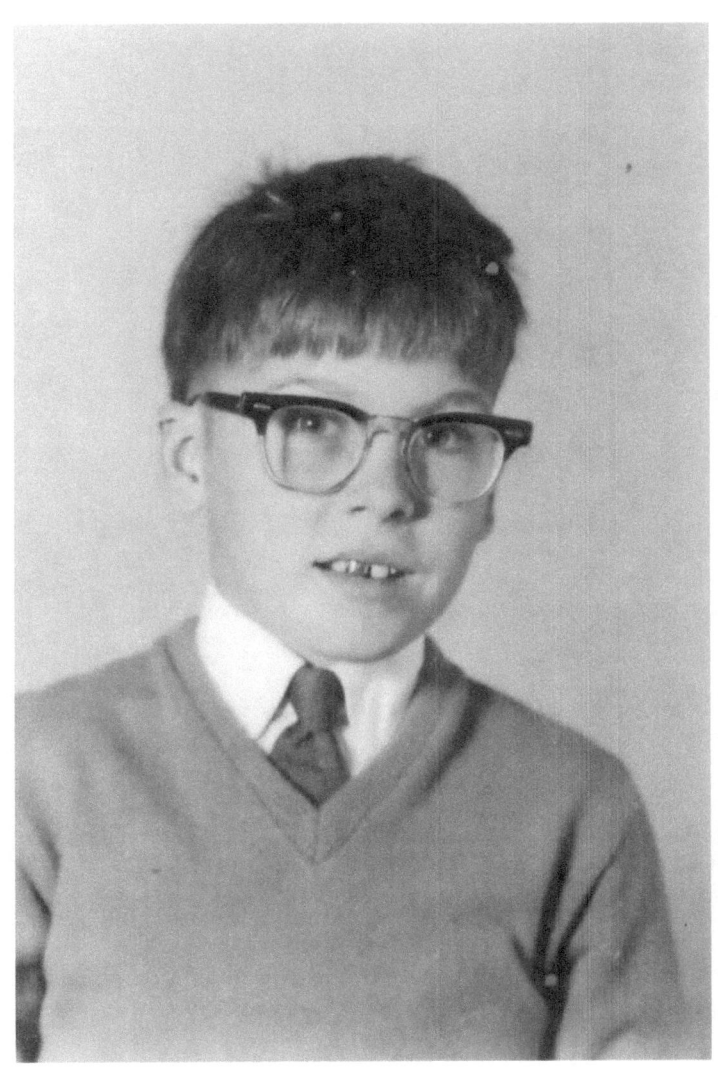
Alabama Bob 1962, Starting at Devine Redeemer

Mom, Bob and Little Billy one week home from hospital hanging onto his big brother at archery tournament

The Grand Hotel, Taiwan, 1967

Griffiss AFB, 1970, Tim with his broken arm
sticking out and Bill wearing my hockey helmet

JJ Scanlon, Official Photo, 1970

JJS Charter Pilot, 1977, flying Grandma Home to Abilene

# CHAPTER 14

# 1967 Griffiss AFB, Rome, New York

**JJS.** August 1967 to New York, Griffiss AFB – Snow Land . . .

**RES.** Dad didn't write much about his postings in Taiwan or New York the sentence above is all he put to paper. Or for that matter, any of his further postings; either because it paled compared to what he did in his early career, or it was Classified, probably some of both. So I'll try to fill in the blanks for those periods from what I know. Dad never told us what he was doing or where he was going. He didn't tell us what he did when he came home. We would just hear the words TDY and we would know that we wouldn't see him for a while. We could judge the duration by the size of the B4 bag he packed big or little. We get hints. Dad lets slip that he wouldn't have a prayer if he went down over the pole. He kids that he thinks twice about taking his pistol with him because as a good Catholic he would be too tempted to shoot himself. He is playing intruder; trying to defeat the DEW line, early warning radar system in Canada and any interceptors scrambled to get him, in an old B 57-A packed with electronics. This is the last A model in the inventory; it goes to the boneyard after RADC is disbanded. Capt. Cunningham blew the canopy on this bird for an emergency landing when the nose wheel gear failed to deploy, standard procedure. The ejection system is one of the old ones with basically a three inch cannon shell under the seat, good thing he didn't have to eject or scramble out. Jim and I go to check out the canopy on the flight line. Dad also flies a lot of T-33 missions to the various interceptor and Administrative bases concerned with the project. This is the Flight Test Division.

Dad takes command of Rome Air Development Center (RADC) he is also the base Operation Officer. Dad's office is in the flight tower. The base is a SAC Base with Nuclear armed B-52s on the ready ramp with Hound Dog missiles slung under each wing. The base also has an interceptor squadron of F-106s and some KC-135s. Snow Land is right. Rome gets 12 feet of snow a year. Dad has to keep the runway clear and you can't use any sort of salt based stuff on the runway because it corrodes the aircraft, they use an alcohol based deicer. Dad shows me his barn of snow equipment: plows, blower's, sand, and tanker trucks. It has a whole maintenance barn associated with it. They do the runways first before they do the neighborhoods. He has more equipment than the city

## The Trip Home July 1967

One day we are ready to go. Goodbye Grand Hotel. Mom has sold all the stuff she doesn't want to the Buy-Sell Men. These fellows have been pestering us since day one. We have a coffee can full of their business cards. They buy the stuff from military families when they leave, appliances and stuff that you bought from the PX. They do it to avoid paying import or sales tax; or to buy stuff that they just can't get otherwise. Mom sells the 1948 washing machine, the piano, the air conditioner heck everything she brought from the states, down to the mosquito nets to make room for the new stuff. Dad has sold his truck and has placed an order for a new bare bones 1967 Chevy Carry All, now called a Suburban that we will pick up in Portland.

We load on a plane late in the afternoon in Taipei on a chartered Continental 707. We land at Tatchikowa AFB Japan deplane and sit in a hanger for a layover while the plane is serviced. It is dusk when we get back on the plane. We are flying the Polar Route direct to SeaTac. We sleep through the flight, best one ever and wakeup for breakfast. We let down in a solid overcast, typical Tacoma weather. We take a taxi to a motel and bring our stuff in the room while Dad goes to get his truck in Portland. Dad had a brochure where you order your truck by mail and check off the options you want. Dad has picked very few options beyond the base model with a 283 V8 and a 2bbl carb. He has upgraded it with a trailering package plus hitch, oversized radiator, 4 speed manual transmission with a granny gear and posi-track but no factory A/C. We are going to live in upstate New York, Snow land. Who needs an air-conditioner?

Dad returns to the motel and finds us glued to the TV, we haven't done a single thing the whole time that he has been gone. Not a single suitcase has been cracked open. We haven't watched TV for two years. Star Trek is on and it's in color. Everything was black and white when we left. Dad looks at us, looks at the TV then looks back at us and says "turn that ridiculous crap off". It's a rainy afternoon, typical weather for this area. We chow down at the attached diner and turn in early so we can hit the road at the crack of dawn. Accommodations are tight in that little room but we're so exhausted

## 1967 GRIFFISS AFB, ROME, NEW YORK

that we pass out where ever we are. Next day we load up. We are limited to one small hard side Samsonite suitcase each. They are a nested set so they are issued by size Mom and Kathleen getting the big ones. How we get all ten of us in the truck with all our stuff is a study in organized chaos. The truck has no seatbelts, vinyl mats instead of carpet, no paneling or insulation in the roof, sidewalls or doors. It looks like the inside of a C-47. We are lucky there are regular vinyl bench seats instead of backwards facing canvas jump seats. Some kids sit on the suitcases and wheel wells.

We're off. First we are going to Klamath Falls Oregon to see Dad's old High School friend and hunting buddy Hub and his wife Molly Edwards. The sky clears, we see Mount Hood and have parking lot tailgate sandwiches with Shasta Cola, stuff we buy at a Safeway whose parking lot we are parked in. The countryside is beautiful big trees everywhere and no bamboo. It's our first look at the good old USA after two years and we are excited about the trip. We are going to New York by way of LA so it's going to be a long hard look.

Hub is a hell of a sportsman he has some property along the Rouge River. Hub puts us all up in his big log house. Hub is a bow hunter; he shows me his 70 pound Bear recurve bow and puts a wood handmade arrow in a tree at 50 yards with it for me. His practice/field tip arrows are tipped with .38 special pistol brass. I'm impressed. Hub has a double-ender row boat that he made himself, it's a Rogue River drift boat, particular to this part of the country. Hub takes Dad and me on a drift fishing expedition down the river, something I'll never forget. Molly drives the truck and trailer downstream and pucks us up. We have our catch for dinner. Staying with Hub allows us to decompress, get over jet lag and get used to not sweating buckets.

We get back on the road after Hub and Dad install some seatbelts in the car. We say our goodbyes early the next morning before the sun is up. We're are making the trip to Los Angles in two days or at least that is the plan. We make it to San Francisco the first day and stay in a roadside motel for the night. It has been a long day for everyone. We have been rotating out rubbing Dad's neck the whole way; there's no center arm rest for his right arm and he is complaining. Dad is determined to fix that when he gets to his Father's house in Gardenia.

When we arrive in Gardenia some of us sleep in the truck and some of us sleep in the house, there is also a loft over the garage where Dad and his brother Bob used to hang out. I find Uncle Bob's old Daisy BB gun in the loft and stash it . Grandpa has his dental shop attached to the back of the garage where he makes dentures and appliances. Grandpa is famous for pulling out his dentures and scaring us with them for a laugh. Grandpa gives a select few a tour and lets us watch him work. Grandpa looks a lot thinner than the last time I saw him; his clothes seem to hang on him now but he is all similes and dismisses my concern. He still sports a pipe but I don't see him light up. Dad is worried that Grandpa is wearing himself out taking care of Grandma like he used to when it should be the other way around. Dad is busy with his arm rest; it will be a plywood box with a hinged lid that he glues some parachute seat foam to, we still have a lot of these in the trailer. He intends to keep his maps in the box. The little guys pick a lot of Grandmas lemons off her trees, all they could reach actually, to make lemonade. They also use up all her sugar, and make a mess in the kitchen. This is too much aggravation; Grandma Scanlon is pissed. Dad comes home to find all the little guys standing in corners of the house their face to the wall contemplating their sins. Dad and Grandma exchange some sharp words ending with Dad telling her that we are not bad kids. I am proud of him for that. We are all split up and are rotated between Uncle Sal's house for the 4th of July and Grandma Nuno's apartment; Jim and I go to Grandpa Nuno's house. Mom and Dad then go on a visiting spree, staying here and there, they have a lot of old friends in LA to see and catch up with. They spend most of their time at the Clarks in Laguna Beach; real nice place to decompress. They also hang out on Balboa island with Mom's cousin Sonia,/Sunny, whose Australian husband Al Larson has become rich cornering the Australian lobster tail market buying futures in frozen tails. He is an interesting fellow. Al was shot down over Italy and captured while serving with the RAF; he managed to escape while being transported to a POW camp and make his way to Yugoslavia and freedom. Al is a very resourceful person. Al and Sonia live in Shirley Temple's old house across from John Wayne's place.

Finally it's time to hit the road. I prep and clean the old trailer for Dad. I pull out the jack stands and set up the overload hitch. We can put our bags in the trailer now so it's not so tight in the truck. Jim and I load up all the bags coolers and food after it's cleaned. The 67 suburban is a couple of feet longer than the one we rode out in. but it still has the same configuration it's a three door unit with four wind down windows, the back is all un-tinted glass. We hitch up and say our goodbyes to Grandpa and Grandma Scanlon both are sad to see us go now that it is really happening. We will be on the other side of the country, too far for a visit. Little do we know that this will be the last time that we will ever see Grandpa Scanlon or Grandpa Nuno. Here we come Route 66 hope it's in better shape than when we left for Taiwan.

We do not get very far from LA before it's apparent that we need to travel at night or get an air-conditioner installed. The Air Force blue paint job with no headliner plus ten 100watt humans, turns the back of the bus into an easy bake oven. We see an ARA sign from the highway and pull over. We had an ARA unit in our last truck. They have something that they think will work in stock; what brackets they don't have they will fabricate on site. The unit looks small to me, smaller than the one we had in the white truck, but what do I know. This place has been here forever on Route 66. The garage is red brick with faded ads painted on the brick; even the driveway and parking lot is paved with brick. There are some huge shade trees that have been here forever too. We back the trailer under a tree and unhitch. Just like the dustbowl folks must have done when they came this way. It's going to be awhile. We only have three folding chairs, Mom and Dad get two, the girls and the babies get the other so the rest of us just mill around. It's hot. We are on the edge of the Mojave Desert. The air is blowing hot, the bricks are hot, but it's nothing compared to humid Taiwan. This is dry high desert, air that will make your nose bleed. There is a hardware store across the highway. Dad goes over there to check it out and comes back with a box of 24 x 12 x ½ Styrofoam acoustic ceiling panels. After the A/C is installed in the truck, Dad cuts some panels to length and pops them in the roof between the headliner bows, the remainder is jammed against the truck's side panels all without adhesive. Not half bad. We hitch up and head east across the Mojave, next stop Palm Springs. The sweat margin is down 20db but I wonder what the long term psychological damage will be from the incessant squeaking of the unsecured Styrofoam panels.

We stop for chow and rest in a date palm forest that goes on forever, way cool with peacocks wandering around. The palms are planted in rows so you can't get lost. The prehistoric looking palms require some investigation and we get a chance to stretch our legs under some shade and take a leak. It's not long before we're pelting each other with whatever windfall we can find under the palms, taking cover behind the trunks until Dad whistles recall for chow.

We are headed to Flagstaff, four-corners and the Grand Canyon. We will not be going to Albuquerque, something happened between Mom and her brother Bill unnoticed by us kids the last time we visited. Bill had married Irene after his first wife, Shirley, passed away and Mom said something to upset her. We are going to cut through the mountains to check on our house in the Springs, it needs some maintenance before it is leased again and this is our one chance to do that. We pass through some sad looking Native Indian Reservations, no fences, some sway backed nags grazing on sparse dry grass, and abandoned pickup trucks, dry red mesas on the horizon. They sold Indian artifacts and souvenirs on the southern leg of route 66 but no such activity or enterprise around here.

We start the long climb up the divide. The drive through the mountains is a treat for us, the air is cool and there is a lot to see with pretty places to camp as we follow the rivers through the mountains. We go up over Monarch pass where we used to camp and fish when we were last here. Finally we follow the Arkansas River down to Canyon City and to the Springs. We paint the house while mom spends lots of time visiting friends, little kids in tow. Then we hit the road, lots of cornfields. We see the steel mills of Gary Indiana belching smoke and flame at night. We then stop to visit Mom's cousin Ray in Cleveland. Next we hit New York going through Niagara but we do not stop to see the falls; it's on to Rome.

## Woodhaven Summer 1967

Dad and I park the trailer in a secure area that the base provides for boats RV's and trailers after we unload. We have been assigned base housing in Woodhaven on Saturn Drive. All the streets are named after missiles. Woodhaven is outside the wire The housing here is mostly small split levels, a basement and an upper story of maybe 1500 square feet with a single car garage, part of the basement. The housing is all the same; some are conjoined to make mirrored duplexes. There are some smaller single story units where the terrain is flatter. There is no fencing. It's zero lot line; we all have a large common back yard. There are towering Sugar Maples and large tame gray squirrels everywhere. This housing is for officers only, mostly Majors and Captains. Dad and our sponsor, Col. Collins are the only Light Colonels in the area. Enlisted housing, E5 and above, is inside the wire; it's more apartment duplex like, there are cinder block two story Barracks near the amenities for the single Airmen. There is a lot more housing in another on base neighborhood called Skyline. There are some bigger units there that we are on the waiting list for. We have more Nelsons for neighbors, they have two boys and three girls, Jerry is my age and Rick is Steve's age but in the same grade as Timmy. The three girls Lisa, Sarah and Yvonne, are Reenie and Gary's age. The crates arrive, steel shipping containers this time which are more furniture friendly to Mom's new stuff. We have left all the stuff that was in storage in Colorado stay in storage. How we cram all of us and our parred down Taiwan stuff in that house is a miracle. There is no room for a car in the garage, it's full of stacked, boxed up stuff that will remain stacked and unpacked. We have Mom's Chinese rugs laid out two deep; her Korean chests are stacked one upon the other. There is no room for her ten person table setting side board and hutch so we leave two leaves out and double park the sideboard and hutch. All the carved teak, jade colored marble inlay, plastic covered velvet cushion, furniture that we can't sit on barely fits in the living room. You have to stand behind the couch or get a foldup chair to see the new Motorola color TV,

## 1967 GRIFFISS AFB, ROME, NEW YORK

our first. We get to watch Dallas play Green Bay in the Snow Bowl. I'm suddenly a Cowboy fan.

Jim and I have a room in the basement of course, across from our room there is a utility room for the washer and dryer, water heater, the furnace and an old fringe. My room, actually a utility closet, is small with barely enough room for our bunk beds and a chest of drawers. There is a small closet to share. Jim has a cassette player. I have no Idea where he got it. He only has one tape, a Stones Led Zeppelin compilation that he plays every night. The hall to the garage and a short staircase to the common area separate the two basement rooms. Steve, Bill, Tim and Gary all share a room upstairs. Maureen and Kathleen are still roomies one a Junior in high school and the other in first grade, sharing a bed. Mom and Dad have their own room of course. It's real tight so we spend a lot of time outside; it's not long before the whole neighborhood knows we're here.

We find the rolling stock, bikes and wagons and are soon doing the block like a swarm of bees. We follow the DDT sprayer like a rifle squad using a smoke screen. The sprayer is necessary because of the proximity of the dump, a huge depression with standing water in the bottom that might have been a gravel pit at one time. The water table is low here, in the loop of the Mohawk River. The lip of the depression is lined with cut branches, beech mostly. It's where the base dumps waste of that type, the area is very wooly. Great place to explore and play. We build a War Wagon out of some lumber, packing crates and wheels off some trikes. The wagon takes two passengers and a driver one pusher and one puller. We always have a ball game of some kind going in the common greenspace under those huge Maple trees.

It was the Maple trees that gave me the idea. The tree right behind our house had a huge branch about forty feet up. During the unpacking I found my old twenty pound longbow. I won an archery tournament in Taiwan; I was the only kid to put arrows in the butts at seventy five yards. The tournament was for thirteen to eighteen-year-olds. I tape a bolt and some para-cord to an arrow and shoot it over the branch. I use the cord to fish a three strand climbing rope from the Alaska haul over the branch that I've tied a block and tackle to. I have three hundred feet of braided army OD climbing rope rove through the blocks. I attach the fall block to a parachute harness that we just happen to have and there you go instant carnival ride. You buckle in a little kid, get two or three big kids on the line who run as fast as they can, only to turn and drop their load with a jerk to a stop, lots of screaming as the branch bounces and swings them around. What fun. I grow a crowd, everyone wants a ride. I'm suddenly very popular, with the little kids at least. It took longer than you would think in retrospect, given all the screaming, for the Moms to pour out and surround me, I'm not too popular with them. They demand that I cease and desist; turns out that they don't like seeing their kids screaming and twisting from a rope forty feet in the air no matter how much the kid likes it.

Jim already has enemies, big surprise, he's the new kid and has pissed off three neighbor kids. He has offered to fight them one at a time. It's like third grade all over again. I'm to back him up. He has told them that. Thanks a lot. What a moron, we just got here. The boys show up. I tell them there isn't going to be a fight and talk them out of it somehow, mainly agreeing with them that Jim is a real pinhead and they should probably just let it go kind of firm like without being threatening. They are the Sellers boys, they like me. I go to their house a block away, turns out they are cool, Tom can sing and play Doors songs on his electric organ; his brother isn't too bad on the guitar. I have some new friends. They are Lt

## 1967 GRIFFISS AFB, ROME, NEW YORK

Col Sellers Boys. And Lt Col Reperman's son. They live on Saturn drive. Jim is forgotten.

Mom gets a new car soon after we move in, a Buick Special Deluxe. The car is a cream colored two door coupe with gold and verdigris brocade upholstery. It has a 350 cubic inch engine with a two barrel carb, a three speed automatic transmission and factory air conditioning. It will do every bit of the indicated 120mph on the speedo. Don't ask me how I know that. It's a little too much car for Mom, she can barely see over the dash, but most cars are a little too much for Mom in the muscle car era. Mom can now go wherever she wants without being chauffeured around by Dad as long as she has a pillow to sit on. I'm thinking I might get a crack at driving it, I'm fourteen and getting the bug; the automatic transmission should make it a snap. Kathleen gets her license but Dad won't let her drive until she demonstrates to him that she can change a tire. Dad makes Kathleen jack up the car pull the wheel and mount the full sized spare then take it off, remount the original tire and put the spare back in the trunk with the jack. Kathleen complies but not without a lot of bitching. She is livid about this pointless unnecessary exercise. All that heavy labor to be right back where you started. One of my friends has a mini bike with a 5Hp Briggs and Stratton engine, just like the one on my old lawn mower. The bike has a centrifugal clutch so it's no problem to operate. Hopefully my old mower is still in storage back in Colorado, I think a lot about making a mini bike out of it if I ever see it again. Turns out Dad thinks I should learn to drive too, but not the Buick; I'm to learn to drive the War Wagon. I'm wondering why the truck, its Dads car, the vehicle he drives every day. Is he going to let me drive it when he's flying? I am intimidated, apprehensive and excited at the same time. It is Saturday.

Dad takes me to a large empty parking lot on base, stops the truck, gets out and tells me to hop behind the wheel. Dad gives me a onetime brief on the controls, then tells me to start the truck. I manage to start the car right but that was the only thing I got right. I kill the motor a couple of times before I get a feel for the clutch and start to roll; that gets me a dirty look. Then I get punched in the thigh because I'm crashing into imaginary cars parked in the spaces marked for that purpose. I'm thinking: Is this how you taught Mom to drive or pilots to fly? If so I'm not taking flying lessons from you. Dad is all over me until I can start, stop and shift up and down without snapping his head and hitting imaginary parked cars. I should not have been so excited about learning to drive, I am being trained to function as a valet. My new duty is to park and shift the cars if they need shifting since the garage is full of stuff.

Its mid-August, time for school to start. I'm to attend Ninth grade at Staley Junior High with Jim and Steve who are in the Seventh grade. Kathleen will be a Junior attending Rome Free Academy, a High School established during the Grant Administration. Rome has been around a long time, the Erie Canal runs through it. Fort Stanwix of the French and Indian war is in Town. The fort was established to guard the portage between Lake Ontario and the Mohawk, which flows to the Hudson River. Staley is located along the bank of the Mohawk River which also flows past the west and south end of the base. It's a pretty campus and a new school with lots of playing fields that are on the rivers flood plain. The good news is that they have the tennis courts here for the High School team, there's no Junior High sports program but they tell me that I can try out for the Varsity High School tennis. There's a bus to school but it's within easy walking distance, around two miles from Woodhaven.

## 1967 GRIFFISS AFB, ROME, NEW YORK

The first day is a trip. Rome is full of Italians. Who knew? Utica, I'm told, is our arch rival, it is full of Poles. The Italians tell Polish Jokes and the Poles tell Italian Jokes; the jokes are all the same. The school is further divided into the surfers and the punks. The surfers wear white socks and the punks wear black and then there are base kids, we transient, square pegs. The New York school system is excellent, you have to pass the Statewide Regents Exam to pass to the next grade. Instruction is rigorous in order to pass finals. They give me aptitude tests and place me in classes where they think I should be. I have a great algebra instructor who really ramps up my understanding and interest in math. I find myself in an after school auto shop class that I enjoy. I try out for the Tennis team and I make it; the season is in the spring but we practice in the fall and do indoor drills when it snows. I'm number eleven on the ladder. They have a challenge system, you can't refuse a challenge but the number of challenges you can make to a single player is limited; it's pretty competitive. I like my team mates, they don't give me too much crap about being the only freshman, base kid, new guy, after I work one rung up the ladder. The days blur into a routine of: getup, get dressed, eat; catch the bus, go to class, tennis, walk the two miles home packing my books and tennis gear because tennis is after hours, eat, do homework, pass out and do it all over again. Jim and Steve go to Staley too. Tim and the rest go to Francis Bellamy Elementary. Maureen and Gary stick together like glue. Bill has to wear his baseball helmet to school. We are all strangers in a new school. Tim gets a surprise. Dad tells him that no one in the family is studying music; someone needs to take up an instrument so Tim is tasked to learn the flute. Why? Because Dad has always liked the flute, besides Mom got a real good deal on one. Tim is handed a real nice Gemvart silver flute and starts taking lessons. Then it snows.

# Winter of 1967

It's been nearly three years since I've seen snow or have been so cold. The kind of cold that makes you want to curl into the fetal position under your blankets, It's only October for Pete's sake. We had exactly one month of Fall. The leaves on all the Sugar Maple trees turned brilliant colors at the first frosts, which I thought was wonderful until they fell and I was told to rake them up. By the way there is a lot of dog doo in leaf piles, I know, so I generally advise against jumping or rolling in leaf piles. Mining the piles and letting everyone know keeps little kids and their dogs from spreading them all over the place after you just raked them up.

With the snow comes snow shoveling. I add clearing the driveway to my morning routine. Dad likes to be at work by 0630 so I'm at work before then, no need for an alarm clock. Dad enjoys waking me up, barking like some barracks corporal; his favorite wakeup is ice cold hands under the blanket while he sings some silly ditty like the "She's Too Fat For Me Polka" or "May the Purple Bird of Paradise Fly Up Your Nose". Rome gets over twelve feet of snow a year and it stays put till Easter. Everyone has dayglow orange balls on their car's radio antennae so you can detect the presence of a car over the plowed snow bank at an intersection. After you shovel the drive you have to hang around if the street hasn't been plowed because the plow will throw a pile of snow across the drive that you had better remove before it freezes. That mix of snow ice and sand requires a pickax to break up. This frozen ridge row greets me when I get home sometimes; I have to remove it before Dad gets home. We get snow days. On one occasion the snow drifted against the house high enough to cover my window. Another day it was forty five degrees below zero and we boys were sure it was

## 1967 GRIFFISS AFB, ROME, NEW YORK

off. Cold enough to freeze flat spots on the buss tires; you can feel them thumping when we roll. We were hovering around the radio waiting for them to call it along with all the other closures. Kathleen insisted that school was on and she was not going to risk missing the bus so she grabbed her stuff and marched out the door. After all, Dad went to work. The bus did not come, it's too cold. The superintendent determines that it's just not healthy for kids to be out and about. Kathleen ends up freezing her larynx waiting for the bus. The Super was right.

A lot of cars are fitted with block heaters; you plug in your car where you park it and it keeps your motor warm and easy to start in the morning. The barracks all have convenient outlets for this but the houses don't, we have a garage, full of stuff. I have to scrape the ice off the cars in the morning until Dad buys a little space heater instead of a block heater and hangs it in the sub, at least it keeps the windows from icing up if I rig it and plug it in.

Turns out there's lots of stuff to do in the winter. We have some old hand me down skates from when we were in the Springs and some old pairs that came in an auction box of stuff. There's a pair of cheap hockey skates and a pair of fancy long bladed speed skates along with some men's figure skates and Mom's old skates that dad bought her their first Christmas in Munich, a pair that she has never used. Jim grabs the speed skates which is a mistake because he is splayfooted. All we need is a rink. We go down to the dump and find the ponding water in the bottom frozen over but with trees sticking out of it along with floating logs that were frozen in place. We start cutting down the saplings, chop a hole in the ice, dip out water with a pail and ice over the stumps. This is a lot of work for a lumpy rink that we can't resurface. When one of the neighbor kids that we lured into our venture falls through the ice and gets a soaking; we abandon the dump, except for skiing and sledding.

We still need a rink. We pack the snow in the back yard stomping it down until we have the outline of the rink we want. The Nelson's get in on it until it wraps around the side of our house across the back of our house and theirs forming a giant L. the snow was two feet deep when we started so we have readymade banks that we pack and build up with more snow. Then begins a nightly routine for me, resurfacing the ice. That rink was pretty ambitious. That's what you get with ten kids stomping around, it takes quite a few cold nights after dinner to get a good buildup with a five eights hose. I like going out at night in the cold as long as it isn't blowing. The air is sharp and clear, like all the bad has been precipitated, freeze dried out; cleaned. It isn't so bad if you are dressed properly. It gives me some alone time. I can turn off my filters and tune in to what's around me instead. After a while I can hose down on skates, so my feet don't get cold like they did standing on the ice in Dad's thermal flight boots. The rink is a big hit; kids are on it day and night depending on the weather. One thing I didn't account for was that having a rink just meant more snow to shovel. You have to get the wet snow off the ice before it freezes to the surface and makes it all pebbly and crusty. At lest you can do it while skating, it really builds up your skating muscles pushing a shovel full of snow. All the snow shoveled from the rink is used to build up some bobsled like iced banked curves into the corners of the rink. We had a flat bottomed shallow plastic tub of a sled that we would place a kid's collapsible chair in/on, load a kid on it, while another kid would push it around the rink. Talk about no seatbelts there were many spectacular wrecks the degree of which depending on the mass of the rider and the horsepower of the pusher.

We get some sticks and a puck then it's on. I find that I can skate circles around everyone, forward or backward. We buy more pucks and make a net. I practice stick handling and shooting. After three months of this I'm getting pretty good. You want to control the puck on this rink; you need to keep it down or you will be hunting for it a long time; there's no boards here just deep snow. Jim of course prefers the slap shot and sprays them all over the place making us duck and yell.

## Spring 1968

Winter turns to spring or a spring thaw. Lt Col. Collins a native New Yorker with a place up on Lake Champlain, shows us how to tap Sugar Maple Trees. We were surrounded by giant Sugar Maple trees that have never been tapped as far as I can tell. There's a woody cane that grows down by the dump that is about a half inch in diameter which has a pithy core that you can poke out to make a spigot of. After the sap run you can just drive them into the tree and close up the hole. I make a bunch of spigots, probably too many, then armed with a brace and bit I start tapping the trees. We have a bunch of one gallon plastic ice cream pails that I hang on a notch I whittle on the top of the spigots. The last thing is to tent over everything with aluminum foil to keep stuff you don't want from getting in the pail, just like Col. Collins showed me. I tap most of the trees in the green belt. Now I need to keep up with it. Once you get started you can't stop because the trees don't. Tapping and collecting requires dragging a toboggan around with the tools first and then the sap. I swap out the pails and put a plastic snap lid on the just collected one and tie it on the toboggan. All that sled dragging makes you thirsty. Fresh ice cold sugar maple sap water is the best thing in the world for that thirst. It takes a lot of sap to make syrup, on the order of forty to one, which translates to a lot of gathering and boiling. I have tied up the stove and have all four burners going. It's a lot of work but I have a lot of help, everyone is looking forward to pancakes and waffles with homemade syrup. We are in a hurry to see what we are going to get; we hard boil a sample and find out that that is a mistake by burning it slightly. The sample doesn't taste too good casting doubt on the enterprise. Some wag suggests that I've tapped the wrong kind of trees. The answer is a long slow boil where you keep adding fresh sap water to the batch. It is also essential to keep stirring and not with a wood spoon like the first batch or it will taste like the spoon or whatever you have been cooking with that spoon. When the run is over we have just under two gallons. It was a lot of work and would not be repeated, just something to break cabin fever and keep the minions entertained.

Spring training for Tennis starts with PT and wind sprints in the Staley gym after school, the high school guys carpool over; I'm glad that I don't have to walk to RFA and then home in the snow for PT but I wonder what I will do when I'm going to RFA.

The Eskimos have many names for snow and now I know why. We are in the black nasty crusty icy pile in the gutter season with surmounting concrete hard mountains of plowed snow. Black ice on the roads. One kid, an eleven year old fell victim to these conditions. He was horsing around with some other kids playing slap and tickle at the bus stop. The bus rolls up as he is wrestling with another kid. He jumps up and slips on this common feature, sliding under the rear wheels of the bus as it's rolling to a stop. The rear wheels crush his hips, legs and gut, he knew what just happened to him. You could see the terror of that in his eyes. He wasn't screaming cuz all the wind was squeezed out of him; he survived to walk again. The Air Force took good care of him. The bus was empty, the two rear tires spread out the load, little consolation. The driver is a wreck; we get a new one. We get lectures about bus safety. The bus won't even pull up unless we are all well away and lined up on the sidewalk now.

Winter is slow to retreat in upstate New York. We have a late start to the tennis playing season and have a few matches called off due to snow and or ice on the courts. We have two feet of snow that Easter. The ice rink is slow to go it's still a thick sheet of mushy ice in the backyard a glacial tribute to our mammoth efforts.

I like hanging out with the team, riding on the team bus to matches. I'm the youngest player at fourteen playing exhibition singles and number three doubles. My teammates relay stories of their various conquests real or imagined with an eye to make sure I was listening. I get a big dose of late sixties jocularity. They give me Camel filter less cigarettes to smoke with them. I like the coach too. He doesn't give a lot of pointers or try to change your game, he just works with what he has and makes us competitive, even with each other with the ladder system, tacked up on the cork board in the gym. Personally I think you have to hit a million balls to even start being good. Coach runs the crap out of us with wind sprints and cone drills on the courts, we do laps after practice. My legs are pretty rubbery walking home with my books, racquet and gear in my gym bag under my arms.

Dad buys himself and Mom new Wilson T 2000s. I inherit Dad and Mom's old Dunlop racquets and my game picks up. Mom and Dad bought the Maxply Forts in Hong Kong. They were big Rod Laver Fans. That guy's left arm is twice as big as his right. Dad's racquet is a medium 4 5/8 with a heavy head while Mom's is light racket with a 4 ½ grip. You could get them like that. The same design in wood using the same materials covered a lot of wants. That's real craftsmanship. I was playing with a tired old medium Jack Kramer at the time. I like Mom's racket, I can hang on to it better with my baby hands, and it's quicker to volley. Dad is base champ, he pulls me away from whatever I'm doing to go play a set whenever he feels he needs the exercise, we routinely play after church. We always play a set after a minimal warm up. You only get two practice serves. Dad never just wants to hit. It's all about winning points. I want to do drills and work on my game not deal with his over the top, spin, chop, lob and drop shot game, this doesn't prepare me to play my matches. I have yet to play anyone with a game like Dad's. The T 2000 exaggerates the English he puts on the ball. It's always best of three and I try to get it done in two by hitting out and not pulling my shots, I don't care about the points. I like to run and hit. I like to go to that mindless place where you are not thinking but reacting, chasing the ball like a pup. Dad is not of the "when in doubt call it out" school, quite the opposite. He always calls the close ones in, in fact he is in the habit of returning a second serve or one outside the line yelling "play it" when I can clearly see it's out from the baseline. It messes with your head, your concentration, and timing to say the least. Dad also insists on "coaching" me while we are playing, which is very bad form, but he is oblivious. You can't get mad, display any attitude, say anything or even make a face, because you know it would not be well received. It improves my concentration; I adopt a detached unemotional self-control that serves me well in my matches. I never bang my racquet on the ground because it's dear to me, I double tape the edges. Dad's English teaches me to keep on my toes and my eye on the ball. It's something we do together.

1967 GRIFFISS AFB, ROME, NEW YORK

## Summer 1968

Its early summer of 68. We have been here a year. It's time to move of course. The good news is that we are not moving too far; the bad, it's inside the wire and triple the distance to the tennis courts. I have made a lot of friends my age in Woodhaven a lot of them pretty girls who I seem to notice more and more these days. It's going to be a pain to go back and forth to see my friends. There is a curfew on base that starts at dusk, which is pretty early at this latitude, in the winter, when I wouldn't be riding a bike. That was still to come. Oh yeah you have to always have to have your ID on you. This is a SAC base, they ae serious about security, Dad is in charge of the MPs it would not do to embarrass him. Don't leave home without it.

    A larger residence has opened up in Skyline at the end of Atlas Drive the main road into the area. We are in the Colonels Circle at the bottom of the dead end road. We are at the end of a spur ridge, the road running on top of it ending in our cul-de-sac, Atlas Circle. The spur broadens upslope away from us to the main drag itself, another ridge at a right angle that descends down to the airstrip. There's more housing on the shoulders of the ridge as it broadens and blends with ridge above headed downhill to the runway. The other side of the ridge drops off sharply to form a little valley broadening as it goes down slope. The ridge on the other side of the valley about 500 yards from us is Married Enlisted housing E5 and above. The valley used to be the small arms range and the ridge we're on used to be the backstop. There's a lead mine down there. You can't turn a spade without finding some slugs. It's sandy in the bottom of the draw, perfect for that. This field is great for playing ball, our favorite game is punt return.

There's nothing past us but a heavily wooded overgrown corps of trees not maples, thick New England woods with briars and skunk cabbage that no one has entered for decades. Below the enlisted housing on the far ridge is a dairy farm with a stream meandering along the edge of the property separating it from the base, that and a barbwire fence. The other side of the woods about a thousand yards from us are the F-106 barns and the B-52 ready apron. All in all a pretty sweet playground even though there are no Sugar Maple trees here.

The house is whole lot bigger. It has a garage for mom's car. From the front it looks like a long single level ranch. It has a full basement, the house is probably just under 3000 sq. feet; room to spread out. The Air Force moved us room by room with stake bed trucks. We are mostly still packed in boxes. The boxes and furniture are in the right rooms so it doesn't take long to get settled. The basement is split right down the middle it has one continuous room toward the street the length of the house, the other half, has one big bedroom, a bathroom, the utility room, stairs and a door to the backyard. The four little guys get the big bedroom downstairs, their turn to be basement trolls. Steve is still stuck with being their supervisor, he's an eighth grader now like Jim. The long room we turn into our kid activity center. The stereo and a TV goes down there. Steve and Kathleen have money from babysitting. I have a job sacking groceries at the Commissary for tips. Steve and Kathleen spend their money on records and I spend mine on model airplanes. It is very pleasant having the White Album, Cat Stevens or Simon and Garfunkel spinning while I build models. Jim and I string heavy fishing line across the ceiling to make a grid then hang our completed models with four pound monofilament from the grid. We eventually have everything that ever flew hanging from the ceiling; all the pilots like it.

# 1967 GRIFFISS AFB, ROME, NEW YORK

Jim and I get a room upstairs that is twice the size of the Woodhaven closet we used to have. Maureen and Kathleen still share a room Kathleen will be a Senior at RFA this fall. Maureen will start the second grade. They seem to get along. I don't know if I would like sharing a room with Gary. The girl's room is next to ours. There is a big bathroom across the hall from my room that we share with the girls and the parents. Mom and Dad have the big room across the hall from the girls. There is a big kitchen/dining room at the back of the house and a big living room on the other side of the partition wall. Lots of room for Mom's stuff.

Kathleen scores a job running a summer archery class/lessons for kids on base. She recruits me to help her. She needs someone to hump the butts. We both shot archery at TAS in PE class. I won a trophy when I participated in a summer class like the one we will be putting on. Dad gets in the act, he buys us both used recurved bows, a 30 pound 1953 Bear Tigercat for me and a custom made 26 pound bow for Kathleen. We each also get a dozen custom cedar arrows with fancy spiral fletching. Dad also lends me his long bow, thing is I have to make a string for it. The long bow is handmade. Dad and his brother Bob made one each out of Osage or Bois De Arc. It is carefully carved keeping the white/softer wood on the outside of the curve and the darker heart wood on the inside. The upper and lower limbs are spliced together and glued It has quite a few coats of varnish giving it a high gloss finish. The handle is wrapped with while elk leather from the monster spike Dad shot in 63. The bow only has one flaw. Dad left his bow on the top of the hay bale he and Bob were shooting at and Bob put and arrow dead center and mid-span into the outside of one of his bow's limbs. Ten to one there were words if not a scuffle. Dad repaired the bow by wrapping it with a thin strip of rawhide that he shrank on Indian style. I have never shot this bow; it must be at least twice the draw of my bow. All that will make it very interesting to shoot. I make a string out of waxed linen strands. I back braid eyes for the ends and serve them with heavy silk thread, same for the center of the string for the nock. I string the bow. The string doesn't immediately pop so on to the next test, the range.

We set up the range in the back of the church parking lot where there is a handy back stop, a bank made from spoil and overburden left there when they graded the site for the parking lot and the Church proper, the next grade up. We set up about a dozen targets for a dozen archers, just about right. It's safety first. I know of one coach that took an arrow in the bicep at Taipei American School plus pin heads trying to arc arrows over the canal levee we used as a backstop into the Chinese village on the other side. Tough with a 20 pound bow. Many tried but I don't know of any damage done. We don't want any of that happening here. The woods are the ultimate backdrop, good luck finding arrows in there. The classes are three days a week so it doesn't get too boring or take up too much of my time. I get pretty good with my bow; I practice snap shots, what you need hunting rabbits and small game. One kid asks how I can do that without aiming I think about it and ask him do you aim a rock when you throw it? Dad's long bow takes some technique to shoot; you have to start with the bow parallel to the ground and swing it up keeping your bow arm stiff while pulling back the sting using your core and gut. If you start with the bow up and just use your arm it is tough going. That bow has close to a eighty pound draw weight so I don't shoot it too much. One day the string lets go at the eye splice when I have it at full draw, quite a shock to the arm and shoulder plus I wacked my leg. I hesitate to make another one the first was a lot of work but put off buying one until the summer is nearly over.

In between archery sessions there is plenty of time for tennis, hanging out at the pool and playing baseball or tackle football/no pads with my buds. One of our favorite things to do is tubing down the Mohawk. Dad would drop us off below Delta Reservoir and pick us up later at the Main Gate. The river winds around the west boundaries of the base in a great loop with nothing but farmland and a few mill races for miles, real peaceful. It would take most of an afternoon to float all the way to the gate, a great way to end the summer. Before you know it school starts. I ride the bus to RFA with Kathleen. We don't sit together. There is already a chill in the air, football weather.

## Winter 1968

I like RFA the teachers are great. Kathleen has nothing but good to say about Mr. French her English composition instructor. Mr. French gets me to read and appreciate the classics, Hugo, Tolstoy, Sir Walter Scott, Twain and many others. I take Geometry and my Math skills accelerate. You can actually use math to understand, define and solve real world problems. I love it. There is only one right answer, got to like that; no arguing but you get credit if you show your work. Chemistry is fun too It has math and then there are the experiments! My good friend David Nelson is a Lab assistant, we engage in some very questionable off site Chemistry; things that go boom.

So… we had our usual backyard hockey game after dinner. We have made another rink since the last one was such fun. Still kind of sweaty, I zipped up in a standard USAF issue artic flight suit, wolf fur lined hooded parka fur backed mittens with trigger finger, fur backed so you could wipe your nose on the back of the glove. When it froze you could just brush it off, zipped on Dad's double insulated flight boots .and started the nightly routine of resurfacing the skating rink. The only one who had a bigger, better rink was Col Jack the Base Commander. Jim has a mad crush on his only child, his daughter DJ, but he has no chance. Dad volunteers us to clear Col Jack's bigger rink when it snows. You have to get the snow off quick before it sticks and makes pits and or lumps on the Ice; we make a lot of trips. It's not so bad if you are skating and have a snow shovel and are dressed for the occasion. Col Jack's quarters were away from the rest of us at the opposite end of the runway, it was a huge farm house with a barn. Col Jack's daughter is a gymnast. She has: a trapeze, parallel bars and a trampoline in the barn along with a MG-A and a Austin Healy Sprite. The Col is cool, he likes us. Normally you clear the ice and put some more water down to resurface the rink. You have to be quick to keep the water from freezing in the hose when you do this. Dad had a pulley system rigged to the eaves of our house and the same system to the Col's barn. I attached a rope to the middle of the hose, winch it up and disconnect it from the bib draining it before it can freeze. I remember this particular night because you couldn't raise the hose spray two feet above the surface of ice before it turned into ice BB's that just bounced off of the surface of the ice, it was probably 20 below. After resurfacing the rink I turned in as usual, nothing special. The Colonel's circle, where we are, is near the east end of the runway, the winds are prevailing from the west. Planes are staged for takeoff in our backyard not more than 1500 yards away. The B-52s were lighting off every four hours, four planes and eight engines per plane add up to twenty four run ups/explosions, nearly constant when the ground crew

## 1967 GRIFFISS AFB, ROME, NEW YORK

are shorthanded. The only redeeming thing is that the six F-106s on alert are in heated barns and don't need to be run up, they just open the front and back doors of the revetments and the planes are away. I had got in the habit of sleeping with the window cracked open just a little from sleep overs at David Nelson's house; it makes you snuggle deep in your covers.

    I wake up, head for the bathroom across the hall and start to get ready for school. Something is wrong. I don't look in the mirror much but this day I do. The left side of my face doesn't seem to be working. I have an unblinking left eye and my mouth winds around to the right side of my face when I try to talk. Mom takes me to the dispensary just down the street. The Doc says that I have Bell's Palsy which may or may not go away. I get a pirate eye patch with an ointment treated gauze liner to keep my unblinking eye from drying out and I am sent to school for ridicule. No one gives me too much crap; I already have stainless steel teeth, I'm a Jock, I play Varsity tennis and hockey although I probably can't hit a thing anymore with one eye, Dad brings a Major to the house, another pilot, to talk to me, he has the Palsy and talks out the side of his face like I now do. The Major tells me that a lot of early aviators have had the same condition, Wiley Post and Howard Hughes to name a few, General Lemay is rumored to sport a cigar as a prop to hide his drooping lip; these fellows froze their jaw sucking oxygen from a hose before the development of a proper mask. I have no record breaking altitude or distance flight to justify my condition. The cigar thing is intriguing. I could work with that. If it is good enough for the General…. After six months or so I slowly regain control of my face till I have about 80% back; I'll take that, and no cigar. Was it the cold or was it that shot I took to the head playing backyard hockey?

# The Big Ice Storm Christmas 1968

It's mid-winter. We're are all home for winter break in fact it's Christmas Eve. Not a good time. Everyone except Jim, Mom, Tim and I are down with the Hong Kong Flu. Jim is down from falling on the ice and turning his face into hamburger the day before. No ordinary ice. We have had a severe ice storm that deposited two to three inches of ice on top of three feet of snow. Last night was a full moon after the front moved through, it cleared the sky and the air turned bitter cold. We strapped our skates on and went skating cross country. It was weird skating up and down the hills where we had been trudging through knee deep snow just a few days ago. The surface of the ice wasn't smooth it was pebbly like skating on frozen peas, hail combined with rain drops that froze on contact, a conglomerate. Jim, skating on his long thin racing skates, managed to punch thorough the ice while going rapidly downhill and fell on his face, the only part of his person that wasn't covered up. It was snowing all day so there is about three or four inches of fresh powdery snow covering the ice. The snow was followed by another clear moonlit night, too cold to snow. The fresh snow on the ice gives some traction for boots and it's the perfect depth for sledding, the underlying ice even makes the shallowest grades fast. We have to do this; I call my buds. Tim and I grab a 8 foot long toboggan and meet up with John and Rick at the top of the steepest part of the hill behind Cols Guthrie and Stahl's house. The 50 yard run has one drawback, there's no runout at the bottom of the hill just a solid line of trees; that's what makes it fun. We play chicken with the trees. The last one to bail before the sled hits the trees is the winner! You are a real looser if you don't bail. After a couple of runs the top of the toboggan and the ropes running down each side, that you hang on to, get loaded with packed snow.

## 1967 GRIFFISS AFB, ROME, NEW YORK

We take turns as drivers if you can call being in the front driving, it takes a lot of effort to warp a toboggan and make it turn where you want it to go without tipping over. Tim and I are on the back for this run, Tim just in front of me. The back guys have to jump first to avoid hitting the jumper in front of you except this time the chickens in the front go first; Tim and I have to wait until we're past their tumbling bodies. Tim's mittens are frozen to the ropes; he's been struggling to free them since we started our run, so I throw my arm around his waist and roll off just before the toboggan hits the trees with a whack and flies up in the air. I roll and hit the trees too but I don't fly up in the air. Tim has stopped uphill of me where we rolled off and he is howling. I grab the toboggan to drag it back up and see one of Tim's mittens still frozen to the ropes. I drag the heavy wood eight foot long Adirondack toboggan up to where Tim is, a real bitch because there is so little traction. Tim is still laying on his back, his right forearm punched through the snow and ice when he jumped so when he somersaulted over his arm he broke it but I don't know this and he can't tell me because he's screaming. I chop his arm out of the ice with my fists which makes him scream more. I free his arm and see that it's not right, but I tell him he will be ok. There is no way that I can haul him uphill on the toboggan. So I grab his good arm and throw Tim over my shoulder in a fireman's carry this brings a predictable response. To make sure I make it up the hill without dropping my little brother I kick a foot hole in the ice for each step I take, this makes his shattered arm, which is hanging down my back, flop around. I keep apologizing and talking to Tim to try to keep him from going into shock; he is delirious with pain when I get to the top of the hill. Rick and John have grabbed the toboggan and run ahead to the house to let everyone know that something happened to Tim.

When I get Tim inside I want to cut his jacket off but Mom doesn't want to ruin a perfectly good jacket so I slip it off, a big mistake Tim howls. When I get it off I see his arm looks like a crankshaft. Dad is sick as a dog but he comes down and collapses in a chair as I am trying to do some first aid. I make a splint out of a magazine and some rags. Tim needs to go to the hospital now. The roads are covered in black ice and snow like everything else except the runway. Mom's car doesn't have snow tires and she isn't good in snow. She would never make it up the long hill to the Dispensary. Dad tells me to put chains on the truck and take Mom and Tim to the hospital. I chain up the war wagon as fast as I can, load up and head out. Tim is a little better after I stabilized his arm, he's not screaming anymore and is able to talk to us so he's not going into shock but he is really hurting. When we get there, there's got to be only three people on duty in the entire building all of them corpsmen, no doctors, after all its Christmas Eve and the ice storm has brought everything to a standstill. They x-ray Tim in the splint and tell us that he has broken both bones in his forearm three inches above his wrist. They are going to set his bones and put his arm in a plaster cast. Tim walks past us with the corpsman on his way to X-ray to check their work. Bad news from x-ray, they have to cut off the cast and try again. Tim goes by in a wheelchair on his way to x-ray this time. More bad news they will try again. Round three Tim is on a gurney now on his way to x-ray, They still can't get proper alignment and Tim can't take any more of this; the plan is to keep Tim in the hospital over Christmas let him rest and allow the swelling to go down. The surgeon will break and reset his arm on the 26th, my birthday. We hold off on celebrating Christmas and my birthday until Tiny Tim comes home three days later, everybody is sick but me anyway. We are sure glad to have him back. Tim's cast runs from his fingers to his armpit, a real monster; his hand and wrist are twisted down and away from his body in such an unnatural way he can't even scratch his ass. When he wears a coat it sticks out the bottom since he can't get a sleeve over it. Timmy usually puts a red sock on it when he goes out. It kinda looks like...well you know what it looks like: Tim has a lot of fun with that when he waves hello. No screws or plates were used to align the break so his arm is never quite right after he heals. The cast finally comes off in May, five months later. Tiny Tim's arm is smelly and withered with white flaky skin, but he is glad to be free of that cast.

## Easter Weekend 1969

Dad has purchased some land, sight unseen, in Port St Lucie Florida. The plan is to drive down there to see it, stopping to see some friends along the way. One thousand miles as the crow flies and thirteen hundred as the trailer rolls, one way. I hope this real-estate deal works out for Dad he has a lot of old friends down there to go fishing with. Dad famously turned down an offer to go in with some friends on some dead orange groves outside LA the property was going for a dime on the dollar. The orange groves were located in a little place called Anaheim. Dad has been trailer shopping and has taken the plunge. He has been taking me with him to look at various models and has traded in the sixteen foot Astro for a twenty four foot dual axle FAN with elastomeric suspension. It has a bathroom and a holding tank, no more porta-potty sack of shit and a shovel. The girls will be pleased. We have already taken it on weekend trips to Thousand Islands, just an hour away from us. The Air Force has a recreational camp there where you can park your trailer with hookups for water and electricity. They have a fishing dock, showers and a bathroom too. You can rent a sixteen foot boat with a ten horse motor and six gallons of gas for six bucks to go fishing or rent a thirteen foot row boat for three bucks. Six gallons of gas only last's a few hours; you don't run out of gas in a row boat and can keep it all day. I love coming up here. We come whenever we can. A boat for a day, I row all over the place.

I love catching Walleye and Pike they get fairly big, folks up here call them snakes. On one foray Dad, Jim, Tim, Steve and I are sitting in a boat, it's raining, all of us are huddled under one GI poncho with our poles sticking out when I see, a real snake, swimming by and point it out to Dad. Dad reels in and makes a perfect cast over the snake and snags it. Wise ass me, I say, "now what are you going to do?" In answer Dad reels it in and tells me to get it off. There is a great dealt of movement in the boat, to see it, get away or both all at once, not good in a small boat. I manage to pin it with the edge of the net and get the hook out with a pair of needle nose pliers, the darned thing had to be three foot long. Ha ha ha.

I learn a lot about trailering. One thing I learn is Dad can't back a trailer to save his life, which makes sense because airplanes don't go backwards. I swear he can't do it unless he has someone port and starboard, with light wands giving him signals. He got by with the sixteen footer by having an offset hitch installed on the front of the truck but the tongue weight on the new rig is too much for that. Dad makes me back the rig while he stands aside and supervises, no pressure there.

Getting the trailer positioned is just the beginning. The truck has to be unhitched and moved away the coupling is a pretty beefy equalizer hitch with a friction sway bar. Then the trailer has to be leveled and unloaded, the gas turned on, the refrigerator loaded, the heater lit, the extension cord run, the water hooked up, the bunks

## 1967 GRIFFISS AFB, ROME, NEW YORK

made up and dinette set up. After all that I throw my pup tent, and a tarp out on the ground with a couple of sleeping bags. I need to set up my tent because that's where Jim and I are sleeping, there's not enough room in the trailer for us. Someone else claims the truck.

    We load up and head out for Florida without a problem we are practiced gypsies now. Not yet across the state line, Dad pulls over and springs one on me, I am to drive through Pennsylvania. Dad has a stiff neck. He always has a stiff neck or maybe he just likes us to give him neck rubs. I can see it though, those flight helmets are heavy and there are hoses and wires tugging on your head as you swivel it around for eight hours. He was flying the day before. I've never driven very far, not even to school. Now I have to double clutch through the Appalachians in an underpowered tuck pulling an overloaded trailer with a live load of ten souls and Dad sitting next to me judging my every move; I think my teeth are going to crack. I just turned sixteen. Anyone who has driven through central PA to the coast knows what a bear the twisting hilly road to and from Scranton is. I spend most of the trip scared, creeping along in third gear, doing 45 going uphill trying not to kill the engine or I'm out of my wits going downhill afraid of jackknifing. The trailer sways and fishtails despite the sway bar. I do the math, this is a half-ton truck with a half-ton live load, not to mention the trailer and all the stuff in it. We are maxed out. It is really hard to keep it steady and between the lines. Dad likes a steady ride, no turbulence, no over correcting, good formation driving. The truck is power nothing just a big steering wheel. What a relaxing Spring Break for Bob. Next stop Annapolis. When we get there I need to see if I can get more out of the equalizer hitch to shift some more weight to the front wheels plus tighten the sway bar. The front wheels are squirming making steering difficult.

We camp out on the estate of Captain Cunningham our sponsor from Taiwan. His place is on the Severn River. The good Captain's elegant wife and Mom are great friends; she has a yellow Jaguar XKE convertible that they jump in and head to town, leaving the rest of us to cool our heels and skip rocks in the river while they catch up. We wander around the neighborhood while the Captain and Dad chat. The surrounding estates are picture book places with white rail fences beautiful estates, horses and barns on one side of the road with mansions docks and boats bayside. The Captain has no boat, he has had enough sea time to know better. When Mom comes back we go to the Naval Academy, a beautiful campus that I'll never experience. I like all the historical stuff, pretty inspiring.

The next day we are off to Old Historic Williamsburg, it's crowded, it takes a long time to find a place to park the rig and then it's a hike to see the sights where we will hike some more. I am bored and impatient wandering around with a group of ten who all have to see everything one at a time. We have to stick together, the place is packed with people so unlike the first Historic Williamsburg. We also visit Mont Vernon. Washington DC is a drive by. I see the Washington Monument from the truck. Not impressed; lots of shabby buildings and row houses.

The next day we head to Shaw AFB where Dad flies to a lot, it's the recon center for the Air Force, there used to be a B-57 group stationed here. When we travel, we don't go from city to city, we go from base to base. We visit Lt. Col. Shakita and his wife Betty at Shaw, Mom's best friend and Dad's Sqd. mate/Navigator from when we were stationed in Japan. They are also Tim's God parents. We burn another day. So far this trip has been great for Mom, she's constantly paging through her overstuffed, rubber band bound, red address book that on end looks like a partially open Japanese fan.

When we arrive at Shakita's, the ten of us are wanting to freshen up, some of us need a shower, we find that their septic system is overflowing in the back yard. It looks like a swamp back there. We can't even pee. After visiting and holding our water we camp at a RV park at Shaw. Next stop is Moody AFB Georgia to visit another round of mother's friends. Mom has been on our host's phone calling around until she scores a place to park for the night on base. In the morning it will be seven days on the road.

The next morning there is a sea change. The troops are surly and dirty. The prospect of adding two to three days and a thousand miles to our itinerary just to make it to Port St Lucie and back to where we are now is depressing. We only have a ten day break, so that would not leave us enough time to return home. Dad makes the decision that we will make a beeline straight home without stopping at any "friends" houses; he and I will share the driving duties. One thousand miles in eighteen hours instead of seven days. Everyone perks up at the prospect of going to sleep and waking up to their own space with a chance to decompress the next two days! Everyone except Mom. She had written a bunch friends in Florida that she was going to be driving through their area and now we are making a liar out of her. Midterm, Final and NY State Regents exams come and I do better than I expected. I am recommended for College bound curriculum and advanced Mathematics.

## Road Trip Summer 1969

Its summer. Dad has taken his annual thirty day leave so we are going to hit the road. What else? We are going to see the Flanagans who are stationed at Kirkland Lake, Canada. Kirkland Lake is an old goldmining town north of North Bay, at the same latitude as

Newfoundland. Major Jack Flanagan RCAF used to work with Dad in the Springs when we were stationed there. They lived on Farragut not even five houses down our common back alley. They have a bunch of kids too. We all went to Divine Redeemer and Church together. Mom and red haired blue eyed Mrs. Bette Flanagan are great friends. Kelly is the oldest, he's Kathleen's age. Kelly has blond hair, blue eyes and fine features. Kelly is notorious for climbing on his garage roof and blowing Oh Canada on his Cornet. Kathleen used to have a tremendous crush on him. Robbie is my age, he has brown eyes and brown hair like his Dad. Robbie is a real jokester, lots of fun. All of us have a Flannigan buddy. We are going to see the World's Fair exhibit in Montreal on the way to Kirkland Lake. It shouldn't be too crowded since the fair was two years ago. The end of the trail is Colorado Springs to paint and rent the house, from there we will see the Lepines of course and some other friends then take Kathleen on campus tours of Fort Collins and CU Bolder. She has always wanted to go back to Colorado and going to College there is her dream. We are also going to see Major Lord in Wisconsin and Col Bush in Omaha on the way there.

 I help Dad hitch up, load and service the trailer in front of the house. The trip to Montreal is a short hop of 150 miles so we get to see the Exhibition that afternoon following an early start. The main attraction is the Biosphere. It is the dawn of the environmental movement. The exhibition grounds are pretty, it's located on an island in the middle the St. Lawrence River, connected to the shore by bridges and a light rail system.. We are back on the road after a short walk through the exhibit and lunch. It's under 400 miles to Kirkland Lake; the days are long at this latitude and time of year, so we press on to our destination. My impression of this part of Canada

## 1967 GRIFFISS AFB, ROME, NEW YORK

is that it's very flat and wet. The highway crosses a lot of shallow lakes with great tall stands of hardwoods and pines around them. Dad drives the whole way on this trip. Before you know it we are there. The Flannigan's have us park in their driveway unhitch and setup the trailer for the night. Mrs. Flanagan has a barbeque going and hamburgers ready to grill. We have a raucous evening trading stories of what happened the last four years.

The next day we go to the Flannigan's lake cabin. Kelly takes us water skiing while the rest stay on shore at a cabin where they keep and launch their boat. Robbie and Kelly are smoking a lot of cigarettes. I don't think much of it, after all Dad smokes, but I ask anyway. The boys tell me that it keeps the mosquitoes away I agree that they are pretty bad here because of all the standing water, swampy almost. We get in the water launching the boat but it's too cold for me; I don't stay in long. Suzie Flanagan is a long distance swimmer I'm talking miles/hours in this lake. I don't see how she does it. Kelly shows me another use for cigarettes by burning a couple of leaches off his legs. Now I'm certain that I don't want to swim or even wade in that muddy bottom lake. Waterskiing is another thing. I get up on two skis for the first time in my life with a little goading and coaching from Robbie; then Kelly gets up on slalom and shows us how it's done. Unfortunately Kelly hits a driftwood log; he's not wearing a life jacket and takes a hard fall. We get to him and fish him out. These lakes are shallow, he's standing on the bottom. I can tell, that fall really rang his bell but he kids that it was only a one leech wreck. We head to a kiosk on the lake by the park for a snack; you can buy burgers, dogs and chips not fries at the park kiosk. The Flanagans laugh at us because we want ketchup with our chips that we mistakenly call French fries, salt and malt vinegar with your chips is the thing eh.

The next day Major Flannigan takes Dad' Jim and me to his place of work, a radar installation, part of the third DEW line, the Pine tree line. We get to go inside a huge radar dome, it looks like something out of a James Bond Movie. There is a large rotating wide beam search array and a huge trainable nodding idiot narrow beam altitude array. At the base of the dome are the office and electronics rooms. Very cool. Major Flanagan shows us the consoles and how they work. Little dots all friends no foes. Better to man a station here instead of the artic. This is what Dad flies against pretending to be a bad guy. After we get back to the house, the boys give me a road tour of Kirkland, I get to see all there is to see on a fairly short ride. We have a big spaghetti dinner after which Robbie teaches me how to play cards. He is very good at cards. He takes some of my tip money; it must be all the time you have to spend indoors up here. I'm not a big card fan.

Were up early and so are our hosts who put on a big pancake breakfast. Its 600 miles to Viroqua Wisconsin. We cross the border at Sault St Marie and get there before sunset. I get to see an old friend, Mark R Lord. I haven't seen Mark for a year since the Lords left Griffiss last summer. Major Lord is working on his Masters. The Lords moved back to Viroqua, where they have a place to stay while he completes his thesis, and what a place it is. I thought Mark was exaggerating when he told me about it. The Lords inherited a huge three story mansion plus a basement and an attic. It's an old place with a carriage house, stable, servant's quarters and entrances. There are secret passageways for servants behind the walls and dumb waiters from the kitchen. The little guys love this place. Mark only has one other sibling, he is six-years-old. Mark torments him whenever he sees him, a game they play. Mark plays the Cornet and the drums; he was teaching me to play a snare drum before he left Rome. The

## 1967 GRIFFISS AFB, ROME, NEW YORK

little guys are having a ball playing hide and seek, they have to be stopped from riding the dumb-waiter.

After visiting for one day we are off to Nebraska and Offutt Air Force Base to see Colonel Bush. Offutt is SAC Headquarters. When we get there I get reacquainted with the Colonel's son Bill. He just got his driver's license. Bill brags that he gets to drive his Dad's car, a Bullet black Mustang fastback 2+2 with a 390 and a four speed. I ask for a ride of course; Jim and Tim want to go too. This car has got to be the hottest ride in the country at the time. Head anywhere east of Offutt and you are on dirt roads skirting cornfields, not where I would drive this car but he's the one driving. Driving real pokey in fact. It's clear that he isn't real confidant driving on dirt roads or this car. I tell him that he should down shift going into the curves and power through in a drift. I have been driving on snow and slippery roads. I am used to handling skids and know not to hit the brakes because that locks up the steering or make any drastic corrections, you lose control if you do that. Bill asks me to show him what I am talking about. I am happy to do that. This car is a lot hotter than the Buick and certainly hotter than the truck. I drove a Chevelle SS with a 396 and a four speed in Drivers Ed. It didn't have dual controls and I was the only one in the class who could drive a standard. The instructor's only comment was "Why do people who drive a standard insist on going through all the gears?" I think I can handle this. I push through some chicanes with a two wheel drift. I do a 90 degree turn with some speed. The car fishtails and straightens out when I step on it and the positrack engages. Bill is impressed and wants to give it a try. I climb in the back with Tim and Jim jumps in the front passenger seat because he is tired of riding in the back and threatening to make some trouble. Bill takes off accepting more speed that he was before when we hit a gentle 120 degree turn to the left going over 50. Bill feels the rear end slip, panics, hits the brakes and spins the wheel away from the direction of the skid. A stunt driver could not have done a better job. I stretch out my arms until I touch both sides of the car as we roll over a couple of times ending up on one side, driver's side up, in a ditch. We had the windows down so there is a cloud of dust in the car. When we come to rest Bill yells "It's going to blow"; he then proceeds to kick the snot out of Jim and I climbing out the driver side window, the only way out of the car. After the rest of us climb out at a more leisurely pace; we assess the damage. Tim has some cuts on his arm from the rear vent louvers but that is nothing compared to the stomping my head and shoulder took during Bill's exit. Jim has road rash from the car's final skid on its side on his right arm cuz the window was down, other than that we are good. We are a solemn group of souls. So… here we are at the corner of cornfield and dirt road, what to do? We look wistfully at the cornfield, you could hide a whole infantry division in there not to mention four sad boys. Fortunately a farmer saw the cloud of dust when we rolled and came over on his tractor to see what was up with that. We are lucky that not much happens around here to attract anyone's attention, making a cloud of dust is something to go look at. The farmer rolls the Mustang back on its wheels with a chain and his tractor. It's is a sorry sight. It's not going anywhere fast or slow. We have totaled the Colonel's ride. Bill goes with the farmer on his tractor to call a wrecker and his Dad in that order, I don't envy him. The Moms come and get us they are happy no one is seriously hurt. Tim is hurting but he says nothing. This event kind of kills the visit so we camp at a facility with hookups on base. It's a big base, I don't know what happened with Bill. Mom and Dad don't interrogate me too much and I don't play up my role. Dad is

# 1967 GRIFFISS AFB, ROME, NEW YORK

glad that he is in TAC and not SAC. Upstate New York is a peach compared to dirt roads and cornfields.

The next stop is Colorado Springs and the Lepines our old next door neighbors from Hill. We are here to paint the house but find time to go visiting and horseback riding, Celeste Lepine who I have had a crush on since we were digging holes in Utah, is working this summer at the Garden of the Gods Stable so we all go to see her and take a trail ride.

We visit college campuses for Kathleen who graduated this spring. We visit CU Bolder which is a beautiful campus and Colorado State Fort Collins. Kathleen has been accepted at both, her heart is set on Boulder, but it will not be this year. She finds that out after we get home. Dad wants her to attend a school closer to home. She will be attending Oswego State located on the shores of Lake Ontario an hour or so Northwest of Rome this fall.

Before you know it's time to hitch up and head home. We stop at Offutt again. Then on to Wright Patterson where I was hoping to see the Museum but we ended up visiting some of Mom and Dad's friends. The next day we head home to Rome. On the way back we start seeing some, then quite a few scraggly looking hitch hikers the closer we get to home. These folks are what's left of Woodstock held 100 miles south of Rome that summer. Griffiss AFB will be the site of Woodstock two.

I don't know why Kathy was sent to Oswego other than the need for the parents to have her close by. More likely the cost. I help Kathy move into the dorm. The campus buildings and dorm are old ivy covered stone buildings, pretty actually. Her new roommate tells us that you can see icebergs on the lake drifting by in the winter, yea. Oswego is on the shores of Lake Ontario. We hug and make our goodbyes. I will miss her, it's all on me now. Before Kathleen was sent off all of us are called into the Livingroom. Dad informs us that he only has enough funds for one kid at a time to attend college. We will all have to help each other out; furthermore he is not paying for a degree in flower arranging or navel gazing. Kathleen makes better grades than I do, straight A's in fact. I am only cracking B's max in math and chemistry. Subjects that Kathleen doesn't like or take. Liberal arts rule.

## Fall 1969

Jim has picked up some firecrackers in Canada with some help from Robbie Flanagan something he has always wanted access to; we went through a lot of fireworks in Taiwan; Fireworks were cheap and available most everywhere at all times of the year. My favorite were cracker balls, they were extra-large. You could shoot them with a sling shot or lay out a mine field across the road, folks think that they have a blowout, ha ha, or just toss them from a vehicle. Every time we would go to the campgrounds at Thousand Islands. Jim told me he was going to cross the St Lawrence in one of the camp's boats, go over to Canada, maybe at night, and get some firecrackers. Fireworks are illegal in New York and on the base too I'll bet. Like always Jim is all wind and no sails to his boat, but it shows how bad he wanted some ordnance. He has no idea what he will do with it, once he gets what he wants; he'll think of something. That something turns out to be torturing his brothers. He couldn't get the big cracker-balls he wanted, but he managed to get some pop bottle rockets, some poppers and some small firecrackers. Jim doesn't dare set off a rocket, even though he is itching to lite one off, the MPs would be on him right away. An occasional pop during the day will largely go unnoticed, especially out on the end of the road

## 1967 GRIFFISS AFB, ROME, NEW YORK

where we are. Jim makes his brothers dance by throwing poppers and firecrackers at their feet when their backs are turned, this makes him happy for a while. After the thrill is gone, he comes up with a new game, a version of chicken that he wants to try out on Tim and Bill. First he needs an audience. When we, the curious, are gathered round the carny barker, Jim explains what's up. He has twisted the fuses of two crackers together. Bill is to hold one and Tim the other. Jim will light the fuse. The first one to let go is the loser which sounds counter intuitive. Jim has not thought this through. He bullies tiny Tim and submissive Bill into playing his game and puts the crackers in their fingers. Jim lights the fuse. Jumpy Bill is quite ready to lose this game. He immediately lets go of his stick of dynamite. Tim is left holding the lit bomb. Unlike Jim, he has thought this out. Tim flings the double charge at Jim. In a stroke of luck, bad luck for Jim, the hissing nasties hit his neck and drop down his lose, open collared, button down shirt, exploding with two reports on his bare belly. Say, thanks Jim; that was a pretty good show. Bill is stunned still trying to put together what just happened. Jim is even more stunned, by the time he recovers, Tim is long gone. He immediately took off running after throwing the charge thinking Jim would be hot on his heels. We are all uncontrollably laughing our asses off; this is the stuff of kid legend. Jim can't tell or retaliate there are too many witnesses and retaliation would lead to full disclosure and confiscation of his remaining arsenal. Jim's enthusiasm for fireworks goes down about 20dB.

When school started, feeling kind of cocky, I try out for the ice hockey team. I am the only kid from the base to do so. It is getting cold and I want to skate. I have already started making the backyard rink. I get to skate on the rink downtown for fee, on ice with a nice surface that I don't have to maintain. I can really get up some speed on it. The coaches like my skating and stick handling I weigh about 160 pounds. I am not the fastest skater but I am not the slowest. My coach tells me that I have the crappiest skates on the ice he thinks that my skating would be much improved with some better equipment. I agree my skates are cheap, poor fitting, old and breaking down at the seams from hard use. I tell dad what my coach says. In a rare act of generosity Dad drives me to Utica; gets me fitted for and buys me a top of the line pair of Bauer Black Panther hockey skates, some gloves and a helmet. My skating immediately and dramatically improves. I really have some wheels. I can outskate 80% of the guys especially skating backwards. They are going to make me a defenseman. I participate in no pad drills, shin pads, sticks, helmets and gloves. We move on to stick handling and shooting. This is better than backyard. I am having a ball playing guys my size, you can be real aggressive and not worry about hurting your little brother. We play no pad, shooting but keeping it low, wrist shots no slap shots, no checking and no contact. Then I am issued my equipment. It was ok getting on the bus in the morning with just my books my skates hung from my shoulder and two sticks in hand. This was before back packs were in vogue. It was ok walking home the six miles from the rink packing the same load. For prior tryout and practice sessions coach brought a bunch of gear, shin guards, helmets and gloves along with the pucks. Now I am issued my personal equipment along with a duffel bag, over 40 pounds of it. Problem is there is no locker at the rink where I can store my gear. I soldier on muscling it on the bus where it sits beside me, the driver gives me sour looks, unpacking and cramming it in my school locker but I still have to pack it three miles to the rink and hump it home the six miles over my shoulder. The duffel doesn't have straps to let you

## 1967 GRIFFISS AFB, ROME, NEW YORK

carry it on your back like a pack. Before you know it is snowing again. I question why I'm doing this. I have stopped carrying my books to lighten my load so I am neglecting my homework. It is miserable walking home after skating your brains out, hungry as hell, hoping they left something to eat, trying to beat curfew. The trip is taking much longer now that there is snow on the ground and it is getting dark earlier. The tennis team is starting indoor conditioning. My thinking is that a racquet and some tennis shoes are a lot lighter and I have a rink at home to play on with my friends. I quit. I just turn in my gear. My mistake is not talking to my coach or my parents about it. They might have helped me out. I am unused to anyone having an interest in, or doing anything for me. I never ask for help with anything. Decades later Dad asks why I quit when I was so good and obviously loved playing. I tell him that I didn't have a ride and I knew better than to ask for one. Dad had no answer to that. My coach is a different story. Coach is furious. I put a hole in his lineup. I blow it off until I have to take PE next semester from him. John Pease, one of my best friends, is in my class so when Coach starts ragging on me John does his thing, clowning around, smarting off, disrupting class and of course I am his biggest fan laughing my ass off. Coach hates us.

Its spring. I am about to walk on the court at Staley to play a match wearing my RFA school uniform, warmups over my shorts and shirt when the hockey coach comes up behind me grabbing me by my jacket demanding where I got it. I tell him to let go of me twisting away. I tell him that I am on the tennis team. He calls me a liar and a disgrace to the uniform and that I should never be allowed to represent RFA. I back off and trot away from him and he gives chase. I easily outrun him turning, jinking, going in circles, looking over my shoulder at him, taunting. He looks winded so I let him catch up. Coach starts to pull at my jacket again falling on me screaming invectives about base kids. My tennis Coach pulls him off me. I am laughing it is so ridiculous. Coach flunks me in PE, the only failing grade I get in High School. When the yearbook comes out I am deleted from the Tennis Team picture and it's no letter for me. My third year playing matches, never missed practice. No A for effort…I think I tried pretty hard.

It's a pitch black winter night. We're trudging home through the snow, the regular gang, me, Jim, Rick, John and Rusty. The bowling alley just closed and we are headed home, it's nearing curfew. We don't want to walk uphill just to hike downhill, if we cut through the woods it's the hypotenuse of a triangle, the shorter course. We can skirt the hill and end up in the same place. We have beaten a trail, we've done this before. I do it all the time, I made the trail. It is the shortest way home, on foot from the Commissary, where I work. As the gang pulls off the main drag and heads for the trail which runs right by the east side of the church, in fact twenty feet from the church. Rusty picks up a handful of pebbles and starts tossing them at a church window, they bounce off with a tink, I smart off of course, and say" if you want to break a window you have to use a bigger rock". Rusty reacts to my sarcasm by picking up a bigger rock and throwing it through a large window with a crash. I call him and idiot and we take off running through the woods down my trail. The MPs are on us pretty quick, they have been watching the trail, not hard to find which way we have been going in the snow. A blind guy with a cane could have followed that trail. They have it and the church staked out. They are heavily armed with M2 carbines, .45 side arms and helmets. There is no resistance from the teenagers. I really don't need this, I am President of the CYO

(Catholic Youth Organization) and the head cook for the Monthly Pancake Breakfast at the church Annex; my sister Kathleen teaches Catechism for God's sake. I'm somewhat innocent.

We are hauled in to the Police Station in the back of the paddy wagon and separated. We are young and naive and the MPs are seasoned and experienced. They separate us putting myself John and Jim in one room while they separate Rick and Rusty for interrogation. We in the "waiting room" talk amongst ourselves leading with "let's get our story straight" never thinking that someone was listening and recording what was said. They shake up Rick, he is quite innocent but doesn't rat on Rusty; he is put in with us. He tells us what happened and we tell him what our "story is". Rusty denies everything and is put back in the same room with the rest of us where we all blab and the MPs listen. We are pretty alarmed at the reaction to a broken window. This is when the Colonel strides into the room in his great coat, snow on his shoulders, he is the MPs CO. Over his shoulder Dad tells the MPs to releases all of us. Then he turns to Jim and I, he tells us that we have 15 minutes to get home and we are to take no short cuts, then leaves. There is no way to get home in under 15 minutes, I know, I am on foot all the time. If I want to go somewhere I walk or ride a bike and this is not bike riding weather. We have a dilemma, be late sticking to the streets or take the trail and be on time. I am so shook up by this time, influenced by lazy Jim who doesn't want to run the long way in the snow and ice just to be late, I make the wrong decision and cut through the woods with him, taking the very trail we were captured on. We get home on time alright but it was a trap. We both suffer the consequences of our poor decisions. Turns out someone or some gang had broken into the church a couple of nights before, spayed graffiti on the walls and then absconded with all the sacramental wine; we were not the perps they were looking for. Thus the strong reaction of the MPs, Dad knowing everything didn't make it any easier on me.

## Pilot Parties

Mom and Dad entertain every once in a while, Mom has always been active in the Officers Wives Club. When your husband gets promoted the wife gets two steps in grade. I get to cook a large amount of Chinese food, served buffet style, .and play bartender making daiquiris. Someone has to man the blender. My daiquiris are pretty potent cuz I don't know what I'm doing, so I manage to get them all pretty loopy one evening. By far the best parties are the toboggan parties when the pilots with their wives come over and join in on our nightly winter fun. They all skate, play hockey and ride our contraptions down our bobsled runs like kids. We make a kettle of hot spiced fortified wine with fruit in it. The kettle is at the top of the hill, just inside the basement door. The sleders have to climb the hill to break the chill.

We Boy Scouts Jim, John Pease, Rick Wyatt and I decide that we need to win a winter camping merit badge. Whether or not there is such a thing is immaterial. I get permission from Dad to camp on base. The spot we want to go is pretty remote. I want to camp northwest of the runway past the ammo bunkers till there is nothing between us and the loop of the Mohawk River. We get our stuff together, two toboggans worth that we are going to drag to the site. Daytime temperature is averaging about 10 deg F. Master Sargent Pease is helping us with his pickup truck. He is TDY a lot to Thailand. He says he is working for the railroad...

laying ties. We pile our gear then jump in the back and we are off. We drive the perimeter road then unload and head into the woods. The snow is chest deep and better we do not have snow shoes or skis so we just break trail shoveling stomping and pushing down brush as we go. We want a good trail for access so we won't get lost in the woods. The snow is hard and crusty from the big ice storm and some warming and melting. It takes a couple of hours to work our way to where we want to go. Then we make a camp site the same way we made the trail but do a better job of it, getting down to leaves on the ground so we can gather some rocks and have a campfire. The walls of the campsite are 5 to 6 feet tall. We pitch two GI pup tents and cover them with a white parachute. Don't all kids have a parachute? Since Utah Timmy hasn't had a birthday in May unless it was under a parachute. We put heavy canvas ground cloths in the tents and unroll double GI artic mummy bags. I start chopping wood, it's easy. No one has been back here for decades. There is plenty of freestanding dead wood. All you have to do is shimmy up a tree and start rocking back and forth. If there is any moisture at all in the wood it will snap off right at the ground. To knock it into logs you just hit it with the blunt end of the ax and it snaps right off. Cutting wood is how I keep warm, everyone else rotates in front of the bonfire. The fire is so hot and your clothes so thick they could catch on fire before you felt it.

 Kathleen and Margret, Rick's older sister, pay us a visit one night, bringing a bottle of cheer with them, Four Roses whisky that we chill in a snow bank. Good thing we made a trail they could find and navigate. We grill chuck steak over the fire and bake potatoes in the coals. The sky is clear and there's no wind; life is good. We camp four days. The morning of the last day it is especially cold and we are especially hungry. John is cooking a pound of bacon on a cookie sheet over the fire. We are going to eat the bacon on some toasted tortillas with warmed up baked potatoes. John is cold and lazy so he is trying to do it while lying in his sleeping bag reaching out of his tent fly to cook. When it is about ready John manages to dump the tray with the grease and bacon on his hands while scrambling out of his bag. John is badly burned and blistered. We zip him in his bag. John rotates sticking his hands in the snow till he can't stand it then, pulls them in his bag and warms them up until he can't stand that, all followed by lots of yelling. Jim and I break camp Rick's Dad is coming out to get us around noon so John has a long wait until we can get him to the dispensary. School starts Monday…

## Spring 1970

I am in love, her name is Cindie, she's s sophomore at Rome Catholic High. Cindie lives in Woodhaven outside the wire. Cindie is a beautiful blond haired blue eyed girl with a sweet, kind, unpretentious personality. She never wears makeup, she doesn't need to. She is one of the girls in our church youth group. I am president of the CYO. The first thing I did as president was organize a dance. That was this fall. I have had the gang over to my house for skating and sledding parties a number of times this winter where we got to know each other better. Cindie has a good attitude, is always up for physical activity and never has trouble keeping up. Her Dad, a SAC Major just transferred here last summer. We were slow to find each other but are now inseparable. The school bus from the base goes to both RFA first then Rome Catholic. We are so goofy, I wait at the curb for her bus at RFA in the morning. The bus from Skyline gets there before the one from Woodhaven. She puts her arm out the

## 1967 GRIFFISS AFB, ROME, NEW YORK

window so we can hold hands while we chat about nothing more than when we will see each other again, usually after school on my way back from practice. I wear a path from her house to mine. We go on dates to the bowling alley or to the movies. I pick her up on my bicycle and ride her around perched side saddle on the crossbar. We are always holding hands, making goo goo eyes at each other and kissing, making our friends uncomfortable. I guess we are like that because time is precious between going to different schools, living in different neighborhoods, sports, and my job. Her house is a little over a mile from mine. I go bike riding or jogging in the snow after dinner which puts me at risk if I am late or forget my ID. I had to go over the wire and through the woods north of the coal pile and the railyard quite a few times. The base has its own railroad and power plant. The railroad just services the power plant now in the past it was used to move heavy equipment and I am sure that it still could be used for that. It has connections to the main line outside of the wire, or 10 foot chain link fence with razor wire on top. There's only a barbwire fence separating the Woodhaven housing area from the base proper where I cross. There are a bunch of jet engine shipping containers in the railyard too, the Air Force must go through a lot of them. The coal is always smoking from the sheer weight of material piled up in the yard generating heat. There is a big rail crane in the yard to handle the coal. There's a few nice coach cars with beds and other creature comforts down by the flight line, spruced up for the ready pilots on alert to hang out in. I don't have to resort to this too often, usually I am through the gate before curfew. I flash my ID; just another base kid.

There is a big Airshow at the base that Dad helps put on, with the Thunderbirds. They are flying the new F-4 Phantom now instead of the F-100 D. There are lots of different aircraft on the ramp to look at, walk around and some you can climb in and check out. Jim and I go down to the flight line for the show. Dad lets us roam around the hangers and tower too. Just before the show, Gary manages to catch a swing with a metal seat square in the mouth. Knocking out his front tooth. Mom gets him a nice porcelain cap, sheesh.

# 1970 Kwan Ju Korea

**JJS.** "I'm assigned Base Commander Kwang Ju, Korea, the only unaccompanied tour I've had, a real waste of time and effort; deep kimchee."

**RES.** One good thing that comes out of this tour is, Dad stops smoking. Dad has been smoking at least as long as he has been flying. That big square ring that he used to tamp his filter-less cigarettes Salem menthols with, and used to propose to Mom, is long gone. Jim stole it out of Dads insignia box, the one he keeps in his armoire. Jim wrapped some tape onto it to make it fit his smaller finger but apparently not nearly enough. He lost it throwing rocks in Delta Lake North of Rome, when he tagged along on a RFA junior class school party, the month before we left. I didn't rat him out. I was standing right next to him when he lost it; I was chewing him out, told him he was stupid for having it. Dad never asked anyone about it being missing. Dad didn't chain smoke, he would usually have one or two after work with his martini. I don't know if he smoked at work, you could do that back then. The Army used to pass them out with your rations. He has one "pine scented" ashtray that he keeps in the house; it never gets too full but it no longer smells much like pine. Dad used to sit on the back porch in Utah to smoke, he would flick his butts into the clover when they got short. We boys would fetch the butts and take a puff or two while posing before gaging and making him laugh. Tim and Steve stripped off their lifejackets and jumped in at opposite ends of the pool at the Hill Officers Club, because they wanted to swim underwater, shouting "look Dad", causing him to jump in to rescue them with his sunglasses on and a cigarette still in his lips. Grandpa Scanlon always had a pipe, cigar, or cigarette, in his mouth. He had a rack of pipes by his easy chair and a pouch of sweet smelling tobacco handy. Grandpa always smelled of tobacco smoke. Smoking surely must have contributed to his heart and health problems. Grandpa had a heart attack and bypass surgery while Dad was in Korea. Dad took some leave and flew back to the States to be at his side and to comfort his Mother. Maybe it was that or maybe it's because Dad bet the Chaplin that he could quit. Dad is so tight he was not about to lose that bet, especially to a Priest.

    Dad doesn't get to fly on this tour at least I can't find any hours logged for this time period. Not flying must have made his time in Korea pretty boring for him. The Air Force likes to have their Base Commanders on the ground, giving orders, signing papers, solving personnel problems not boring holes in the air. He has to deal with everything from Operations, Command Staff and personnel to gas being stolen from the cars parked at the Officers Club. Dad is able to take some time to go bird hunting, mainly Pheasants in the countryside and on offshore islands. He does quite well. He took his Winchester Model 12 12ga. pump with him and left the rest of his arsenal stateside with us, stowed in the trailer. They have a skeet range plus a rod and gun club at the base. Dad likes competitive shooting. He shot with his MP's in New York against the local police pistol teams and did quite well. They all had to shoot the same Ruger standard Mk 1 pistols, the one based on the old Japanese Nambu action. I had to clean this pistol once a week after his matches. They are a pain to tear down and reassemble. Dad has the Small Arms Expert Marksmanship Ribbon.

Dad keeps up his tennis, playing daily. His match is interrupted one afternoon; seems his attention is required immediately concerning an incident that just happened at the front gate. The base hires a lot of local civilian help to run everything from the mess hall on up as is common overseas. These folks are organized and are on strike at the time. Kwan Ju is out in the sticks in South East Korea. They are getting paid less than the civil servants in Seoul and want to be paid the same for doing the same work. The strikers have been blocking the gate to all nonmilitary personnel keeping them from going to work. Seoul can't have this so they sent busloads of scabs to bust the strike, feed the troops, mow the lawns, clean the bathrooms, run the BX and shuffle paper. The strikers wearing helmets wielding shields and clubs aren't having it, they attack the scabs trying to force them back on the buses. The ROK guard at the main gate in the forward fighting position strongly sympathizes with the strikers and cycles the bolt on his M-16 to assist them. An American MP sharing this position with him, as is common, drew his sidearm and shot the ROK guard, stopping him from firing on the scabs and their buses. The scabs had come armed with helmets, shields and clubs too so there was a big brawl until the scabs got back on the bus and left. The strikers got paid scale and went back to work. That's how they do it in Korea. The ROK was only wounded and recovered. The MP was court martialed, cleared, thanked, and sent home.

Dad is a whistle blower. Seems that a Col., our neighbor a NFO (nonflying officer) was taking kickbacks from a defense contractor. From what I know and now do, it probably was for falsifying radar intercept performance reports. Our neighbor's daughter is Kathleen's best friend, she is dating an Airman. Its 1970, one year after Woodstock, which was just down the road. They try to get Kathleen to smoke pot; she turns whistle blower. Like father like daughter. I am conflicted, the Col's son is an Air Force Academy Senior he has a green Corvette Stingray with scavenger pipes, and a 454, way cool. Sally hangs out at our house and goes to the skating and sledding parties. I have seen Sally in her bra and panties, flirting with me, a confused adolescent, while she ironed a skirt, she knows I have a steady girlfriend. She is the youngest in their family so she treats me like her little brother. I love Sally. She drives a VW bus and has an older boy friend in Maryland where she went to collage when Kathleen went to Oswego. We sing along to songs on the AM radio and seat dance when we go on local road trips. Sally's Mom has a Mustang with a four speed; she is cool too. The girl will be mustang Sally one day it's just a matter of time.

RADC is disbanded Dad is sent to Kwan Ju Korea some of the group is sent to Richards Gabor on the Kansas Missouri border, the rest go to Offutt outside of Omaha Nebraska. My friends and I are smeared, east to west from New York to Nebraska. Kwan Ju is an unaccompanied tour and we can't stay where we are if Dad isn't stationed at Griffiss. We are turned out of Base Housing. We head back to Colorado where Dad still owns the house outside of Ent AFB. Dad has hung on to the house so Kathleen can attend CU as an in state resident at Bolder as promised. At this point in time Colorado would let Military families attend state institutions at in state rates if they owned property and paid taxes in Colorado. This is my Senior year in High School. I leave all my friends 2000 miles behind me.

We crate up our lives and leave our home in separate groups. This is the first time that we will travel this way. Dad is leaving first. He is taking the truck and trailer to Colorado Springs and is going to park it at our old friends the Lepines. Kathleen is to fly out with everyone except Mom Jim, Steve and myself. They will stay in the trailer until our stuff arrives; the rest of us will follow in the Buick. We three boys are to do the packing and heavy lifting for Mom. We also have to clean the place after our stuff is moved out. We will be there to help Mom on the road if she should need it. I still don't have a driver's license so Mom plans to do the driving.

It's a sad day when Dad laves. Summer is just a time of the year when you move I guess. I'd hitched up the trailer for him. It's packed with everything that Kathleen's detachment will require for a month or more. When Dad leaves he is wearing summer Kaki short sleeve uniform with his blue belt and overseas cap, he's going to a Command Retirement Ceremony before hitting the road. He gives all a hug and kiss. I'm going to miss him; he is gone a lot but usually no longer than a week. Dad's stuff has already been shipped out. He's not taking leave before he reports for duty but he has 30 days to get there and plans to fly out and visit his Mom and Dad in LA. Don't know what will happen from here on except that it will be different.

Kathleen is next to go she has to get Tim, Bill, Maureen and Gary to Syracuse first on the bus and then by plane to the Springs where the Lepines will pick them up at the airport. We find out later that things didn't go too smooth for Kathy. They almost miss their connection at O'Hare and have to all run across the apron to the awaiting plane abandoning their checked bags. This was back when you could do stuff like that. They all eventually got their stuff but this didn't help the stress level. Kathleen is on the glide path to her dream; she is back in Colorado and will be eventually attending CU Bolder. She leaves a week after Dad. I don't see Beta flight and hear about their troubles until two weeks later.

Mom and Dad have arranged for me to have a goodbye dinner with Cindie and all my friends at the Officers Club. We are in coats and ties. The girls are all dolled up. We eat, talk and laugh about things we have done together. There's not a dry eye as we all solemnly swear to get together again. Cindi takes a cloth napkin and writes the poem below about us and what we mean to each other she then passes it around for everyone else to sign and add to what she had written; then she writes a note just for me also transcribed below. We linger as long as we can, then we go our separate ways.

## Our Second (and last) Reunion

Everybody Was There
All The People We had Ever Known
And We Sang And Danced
And Ate and Drank
And For The First Time In Our Lives
Everyone Was Really Happy
Not Just On The Outside
But On The Inside Where It Really Counts
Then Someone Got The Crazy Idea

That Everyone Ought To Get Married
Everyone Thought He Was Kidding
But We Weren't
And Laughing We Ran To The Car
And Drove Off Into The World
The World That Had Once Separated Us
But Would Never Be Able To Again
For Now, We You And I, Are One


*Bob,*

*You have to be the greatest guy I have ever known. You probably taught me an awful lot of things, things I never would have learned if it wasn't for you. I love you more than anyone else in the world and I'm going to miss you a lot. I'm not going to say goodbye, because this isn't the last time that I'll be seeing ya, at least not if I can help it. Someday we'll meet again someway.*

*Love always,*
*Cindi*


We spent the night on the floor, so there's not much to throw in the Buick when we hit the road the next morning. We're on the New York Turnpike headed west. We're all gassed up, I've checked out the car, the sun is shining and the tires are humming. I didn't sleep much last night, thinking about how much I will miss my friends, especially Cindie who was kissing me and crying outside club before her Mom came to get her. I was pretty sad too…

I'm in the front passenger seat with all Mom's pillows except the one she is sitting on to see over the steering wheel. I nod off like I usually do when traveling, it's a talent, just wake me when we get there, that's the best way to travel. I'm not out long when I wake with a start there is a lot of racket and the car is bouncing and pitching. We are off the pavement at a bend in the road, plowing across the bar ditch and heading for some trees; Mom throws the wheel over and hits the brakes. We spin out of course and come to a stop before we hit anything bigger than tall weeds. I am completely awake now. Mom fell asleep after less than an hour on the road. She obviously cannot be trusted to drive cross country. I make my first temporary head of the house decision and tell Mom she is done. I'm seventeen and I have no driver's license but I drive all the way to Colorado. I know the rout because we just drove it last summer, lots of corn. Mom for the first time I can remember, doesn't argue.

I make the trip in two legs stopping in St Louis. Mom is navigating. Her navigation was never up to Dad's standards who always wanted waypoints and to know when to change course thirty minutes in advance; now I share in his frustration. We get off the highway too early and get lost in East St Louis during the wee hours of the night, absolutely the worst part of town at the time. Mom is trying to navigate to a friend's house where we can crash, to save some bucks. I would be grateful if she'd let me just pull over, release the seat back, and sleep in the car at this point. We pull over to a succession of pay phones and call her friend for directions, who understandably, knows nothing of this part of town. We finally find the place. That hour and a half hunt cost 100 miles further down the road by my calculations. A Motel 6 would have been fine for me but you can't get too mad at your Mom and you cannot win any argument with her. I slept like a rock and remember little except a nice breakfast, I don't drink coffee, ick. Mom stayed up all night, chatting and drinking coffee till dawn so she snoozes most of the last leg. I drive straight to the old house on the corner of Meade and Uinta and crash on the living room rug; I'm home.

We drive out to the Lepines the next day to see Kathleen and the little guys. The boys are running around the place with the two little Lepine boys. The little boys are ten years younger than their siblings. Their house is still out in the middle of nowhere on the shoulder of a foothill south of the Air Force Academy. Emery is still working on his house. He's an engineer so everything is a work in progress, he is very smart and good to talk to. Emory's soft spot, which I can't understand, is Citroen's he has three of them.

Everyone wants to get settled despite the house being empty. The moving van is about another day out. In the meantime Mom is anxious to pull her stuff out of storage, stuff that's been there for the last six years, to make everything twice as busy. After dropping off some kids at the house she takes off to see to about getting a phone hookup and utility services turned on. I hitch up the trailer to the war wagon and head for the house thinking that I really do need to get a driver's license at some point. I tow the rig to the house and pull up out front, but it can't stay there, city ordnance. I look at the old carport and there is no way that I can get the rig in there, we used to use a come-along to get the old sixteen footer in. I need to put the FAN in the backyard. Without permission I tear down a section of the basket weave fence and pull a four by four fence post with a come along. I throw some of the fencing in the gutter to make a ramp, then circle the block and back that sucker into the backyard with some help from my brothers stopping traffic on Uintah. The boys have found some nails in the basement and put the fence back up while I level and hook up power and water to the trailer. At least we will have some beds if you want one and a table to sit at when you eat. There is food in the trailer too. We move all the lawn furniture in the trailer to the house with the black and white TV and have our first meal in the house together, spaghetti, I cook. It's nice. We are home.

The storage stuff comes first, a bunch of miscellaneous claptrap we didn't need for six years and can't use now, like a piano and an organ, the industrial ironing machine. Right away Mom notes that the kiddie surrey is missing, the one that her Dad gave us, and has a fit going through the stuff piled in the family room and the old manifest to see if more is missing. They put a numbered sicker on everything they itemized and items are missing, boxes of treasure that someone else packed that Mother has no idea of what went in the treasure box. Most of the stuff is left in the boxes and we boys move it directly into the basement to make room for more useful items.

The Mayflower moving van shows up the next day. It took longer than expected because when the truck left Skyline they had stuff sticking out the back and couldn't shut the doors. They had to wait for a larger van to become available and transfer the load. The movers put the stuff in the house and the nine of us put things in their place in short order.

I have a lot of time on my hands. I don't know anyone and don't have much to do. Cindie and I exchange two letters a week, Cindie's are penned in white ink on black paper. I am determined to go back to NY, I'm heartbroken. To get my mind off things I take on a project. The basement is roughly separated in quarters half of the basement is lit and has a concrete floor while the other half is dirt with no power. One of the dirt floor quarters had the furnaces and associated chimneys. The other quarter is clear we use it for a BB gun range. The problem with the range is the low ceiling, the dirt floor wasn't excavated to the depth of the other quarters. I want to dig it out about a foot and a half, pour a floor and put a ping pong table in there. Mom is onboard and will provide the ready mix and steel mat, we will do all the work Mom is ok with this because it will keep us busy and tire us out, something she is glad to pay for. The boys and I knock it off in a week and then we need something else to do. Timmy turns out to be an excellent shot. He can hit anything with anything; he proves this over and over again on the BB gun range. He has Uncle Bob's old silver BB gun. We have an old 2 inch tetherball pipe that Timmy uses as a bazooka tube, perfect for launching sky/pop bottle rockets. We are back on the roof, we load him and fire him up just like Combat on TV. He puts a round right thru a pickup window going down Uinta, an incredible feat, boom goes the rocket, brakes squeal, we skater. A classic ambush, the element of surprise, we had the high ground, no killed or wounded and best of all none taken prisoner.

## Eleven Mile Canyon

What I want to do is hike up Eleven mile Canyon, camp and try to live off the land. The idea comes to me when we are assembling Dads gun case, cleaning and placing his firearms in it. Dad took all the guns with him in the trailer except for a Model 12 12 gage that he shipped to Korea. With the guns we find half a pink split long bowl toilet seat. It has an engraved silver plate mounted to it that says, "Good Luck from All Your Half Assed Friends at RADC". It has over one hundred signatures on it. You never get an outstanding, a superlative, not even a good, from the Col. everything is half assed or not half bad, somewhere in the upper half of the lower third. We went fishing and camping at 11 mile canyon when we were stationed here last. A crushed granite road with several bridges had just opened up the narrow canyon and its river basin to folks. The same record flash flood of 65 when we left for Taiwan, took out all the bridges and wiped out most of the road in the canyon rendering it inaccessible by vehicle. The park service placed barriers across the entrance, the last bridge in the canyon; not even motorcycles are allowed.

Somehow I talk Mom into letting me, Jim, and Tiny Tim go for a two week camping trip up the canyon. A two week trip requires some planning. We have to take our shelter halves and the canvas tarp, this alone is a load. Tim is only twelve and weighs 70 pounds, less than my pack so we can't count on him to carry very much very far. Each of us has a poncho and a sleeping bag. All we have is a WWII radio pack board, a trapper pack frame and little GI back packs, web gear, Musette and duffle bags. They will have to do. We pack stacks of tortillas cans of bully beef, cooking oil, salt and pepper, aluminum foil and beans to supplement the trout we think we will catch and the rabbits we count on shooting. We take our bayonets, entrenching tools, a hatchet, and canteens hung from web belts and suspenders. Spare clothes and matches in plastic bags are stuffed in where there's room. I'm taking Dads High Standard .22 automatic pistol. I am a fair shot with it and it's easier to pack than a rifle, a .22 should be plenty for anything we may run into I have 100 rounds of .22 long rifle for it. Jim is taking the Annie Oakley Remington pump .22 with about 500 rounds of .22 shorts. He's a bad shot so I don't think anything of it at the time. Jim says he gets more bang for his buck with shorts. I think that they are only good for carnival shooting galleries. Each of us packs a rod and reel with some lures leaders and hooks. I grab a light windbreaker, and wool shirt, like Dad wears, and shove an old felt wide brimmed hat that Grandpa Scanlon gave me the last time I saw him on my head and we are off. This will be a blind drop and pick up. We are hiking as far as we can up an inaccessible canyon at the end of a long road that no one travels. We have no communication capability. I tell Mom to just pick us up at noon, in two weeks, at the same place she drops us off, where the road in necks down at the barricaded washed out bridge.

We hike until we are squeezed against the river by the canyon wall and have to ford it to get on the other side. We have been looking for a place to ford for some time. We double back to a place where the river is wide but shallow and not moving too fast. I pull off my boots and pants then cross first with a rope and no pack, can't afford to lose anything to the river gods. The water is ice cold and up to my crotch, a little deep for Tiny Tim. The rocks are slippery so the rope is a good thing. I tie off to a tree on shore and return for my stuff, you have to carry it over your head. Jim strips down and gets himself and his stuff across. I go back and wade over with Tim, then I go back and get his stuff and untie the end of the rope; Jim reels it in in for me. We don't want to do that too much it's a real time burner.

After we dress, Jim want's to lock and load in case we jump a rabbit; I don't see anything wrong with that until he takes a shot at nothing. He grins and says "test fire". I think he has scared off anything nearby for no purpose, Jungle Jim.

There's a pretty good camp site about 100 yards from where we crossed, no one will be coming upstream or downstream from us. I dump my stuff and take off to retrieve some stuff we had cached. When I get back Tim has made a fire pit and is gathering firewood. Jim is resting on the tent floor, lounging against his pack and playing with his rifle.

I have to goad Jim to help set up camp, he's the grasshopper I'm the ant. We set up the shelter halves at right angles from each other then use the floor canvas and some poles to make an open sided shed roof over the two shelter halves which form two sides of our enclosure. We cut pine boughs to put our mummy bags on. It's getting dark quick like it does in the mountains, in a canyon; good thing we have a fire going. We divvy up a can of navy beans and a can of corned beef between our three canteen cups then put them on rocks around the fire. We wait. When its warmed up and stirred up we toast tortillas with our beans. Nothing ever tasted better after the day we've had. So far so good.

    You'd think it would be quiet out here but there's plenty to listen to, the wind in the pines, the gurgle of the river against the white noise of the rapids up stream. The rattle of aspen leaves. We listen for critters moving around, our senses heightened. We finally get tired of feeding the fire and turn in. During the night we hear a big cat snarling, trying to spook something to run. It makes me uneasy but I have the pistol under my roll of clothes that I'm using for a pillow.

    When I crack my eyes open early the next morning I want to lay there because I'm still sleepy when I hear some rustling, it's a cotton tail hopping around near the edge of the camp. I ease my pistol out and nail him. Lunch. That wakes my fellow campers up.

    I clean and wash the rabbit in the river. Then I oil, salt and pepper him, spit him and put him by the fire for a slow roast. Our camp is right by the river so we go try our luck with our tackle after toasting some tortillas for breakfast. We are bad fishermen. We're doing something wrong. They are not going for the Mepps spinners or any spoons we cast. We can see the fish in the crystal clear water. Can they see us? I remember that I've been here before. I turn over some river rocks and find what I'm looking for. Grubs that they feed on, fat black rock worms you find in shallow water along the shore, under rocks. We change to a Texas bait rig and catch a couple of keepers. Now we have dinner.

Next day we have tortillas and leftovers for breakfast. The plan of the day is to hike around, explore and maybe get another rabbit. We are going to go back to where we stopped the first day and try to work our way further along the river. After a couple of crossings we come onto a wide draw that tends north. We take a break in the shade of some trees before heading uphill. After a while a chipmunk chatters and jumps on a rock next to us. Tim and I are watching him hop around checking us out. I toss him a little piece of the toasted tortilla I'm snacking on and washing down with water from my canteen. As the little creature hops closer to sniff it, I'm startled by crack of a rifle behind me as Jim blows him away. What a jerk. I ask him if he is going to eat what's left of him and thank him for spooking every animal in the draw and making my ears ring. He shot Chip what's Dale goanna do now?

The hike up the draw leads to a saddle and a wide valley. Here I have some luck, I see a cottontail trying to be invisible and not bolt, he should have bolted. After I shoot him and two big fat grey squirrels we head back. Jim shot at some squirrels and some birds' maybe but he didn't hit any of them. Tomorrow we plan to fish some promising spots that we've seen on our hike up to this place. I field dress the game and we head back.

It's starting to get dark as we hoof the last stretch to the camp when we start smelling something skunk like that gets stronger the closer we get. As we get to the camp were happy to find that it's not the camp that stinks. The only thing that has changed is the weather, its warmer and the wind direction has changed. It's enough to gag a maggot. It's too late in the day to do anything about it, that's assuming that anything can be done about it. We light a fire which at least overlays the stench. We are sure that a skunk did not spray our campsite, why would it? They don't do that unless something was after it. The smell doesn't seem to be localized. It's misting and wanting to drizzle so we put all the fire wood we can under the fly and start improving the tent making walls to the fly with two of the ponchos and more poles. I skin the squirrels, salt them down and roll up the pelts. Tim wants to cure them, make a hoop and stretch them out to dry later. Rabbit hide is too delicate to cure. Skinning a rabbit is like opening a bag of chips; squirrels are tough especially if you want to get the face, paws and tails. I quarter the rabbit, salt and pepper all the cuts and wrap them in foil with a little oil to roast on some low coals with some more beans in a canteen cup, a hobo dinner. It take a while to cook and we are tired from our hike so we turn in early, just after we eat.

The next morning it's raining, not so much to cause runoff, just kind of steady. Our plans are bust. The fire is out. We have a cold camp. You want to just lay snug in your bag and look at the rain. The stench has either abated or we are used to it; no, no one could get used to that. About mid-morning I get up; just can't lay thee any more. I throw on the third poncho, grab my pole and go out to try my luck. Some sage told me once, that fish like changes in the weather; I'm going to test that. I give up after whipping the water for a couple of hours. I go back to the tent, strip off whatever's wet and climb in my bag. All I can do is think about all the stuff I came out here to get my mind off of. That dear Bob letter from Cindie was a stab in the heart. I miss my girl, my teammates and all my friends. I am going to be a Senior new kid in in a new school in less than 30 days; it makes a pit in my stomach. All three of us are in the same boat and feeling lonely but closer than ever.

It's raining the next day too but stops before noon. This is getting old, you can't even go out roaming around when it lets up without getting all wet going through the brush or walking under the dripping trees. The plan is to start a roaring fire with the dry wood we have left, so we can feed it wet wood without it going out. We need to dry ourselves and our gear out. This keeps us busy. Being proactive always improves morale. When you concentrate on the immediate both the past and the future fade. The last thing I want to do is obsess. It never helps. I'm not obsessed but I sure am interested in driving around the war wagon. Mom has let me know that it will be my job to drive us back and forth to school. I'm to take Jim, Steve, Gerard Gasparini, an old friend I'm supposed to know from church now grown up, and Mohamed, a Saudi exchange student, living with Gerard. Gerard and Mohamed are to pay me a quarter a day each for curb service/gas. I haven't met Gerard or Mo and don't know where they live, the school is, or where to park. I need to get my driver's license first. I'm sure there's a book and a test for that. For the moment I need to catch a fish or shoot something.

The wind has been nonexistent since the sky rained itself out. The smoke from the damp blazing pine fire has been going straight up in the air. I'm cleaning and loading my pistol that I've managed to keep dry, I've already checked my rod and reel. My mini tackle box is in my breast pocket. I just need to pick up my web gear to be off. When there it is, the stench. It's faint at first but the stench gets stronger as the breeze freshens and bends the rising smoke of our fire over in an arc. I put on my gear, holster my pistol and head up wind leaving my brothers and pole behind.

As gross as it is, I have to sniff it out. I follow the river until the stench goes away then I backtrack to a narrow ravine with water flowing in the bottom that leads upslope. I work up the ravine about

300 yards and come upon a dead elk swarming with blow flies and maggots. Mystery solved but at what cost? I go back and tell Jim and Tim about the dead elk, they want to see it. I'm not interested in seeing it again so it tell them where to look about five hundred yards away adding you can't miss it. Just follow your nose.

We are too lazy to move the camp and agree we like this spot and we would be just chasing the wind unless we moved a mile where there is another spot we like, next to some deep pools in the river we have been fishing in. In the end we hope the wind changes direction, understanding that our comfort depends on which way the wind blows.

We eventually run out of hard rations, the tortillas are long gone flour first then corn. The flour ones are prone to burn; I like the corn ones, they are slow to burn and toast up nicely. We are out of corned beef, beans and sardines. It's been fish or die the last couple of days. Jim has now a knack, for lying on his back, with his rifle on his knees, shooting pinecones from the trees. That, of course, will attract game for miles. I haven't seen a rabbit moose or squirrel since he has taken up cone capping. He has 500 rounds to burn, we're supposed to hike out tomorrow. When you're hungry, it's frustrating to see these large trout swirling in the clear water ignoring your fly, bait or lure. I wish I could shoot them. I have a chant: fishy, fishy, fishy fish, on thee to dine is what I wish. The chant doesn't help. Working along the river, I spot a huge trout in a pool apart from the river. He must have washed up in there during the rain when the river came up then was trapped when it fell, there's a big pink carp in there with him too. I'm going to catch and not release both.

I bring my catch back to camp after cleaning it. The trout looks great around 24 inches long but the pink carp is sort of disgusting looking, I've been told that they are bony but Orientals and people from Oklahoma smoke them and eat them. We have trout, all of it. That night we burn most of what we have left in the way of firewood and settle down for a last night in camp. It will be good to get back and see everyone.

The next morning the carp looks even more unappetizing, so everyone passes on it, even though our stomachs are growling and we have a long hike in front of us. We're talking and joking about what we are going to eat first when we get home. Tim cracks me up when he says that he wants a lard and pickle sandwich, we all have so very little fat on us. That gets me craving butter, which we never get, it's always margarine. Something developed to fatten up turkeys but the turkeys wouldn't eat it so they put some yellow dye and some salt in it then market it to the public as good. It doesn't take long to break down the camp. The fire is out. There is less to carry back. We load up and leave Camp Dead Elk behind. Mom is supposed to meet us at noon where she dropped us off, we have over six miles to cover but we are going to stay on this side of the river. On the first rest stop I take a good look at my brothers and hope I don't look like that. They are grimy, their faces are black with smoke and ash; it makes the pores in their faces standout. Their hair is matted and greasy. We haven't touched a bar of soap for the entire time except to clean the dishes and we used sand for that mostly.

We get to the rendezvous on time but Mom is one and a half hours late of course. Mom is happy to see us but we are a fright. Mom is worried about her car seats, she's brought towels for the seats. We ride back to town with the windows rolled down all the way. Mom is chatting away about something despite the wind noise I don't know what she is saying; I'm just glad to hear her voice.

## New Schools

When we get home the big event is, Kathleen has a job as a motel maid and is registered to attend CU Bolder Cragmore Extension Campus. My getting a driver's license turned out to be no big deal and before long it's time for school to start. Kathleen got a taste for college at Oswego and it changed her outlook on things. It made her question the things her Mama has been telling her for years. Mom is in the habit of still treating us like little babies, telling fairy tales to get around uncomfortable subjects.

Then school starts for me, Palmer High, right downtown. There is a statue of General Palmer in front. Famous alum: Elvira, Mistress of the Dark... Picking up Gerard turns out to be a good thing, he's on the tennis and gymnastics teams things Steve and I are interested in. Gerard is a Junior he holds court to us three newbies telling us what's what; he has an opinion on everything. We get a very concise briefing by the time we get to school. The Colorado School System is not as good as the New York system. They let folks slide here, the bar is pretty low. If you are in athletics you get a pass. And the corollary, if you show an interest in a subject, it's easy to excel. I like my History teacher, he was an aid to Patton. My physics teacher is very good as well as my Analytical Geometry Instructor but it's not the same without my nerdy science geek friends Dave Nelson, Rick Wyatt and John Pease from RFA. Even Tennis is easier, I go to tryouts and make number one on the roster; in New York I would be lucky to be playing number five. I team up with Gerard for a doubles team and we beat everyone. The Inter Scholastic Tennis season is in the spring, fall is just for practice. We practice on the courts at Prospect Lake Park. I drive fellow players to practice for a quarter. The fall season ends in October. The only other way to make money is to work concessions at Academy football games. Celeste Lepine gets me a job there, I am the low guy so I have to lug Cokes at first which is real heavy, I then graduate to hotdogs, better but not by much, lugging a warming box around. The coke rack gets light but the dog box has more dead weight. It's a real workout going up the stadium stairs with a load of cokes.

I've been asking around about playing hockey and finally talk to the coach about trying out. Coach is skeptical, most of his players he has known as Pee Wees and coached their older brothers who played for him as Pee Wees. I go to a no pads practice at the Broadmoor World Arena. Practice starts at 0430. Ice time is dear so we skate when anyone in his right mind has no desire to. When we don't have ice time we play soccer. I meet the team, a lot of them play football. The football coach is also the hockey coach. Coach likes my skating, I'm not in the front and not in the back of the pack in the drills. I can skate backwards faster than most so by the end of the session I am playing second line defenseman. I'm partnered up with a tall lanky guy named John Carlyle. There are two players on each defensive line. John and I are destined to be best friends. We get to where we can read each other on and off the ice. Now after all that I get to go to class.

I love playing, the early morning practice, watching the sunrise color and walk down the mountains, the uniforms and the games, away games in bad weather. Mom and my brothers get to see me play. I inadvertently start a bench clearing brawl in one game. I'm right on this guy's tail when he tries to freeze the puck against the boards behind his net, a rookie move and I punish him for it. I'm going to make him blow snot bubbles. I put a knee between his legs and lift him up as I'm throwing my stick arm and shoulder across his back. He hits the boards and glass with a bang and I bounce off him. As luck would have it he kisses a seam in the glass that splits his lip and cheek, making him squirt blood all over the place. The goalie happens to be his brother; he comes charging out of the crease with his stick raised and hits our winger who is closer to him than I am. The benches are clearing, players are charging past me as I'm drifting slowly backwards from the melee. People in the stands are trying to get on the ice, it's their brothers and sons out there. Coach grabs me as he is going in to help break it up, he tells me to get off the ice and in the locker room, quick. I have to sit out two games for a legit hit. This contributes to our leading the league in penalties that season.

Before you know it's Christmas break. Organized team practice is banned during the break so we skate on the lake at Green Mountain Falls. After a long scrimmage game in full pads, I load up about half the team plus their gear in the war wagon and head down Ute pass. We are laughing because we are packed in like sardines and trying to strip off some gear; we're all hot and sweaty and on a post-game high. Heading down the mountain I see flashing lights in my mirror. I'm seventeen and I just got my license a little over three months ago. I've never been pulled over. The guys are giving me shit as the State Trooper walks up to my window, teammates are always there for each other like that. The Trooper asks me for my license, then to step out and walk back to his cruiser. I haven't had an opportunity to take off my gear so I literally roll out, still in my ratty taped up practice shorts, holey sweaty ten year old number 14 Palmer Terror Eagle Beak jersey, and equally holey socks with my shin guards still taped on underneath. The trooper tells me to walk back and get in his car; it's cold. I squeeze into the front seat like he tells me while the trooper radios a check on the truck and me. The trooper tells me that he pulled me over for speeding and weaving but he doesn't believe that I am drinking before noon. He pulls out his ticket book and starts writing down my information while I try not to look at him and what he's doing, I'm checking out the gear in his car and listening to the radio when the trooper says "your friends must think they are pretty funny". I look up and out the windshield. I am horrified to see my buds lined up on the road side of the war wagon, spread eagle in various stages of undress. The twins are naked except for their hockey shorts, suspenders and tennis shoes no socks. John is only wearing his red button bottom full length long johns, a watch cap, gloves and army boots. The others are just as bad. We're pulled over on a major four lane highway, it's a downhill 7% grade mountain pass. Everyone is going slow to begin with, so it doesn't take long for a huge traffic jam of gawkers to accumulate going both ways.

I'm not sure but I think that the trooper was trying hard not to crack up; he turns to me as he is tearing up what he was writing and says "tell your friends to get back in the truck, get out of here and keep your speed down. He lets me go in the interest of public safety and a good story at the station house.

Christmas means another birthday for me, I turn eighteen; time to register for the draft and start drinking. There's a war on. I register downtown and go home to wait my letter. The lottery system has been in place for a year or so, you are drafted in the order of your birthdate drawn from a bingo basket. The drawing will be held in July, a nice graduation present for you graduates that haven't selected a career path yet.

I receive a letter instructing me to report to the Selective Service office in Denver for a physical. Mom and I drive up in the blue bomb on the appointed day. It gives us a chance to talk. I'm conflicted. I just assume that I will have to go. Dad has given me mixed messages. He has some distain for the current conflict, he tells folks that "he was in the one we won", Nam is just a place where you get your ticket punched. Then next it's "don't you want to see some action? The Military has been your bread and butter after all". He really doesn't want to see me as a ground-pounder; he joined the Air Corps to stay out of the mud. He'd heard enough from his Dad about that to make up his mind.

Mom has been working tirelessly on a NROTC appointment for me, writing our congressmen constantly; along with a lot of other Moms, the best I, she, could do was a two year scholarship, after I complete the first two, and if my GPA is acceptable. My only hope is for a commission is the Navy, they will accept Line Officers with vision correctable to 20/20; I'm resigned to my fate for the present. When I get to the induction center it's like Candy Land, you follow the colored footprints on the floor. First there is a short interview, I'm asked if there is anything that would prevent me from serving; I tell him about my Bell's Palsy, he is unimpressed and passes me on to the next color footprints. A doctor checks our ears nose and throat then it's off on another set of color footprints. In groups of a dozen we are led into a room and told to drop our pants and underwear. The doctor stops in front of each of us and asks us to cough then he asks us all to bend over as he walks back behind us. We are just a bunch of dicks and assholes to this guy. They determine that I am 1A prime American fighting stock. I'll get my card in the mail.

When I get out there's quarter sized snowflakes coming down; I have to knock three inches of accumulation off the truck and of course the glass is iced over and the wipers are stuck. Really didn't expect this or to be in there so long. We start to roll after I do some scraping; conditions do not get better down the road, we have over sixty miles to go. By the time we get to Castle Rock the snow must be over two feet deep. The plows have not been through; you can't see the edges of Interstate Twenty. A white knuckle ride the whole way. Not another car on the road. It was the first time that I have ever known my mother to be wide awake and silent at the same time, now going on two hours. We make it. It's going to be a long six months.

Jim was on the swim team in the fall; they made him a breast stroker because of his splay footed stance; good for a frog kick. Jim is bored of swimming laps and wants to play hockey. I think that is a bad idea. The team has a few hopefuls that Coach lets hang around to help with the equipment, he lets some of them join our drills and skate laps. Jim is one of these. He is a sophomore, he may have a couple of years in him so Coach gives him a shot. There's only two weeks left of the season. That will make three sets of brothers on the team.

Jim gets on the bad side of everyone right off the bat. We are having no pad drills. Our star center Eric comes down the ice on Jim who is defending. Eric puts a move on Jim and easily goes around him. Jim trips and falls trying to follow Eric. Jim slashes at Eric's skates with his stick as he is going down, tripping him. Not good. Cheating, we're only wearing shin guards, in frigging practice, he's the team captain and your teammate. Eric kicks his ass, our goalie joins in. I watch mesmerized. Coach slams the stick I loaned him on the boards, shattering it, then he breaks it up. Dumbass. I'm real embarrassed. That's Jim.

The season ends. I take up tennis again. We practice in the fall and play matches in the spring. The Hockey team takes up "Woodsies" weekend night, beer drinking, hill climbing, bonfire burning events. You can buy and drink 3.2 beer if you're eighteen, better yet, if you have a Military ID you can buy the real stuff on base, in the same store that I used to buy model airplanes. I'm real popular or could it be the 6.0 Colt 45 Malt Liquor tallboys. I have the party bus and all the party people with me.

The parties are in the foothills and some are quite big with lots of goings on mainly trying to get up a steep hill on bikes or in jeeps without stalling and falling back down the hill. Saw some specular fails, jeeps rolling. I thought John was going to kill the both of us one night, when he tried to take a hill in his purple metal flake 48 Willy's. We stalled, rolled backwards then sideways, with no rollbar. Pretty dangerous, sort of what folks do out in the desert these days. All this is at night lit by a flickering bonfire, kind of tribal. I picked up a case of Colt 45 one night; you don't need a cooler, its beer-cold out at night, then I picked up the guys. We run out of seats and seatbelts so Kevin crawls in the back with the beer. I drop the tailgate to let him in. They start drinking as I start rolling Keven is handing out the beer. A case is just about right for eight guys for the night. Everyone but me has finished their first as we roll up to the party. It's dark out with no moon. I pull the sub off the jeep trail nose up a steep embankment, goosing the truck and making my passengers squeal and hoot as I make the wheels spin in granny gear. I stop and we hop out. I walk to the back of the truck, raise the hatch and drop the tailgate. Out rolls Kevin like a sack of potatoes; he hits the dirt and rolls a couple of times. We try to check him out in the light of the bonfire. He's out cold. At first I think that he must have hit his head when I was skidding the truck around, he didn't have a seatbelt, or maybe he was goofing us when he rolled out of the truck, and hit his head. I've seen Kevin out before. A player from Cheyanne stomped on his skate hard enough in a game to break the blade off the boot. The next stride Kevin took with that bladeless boot caused him to cartwheel right into the net.

Kevin knocked the goal loose and was left hanging, out cold, from the top of the net by heel of his other skate. Rick tells me Kevin's back is bleeding, he has rolled over some broken glass. Everyone is very worried until they realize that he has guzzled over half a case, in under an hour, and has passed out, well pleased with his joke on us. Well nothing to do but head into town and get him sewn up.

## Spring 1971

Tennis season starts, Jim wants to play too. Jim came to the first practice last fall so Coach could see him play. I like our Coach, Lloyd Samuelson, he's a straight shooter, a very fair man, someone I can understand, he is also the art teacher and I like to draw. We pair off and start hitting. Palmer plays at Prospect Park, we only have one old court on campus. Halfway through practice. Coach pulls Jim over and says "Jim, we are going to revamp your entire game". Coach tells Jim that he particularly dislikes his under slice serve and that is where they will start. He hands him a junk racquet then leads him over to two cones he has set in the grass. Coach tells Jim to throw the racquet at one cone then run and get it and throw it at the other cone. He is to repeat that till he learns to pronate. Gerard of course is beside himself, Jim is often the butt of his jokes on our daily commutes. Gerard can run intellectual circles around him. He loves to goad Jim and make him lose it. Now Gerard teases Jim mercilessly about that. Welcome to the team.

    I play number one singles and number two doubles with Gerard. I am somewhere in the upper half in singles but Gerard and I are unbeatable in doubles. Coach plays us number two to put some consistent wins up. Our lineup is pretty shallow so you do what you have to. We take on the state girls champs at the end of the season and trounce them. Jim gets to play exhibition matches so at least he gets some competitive experience.

Just a few months into the season, Coach Samuelson gathers us around him after practice for our usual talk. When he is finished with the more mundane subject of dates and schedules, he says" someone on the team has to get a Tennis Scholarship and it's going to be you Bob". I'm elated, he tells me he thinks that he can get me on at Adams State and would I like that. You bet. This is perfect, I can get in my two years toward a NROTC slot there, earning my way as a tennis player; maybe I can attain a slot before I'm called up. I can be a bench warmer/practice team member for their soccer and hockey teams. I am over the moon, all that hard work, destroying a pair of Jack Purcell Converse tennis shoes. a month is going to pay off. Graduation is in less than ten weeks. Less than a week later, Coach, coming home from the grocery store, surprises a burglar in his house who then shoots and kills him while he is still holding his groceries. Lloyd Samuelson's murder remains unsolved to this day. The English teacher is kind enough to step in for Coach, but he's not a player and he doesn't know the league, the players or the coaches in high school much less the College circuit. He does his best but he is limited to making phone calls and arraigning transportation to matches, sometimes he drives us himself in his Volvo station wagon. My plans are adrift again; can't feel sorry for myself, I feel sorry for Coach, he didn't deserve that. I loved that guy. He took an interest in me.

I am bummed all around I miss 20 days of school my senior year. The Colorado school system seems to let jocks slide, you just have to wear your jersey on game day. Jim gets a mercy letter in tennis from the English teacher. I get a letter jacket; I share it with Jim, it means more to him than me, I'm graduating. I am so... don't care. I get hammered at a Woodsie the night before I take the SAT. I crack a 650 in math and a 750 in reading and writing. I wonder what I could have done without a hangover. It doesn't matter Dad has already told me he doesn't have the money to send me to school. The Lottery is in a couple of months so I will just have to wait to know my fate.

Kevin and Rick have motorcycles, I want one too for graduation. Dad will want his truck back when he gets home plus it burns a lot of gas. I will need something to get around in. Dad is still in Korea and will be there until July. I pester Mom and she pesters Dad; in what he will later call "a weak Moment" he agrees to pay for a used bike I have found for $500.00. It's a sweet blue and white 69 Honda 350 Scrambler with knobbies. My unstated plan is to ride it all the way back to Rome as soon as I graduate. My stated plan is to start taking classes at the CU extension here in the Springs; I will use the bike to commute, and it's cheap to operate. Dad always asks "what's your plan?" So I have to have one. It's a used bike, it needs work, a new battery, a tune up and cam chain tensioner adjustment. I put megaphone pipes on it and a sissy bar. The electric system on these bikes turns out to be very weak. It has an electric starter but you usually end up kicking it. I work on my bike along with Kevin's 450 in the basement area that I improved. Yeah I wheel motorcycles through the house; we are all out of control.

Mom hasn't ever been separated from Dad longer than two weeks going on twenty years now, about the same length of time she has been separated from her Mom, Dad and brothers. One weekend home Kathy tells me that Mom is going through Menopause; that explains the mood swings. Mom is frustrated, she can't make us do much of anything without the threat of use of Dad. Kathleen was a big help to Mom and now she's ineffective. Mom has been depending on me since she put us in the ditch an hour out of Rome. She went crazy and had a crying fit a few days ago because no one wanted to go to a spaghetti dinner at Divine Redeemer. She had told her friends that she was bringing the whole clan. Her presumption was not well received and she had no takers. As it is we never have dinner together like we used too. Mom has discovered Swanson TV dinners and TV trays. We cook our own dinners and eat it in front of the TV. There are stacks of them in the big freezer in the basement. My friends come over just for that, Friday nights for Creature Features.

Jim has also been a problem for Mom. It's a good thing that Kathleen is busy, those two hate each other. With her gone most of the time to school and her job the hostility around here has dropped off 20dB. She never comes home unless it's to sleep. She has escaped permanent babysitting duty and is never going back. Kathleen and Jim have a knock down drag out fight that I witness. He was running slot cars and she was trying to watch TV. She told him to stop and he snarled back with invectives and curses then turned back to his slot cars. Kathleen leapt off the couch, snatched up the figure eight track, swung it around and smashed Jim in the face with it before he could react. React he does. He goes WWE on her picking her up over his head with both arms, spinning around and smashing her down on a coffee table, which splinters. Kathleen recovers, getting up screaming and cursing as they face off. Stephen, always ready to assist Kathleen against Jim in every way short of actual help, hands her a steel Zither that she slings at Jims head. Jim ducks the three pound instrument. It shatters a sliding glass door behind him. Ok now I got 'a break it up, they are making work for me. As I get between the cursing combatants Mom comes down the stairs, she is overwhelmed, starts crying, I tell her that I'll take care of it and she goes back upstairs after trying to shame the shameless. Nether

Jim or Kathleen are ashamed in the least about what transpired and would gladly go back at it if I wasn't there standing between them. Kathleen finally goes back upstairs to her room, she's got' ta be hurting. Jims face has a welt from his left ear to his nose which is bleeding, and cuts across his eyes. I get him to help me cleanup and temporally repair the door. Giving him something to do takes the edge off his fury.

Just this spring Jim spilled something sticky on the kitchen floor and didn't clean it up. Mom asked Jim if he spilled it. He lied and said that he didn't. At least three kids saw him do it and ratted him out. Mom was making the liar get down on his hands and knees, on the floor, and mop the spill up with a sponge. He was doing a half assed job. Mom pointed out that he was doing a poor job and didn't get all the splatter. Jim started saying things that you should never say to your Mother. Mom lost it and went to kick his ass, conveniently on the floor pointed at her. Jim sees it coming. He sweeps his arm back behind him and under her leg, flipping her on her ass. Mom makes the mistake of sticking her arm out to break her fall and breaks it instead. She snapped the bones of her forearm like dry twigs. I have never seen my Mom hurt before and it pains me greatly to see her suffer. It's all I can do to keep from attacking Jim but that would not help Jim or my Mother. I use a magazine and an Ace bandage for a splint then take her to the dispensary in her Buick, she can't get in the truck with one arm. Seems like I have done this before. No one ever tells Dad. The cast is off way before Dad comes home. Mom knows that it would be the end of Jim, he'd be thrown out on the street where there would be wailing and gnashing of teeth. Jim is sorry he broke his mother's arm and tells her so, she is his shield, who always sees the good in him and protects him from Dad.

Jim wants to drive. He turned seventeen in March, He keeps taking Mom's Buick on joyrides.by himself. He's supposed to have a licensed driver with him when he's driving; he only has a learners permit. One day he steals the Buick after school and we don't see him for three days. I think that the only reason he came home was that he was low on gas. When he does return, it's with a smashed taillight and bumper on the Buick; Mom's Special isn't so special anymore. I later find out from my buds that Jim got drunk using my draft card and backed into a light pole in a Pizza Hut parking lot, trying to impress some girl. He headed for the hills, slept it off in the car and hung out, feeling sorry for himself till he got hungry and had the balls to come home and face the music. Mom and Insurance cover the damage and Jim's ass. Dad never hears of this either. Not satisfied with that, he moves over to the Sub after I get my bike, starts it in gear, in the carport, in reverse, and takes out the left rear tail light and dents in the end of the quarter panel around the lens. More work for me. I get a new lens and install it, intending to fix the dent later.

I'm no angel either. John and I have been after these two girls all spring. John has the hots for Sue who wears a mean mini skirt and has legs for it and I am crazy about Sandy one of the Pom-Pom girls. Sandy is a beautiful brown eyed Greek who is full of fun, she knows that I am mad for her and teases me about that with her exaggerated flirting. The four of us decide to leave school early on Friday, drive to Denver and ride the old wood roller coaster at Elitch's before it's torn down. I bring the truck and gas to the party John takes care of everything else. I pick up John and a large cooler then we swing by and pick up the girls. John has thought of everything. He has beer, champagne, sodas chips and sandwich stuff. Sandy climbs in front with me, she wants to shift for me. Making jokes about playing with my stick she cracks herself up. We are off to a jolly start. We are on the highway headed north, John cracks some champagne and hands it out. We toast our little adventure. We knock off the bottle, Sandy gets a little goofy after a while; she starts poking and tickling me laughing when she makes me jump. Before I know it she is in my lap, knocking the car out of gear in the process. The truck is all over the road. We are all laughing; that is until I see flashing lights in the rearview mirror. The trooper walks up and asks for IDs, the girls don't have theirs, John and I hand ours over. The trooper looks at John and says "I know you; wasn't your Dad the state Alcoholic Beverage Commissioner?" After John answers in the affirmative he looks us over real close then says "I knew your Dad he was a good man", then he tells us we can go. We are pretty sober now. John tells us that his Dad used to get him and his brother Charlie to buy beer and then his Dad would bust the vendor for selling to minors. We ride the rides till dusk then we hit the road, John wants to take the Rampart Range Road back to the Springs. We head up to Evergreen, gas up and hit the dirt road through the mountains south. Driving on that road at night with no moon is like riding a rollercoaster in a tunnel. It's midnight. We are all tired but only half way there. John suggests that we pull over and camp. Something tells me that this was John's plan all along. The girls are wise to him. They tell us to go start a fire, its cold out; they will wait in the car until we get it started. We get out, unload the cooler and get busy. We get a fire going. The girls throw out a ground cloth and two sleeping bags then we hear the sound of the door locks going home. They laugh and point at us with our sad faces in the fire light, we at least have the food and the beer. After cleaning out the cooler we say good night to our hearts desire, a kiss through a cracked window and turn in. The next day we slowly make our way back to the Springs. We are all in big trouble. Our parents have no Idea where we went or where we are now. They must be going crazy. John tells me that we should buy the girls breakfast and talk over what we should do. After breakfast the girls tell us that we should just leave. They will call their parents to come and pick them up at the dinner; that will give them a chance to defuse the situation. John and I get lost. We take the truck to a motel that his Mom owns and where he lives. His Dad has a garage and a collection of Edsels in the back. We decide to hide out there and do some body work on the truck where Jim bashed in the tail light. That should take our mind off things and occupy us for the better part of the day; fun with Bondo and sandpaper. We are concerned that the cops may be after us for abduction and want to lay low. We are really counting on the girls. The story is, we got lost in the mountains, didn't know where we were and we're running out of gas so we decided to pull over and wait for daylight; which is true, John was the only one who knew where we were. John's Mom discovers us just as we are finishing up; she tells me to call my Mom.

Sandy's Dad will probably never let me near her again. I need to get out of town. I am going to use my graduation money to ride my bike back to Rome the day after I walk. I have been talking to Rick Wyatt on the phone. His Dad is stationed at Richards-Gebaur, Kansas City Missouri. Richards-Gebaur is where F 102s go to die. Rick's Dad is an electrical wiz, a flight engineer, a Chief Warrant Officer who I have a lot of respect for. He ended up there when RADC was disbanded. I don't need a map. Mom and I drove there in the Buick, in a road closing prairie snowstorm one weekend last fall to attend his sister Margret's wedding. That's when Rick and I hatched our plan. He bought a Honda 350 too, an older model like the one Elvis rode in Roustabout. Rick's Dad has been going over his bike with a fine tooth comb. Kansas City, at 600 miles is a third of the way to Rome. This will be a good start. It took two days to drive from Rome to the Springs; I don't know if I can do that on a bike. The longest trip I've made to date has been a 120 mile trip into the mountains with Kevin McNamara and I broke down. Kevin brought me back to town the last 30 miles on the back of his bike. I managed to get my bike running after going back for it with some tools and a new battery. The sissy bar will be good for tying my stuff to. I can also tie a roll to the handle bar cross piece like Then Came Bronson. I will have to pack light, that's for sure. I skip prom but walk across the stage and get a roll of paper with a ribbon around it, they will mail the real one later when they are sure I passed and owe no library fees. I'm not waiting around. In the morning I pull a pair of painter's overalls over my jeans, tee shirt and long sleeve flannel shirt. I grab my heavy leather rough out jean jacket and the helmet I painted to match the bike. It's white with a blue skunk stripe and reflective American flags on both sides. The bike is packed and I said my goodbyes last night. It's a long way to go and it's the crack of dawn so I head out.

The weather is great. I take Hwy 24 to Interstate 70 with no changes until Kansas City. The speed limit in Colorado is 65 mph which the bike seems to accept ok. I'm running knobbies, which tends to make it squirm in a turn and the gear ratio on the scrambler model is a little on the low side compared to the street version of this bike. Things change when I get to Kansas where the speed limit is 80 mph which of course every one exceeds. Eighty is asking a lot from the bike but I have to pick it up to keep from getting run over. I have no faring so I have to lay on the tank and draft behind tractor trailer rigs to keep up. After more than ten hours I'm there, exhausted, suffering from mild hypothermia and there is a buzz in my tail that won't quit. Rick's Mom and Dad feed me and let me pass out. With the heath of an eighteen-year-old I'm up and at'em again the next day.

Rick want's to check out my bike and go riding of course. My bike is pretty tired and needs maintenance. Rick also has an Olds 88 coupe that we go tearing around in instead; he has a lead foot just like his sister and his older brother Mark who drive a two door Ford Falcon and a MG-B respectively. We go tearing around and talk about our options now that we have graduated. It's a big help to talk it out with someone who is in the same boat. Rick is determined to return to Rome. The Wyatts were stationed there a lot longer than we were so his roots are much deeper. Margret and his brother Mark are up there so he has a place to land. I would be alone and overextended. I would have to mooch off friends. I jealously guard the illusion of my self-reliance so I could never do that, too proud. There are no prospects for me there; I've few resources to enable me to endure once I got there. It's a long ride on a bike given my recent experience. I share that with Rick. I can't even be sure that my family would still be in Colorado after the end of the summer when Dad's Kwan Ju posting is up. Cindie, my true love, has moved on for more than six months now. I may be in some boot camp this fall. I'll enlist in the Navy, take some tests and see what happens, before I'm drafted, everything keys on the lottery in four weeks or so. Rick and Mr. Wyatt talk me into hanging around for a while instead of continuing on right away. He helps Rick and I work on our bikes while we talk it over. Turns out I need a new cam chain tensioner and chain plus a full tune up. The parts eat into my gas money. The Moms have been talking, they want to pour water on the idea. I hang out for a week. Mom needs me back and finally guilt's me into heading home to help her.

I head back the way I came after having breakfast with my hosts. I will miss them. Rick will call me if and when he gets established in Rome. His lottery is next year. The trip back is good, the bike is running great. I don't push as hard as I did on the way out. I misjudge my fuel stops because of a headwind and run out of gas 6 miles short of Simla. There's no fuel gage on the bike. I push my bike to the nearest station because I don't want to leave it on the side of the road with all my stuff tied to it. It's late in the afternoon and the traffic is very light. I strip off most of my layers during this hike which is a continuous uphill grade when you are headed west on 24. I am pretty pooped when I finally get to a gas station and have to rest, my legs are shaking; I'm very thirsty and need to cool down and get something to eat. This evolution cost's me a lot of time. After a large coke and a couple of candy bars I perk right up. The sun is dropping behind Pikes Peak before I roll through the carport and park in the yard where I usually do leaving the carport for the cars. The Buick is missing, someone is out. The megaphones on the bike announce my arrival.

Mom is real glad to see me. Tim and Steve have had a spat and Tim has moved out to the trailer. Tim is real popular at middle school, he sings in the choir, plays the flute and has a lot of friends mostly girls. They think he is cute and funny. He and his friend John Schable have taken to partying in the trailer since I left. The next night I catch the both of them staggering drunk. I make Tim walk around the block; its two o'clock in the morning, he has to stop every once and a while to puke. At least he isn't puking in the trailer.

Bill too decides that the corner room is too small to share with Tim, Steve, and Gary, The politics are working out where he is the low boy on the totem pole, Steve babies' Gary who shares his bed. Tim is wittier and older than Bill. Steve is fastidious and Bill is a slob who doesn't want to take orders. Tim usually goes along with Steve. Bill wants a room of his own. The boys have trundle beds; you have to pull out the wheeled lower one to go to bed. The room is so small that four beds abreast span the whole room and leave little room to maneuver or put other stuff. The basement is out, Bill hates it in the basement. Jim likes to lock Bill in and turn out the lights in there and make him cry whenever Jim catches him or anyone else in there. There are no windows in the basement; it's pitch black. Terrifying for poor Bill.

There is a large closet in the family room, just about the length of the boy's room that it shares a party wall with. The closet is deep and has some sturdy sliding solid wood doors. We empty the closet and put what was in it in the basement, a couple of trunks and some unused clothes. We have an old two piece Japanese clothes chest in the family room that we help him slide in the closet. We place some boards and a twin mattress on top of the chests. Bill can keep some of his clothes in the chests and hang the rest at the foot of the bed. Bill has the upper shelf for his other stuff. Sheets and blankets go on, then Bill crawls in and shuts the door. It's a couple weeks before mom knows that Bill has moved into a closet but by then he is well established and much happier, she leaves him be. Makes me wish I had thought of it first.

Kathleen talked me into taking some hours at Craigmore; it's close and cheap only $50.00 a class. I am late signing up, waiting on my SAT scores for admission; they arrive with my diploma while I am headed east. Course selection is limited I could only get an opening for Biology 101, three hours of class and one hour of lab. Something no one else wants. I'm really not interested in Biology but I sign up anyway, mostly to be able to tell the Old man what my plan is, which he is sure to ask. What I want to know is what his plan is.

1970 KWAN JU KOREA

## The Last Posting

We are all waiting to hear what Dad's next and probably last posting will be. Most of us like where we are and hope to stay put. We have a house here and have just made new friends. We are barely settled in from our last move. Not Mom; she says that Dad promised her that they would end up back in LA with friends and family when he retired. Dad has a lot of choices/input in the way of requests concerning his last assignment so hopes are high. Hopes are soon dashed. Dad has picked Dyess AFB Abilene Texas, 45 miles from Coleman, where he started flying. Crap, why there of all places? He later tells us his reasoning "The flying and hunting is good there and people are real friendly." We will not see Dad till mid-August. Mom has less than five weeks to buy a house in Abilene and move all of our stuff down there. Dad has already lined up someone who wants to rent the house; it has to be empty and painted before the last week of August so they can move in. The trailer has to be ready to roll too. I am to help Mom and paint the house for a few bucks while I am going to school.

Wasting no time I load Mom in the Buick; just the two of us hit the road to Texas, Kathleen is stuck with the rest of the tribe. Mom and I leave early in the morning and make Abilene in one day staying in a motel for the night. We are staying on the north end of town. Abilene Christian Collage is the first thing that we see in the morning when we start driving around and as luck would have it there is a relators office right across the street from the campus. We pull up. The realtor is a real nice lady with the thickest drawl I have ever heard. She drives us all over town briefing us on the different neighborhoods the north and south side. They like the Air Force in Abilene; the base is located on the west side of town. The town is split north and south by the railroad that runs through the center of town. There is a great rivalry between Abilene and Cooper high, the north and south of the tracks high schools respectively. Football is very big in Texas we are told and Cooper was State Champion last season. I notice right off that you see almost no one outside walking around. It is hot and humid here, working under the dash to replace the signal flasher and checking under the hood has left my tee shirt sopping wet. As it turns out Mom likes a place just a few blocks from where we started, it's off Washington Blvd, in Radford hills, a fancy neighborhood where a bunch of ranchers and oilmen live. Billy Sol Estes lives nearby on Lytle Lake, actually a manmade tank, his daughter Joey eventually gets a house next door. The house is a split level 3000 square foot 4 bedroom 3 bath, red brick house with white iron work accents on a corner lot. It has plenty of room for Mom's stuff. There are no other houses on the block, across the street and back ally. We will be the only Catholics in a church of Christ neighborhood. Mom makes a cash offer around 32K that is accepted; the deal is done in short order and we leave for Colorado the next day with the keys. The kids ask my assessment when I return. I tell them it's flat, red dirt country, humid and hot, oh yeah there are these large fly looking insects that make a constant annoying noise in the trees called cicadas. I see very little joy in their faces. As for me, I am going to try everything I can to stay in Colorado. The lottery will be held in two weeks then I might have some certainty. Until then I have my friends my classes and my chores. I have the rest of the summer to find some sort of work and a place to stay until I am called up or enlist.

John Carlisle and I go to Abilene in the Buick to open up the house. We need to get the utilities turned on, clean it up, paint, and make some minor repairs. The carport/garage has no doors so we have to buy and install some so Mom can secure stuff in there. We leave late in the afternoon because we want to drive through Texas at night when it's not so hot, my idea. When we get to the first fuel stop, John jumps out to hit the head. I am still pumping gas when he comes back practically beaming, I'm thinking he must have really had to go bad but instead he says "Bob, They sell beer in the gas stations here and it's 6.0!". It turns out that Texans think it's a God Given Rite to have a cold one on the way home after a hot day working in the sun when you only have a two fifty five air conditioner in your pickup truck; that is two windows and 55mph. We drive on but get hungry south of Lubbock and pull into what looks like a local dive, there are a lot of pickups parked out front so it must be the place to be. The joint has a red sign outside that says Steak and Beer, what else do you need to know? We park and go inside. I open the door and John goes past me then dammed near busts his ass slipping on the floor. Good thing he can skate. After he recovers he says "what's this shit on the floor?" It looks like sawdust to me, everyone in the place looks us over when John blurts it out. And there we stand, wearing cutoff jeans, tee shirts and climbing boots, our hair is long and would be way over our collars if we had one. We are both wearing tinted wire frame glasses, round John Lennon glasses for John and rimless 1920's grandma wireframes for me. Hippies. After a while they go back to what they were doing and we get a table. Most people in Colorado don't like Texans who come up and buy property in the mountains with their oil money or are a lot of the tourists that folks who run the motels and the tourist traps depend on. There was one kid at Palmer who transferred from west Texas. He joined the football team of course and was always bragging about Texas and Texas football. Everyone teased him mercilessly in return, mocking him and giving it right back. I kind of liked him because he took it good naturedly. He was a good player. Everyone in this place is dressed up to the nines western style with big hair, big hats, boots and jeans. They have big belts with their name on the back, secured with even bigger buckles. The waitress tells us that there is going to be some boot scooting tonight in an hour or so, after the band, Tony Douglas and the Shrimpers, finishes setting up. The steak and fries are great, the people are friendly, it would be cool to check out the band and the dancing but we have somewhere to go. We leave. It's a full moon. I turn off the headlights. I floor it and peg the needle. We ae heading south and downhill fast to Abilene.

When we return, I watch the lottery on TV. My number comes up 362; they match your birthday with a number pulled from two separate drums. They will call up almost every other eighteen-year-old in the country born that year before they get to me. I was certain that I would have to go, now I have options. I haven't even thought about that.

Mom really blows it when Dad comes home. She is late for everything; she runs on Franny time. She knows when he is to arrive and had planned for all of us to meet him on the tarmac but couldn't get it together. We are still milling around in the carport loading into the sub waiting on her when he rolls up in the back seat of a staff car in uniform. Dad is very disappointed in the lot of us and lets us know it. This sets the tone for his return. We get a dressing down instead of hugs and kisses.

Dad is unhappy with me in particular, the way I look and dress, my bike, the condition of his truck, the Buick what I have been doing with my time and the house in general. We have not been keeping the log books for the rolling stock up to date, gas mileage, oil changes, repairs etc. He is unhappy with the Bondo and primer job I did on the rear quarter panel of his truck. He thinks that the steering in the truck is sloppy too. I tell him about my recent tune-up and the repair job on the steering column turn signal indicator switch, you have to pull the steering wheel for that. I tighten up the worm gear in the steering box taking out some of the slop, but he is still not happy. I tell him that Jim has been the one driving the truck, I have my bike and I think that the truck just needs new tires and shocks. That is backtalk and not well received even if true, besides it will cost money. It is getting close to that day Dad has been telling me about. That day is: "Plate Breaking Day", that's the day when they break your plate, you have to go out and get your own, and what goes on it from now on. I get the tell, along with the Bum, and Digging a Ditch predictions for my future. I'm told that my services are no longer required. I split with everything I own or want in a duffle tied to my sissy bar bag and rolled up in an old army poncho, bungeed to my bike's handlebars. I don't need this shit. Things are tough enough already, life's a bitch for sixty years, if you struggle and are lucky to live that long, then you die. Mom is upset with this turn but everyone's plate is full for the present and no one worries about this stray dog. Decades later I find out that Dad had sent me a very nice letter thanking me for what I had done in his absence, offering his support, and asking me to accomplish a detailed list of things before he came home. Dad had put the letter in with a package he sent to Mom as packing to "save postage" thus the misunderstanding. I never got it.

 Dad hitches up the trailer and heads south with Tim, and Steve along with the bunnies Bluebell, Raquel, Jack Stewart, Sugar and Earnest T Bass. Its hot crossing the Llano Escondido, the land of the Comanche, the air conditioner is not keeping up with the heat load, there is general concern for the bunnies as they cross the Canadian

River, a muddy red delta. Tim and Steve are gob smacked, they just can't believe Dad actually chose this; they are trading the beautiful, cool green mountains of Colorado for the hot, flat red dirt, arroyos, cactus and Mesquite of the high plains for a last posting. Odds are they will be stuck here, the rest of their lives; they're all depressed no matter how much Dad keeps, pumping them up, telling them how great it will be living in Texas. He repeats: "The flying is great, so's the hunting, and the people are real nice."

Mom follows on two days after in the Buick with Jim, Maureen, Bill, Gary and Midget, after the Mayflower leaves. Dad left early to report for duty, he has used up most of his leave attending to his parents in LA. Grandpa is not doing so well, he is taking care of Grandma. Grandpa Scanlon Passes November 21, 1971, God rest his soul. Mom, Jim, Kathleen and I take care of packing and cleaning up before they leave. I get some money to paint the house, Kathleen is camping on the floor with a friend until the dorm at CU Bolder opens up. I'm sleeping on the floor too. Kathleen has achieved her dream. I have no idea what to do...

I had been hanging around more and more with my best friend John Carlisle. His Mom has recently purchased a big house a few blocks from ours so they no longer live at their motel, the Glen Russ. She Charlie, Mary and John have moved there, older brother Bob lives in Boulder where he graduated and now teaches. Bob is an accomplished climber and is in Rocky Mountain Rescue. They all know me, my situation, and that things are rough for me at home. John offers me a spare room while we do some renovations on his Mom's house. I take up their offer. I guess they feel sorry for me. I can't decide if I am a moderating or instigating factor in John's behavior. Certainly the stuff we do is best done with a buddy like climbing in Cheyanne Canyon or four-wheeling on the Old Gold

Camp road. The hill climbing and the woodsy parties. We also get into a lot of mischief best not recounted here. John and his family are car mavens like me so I'm right at home. We fetch some old cars that John's Dad had stashed here and there in the mountains, which always requires two drivers, experimental and chase. Driving those old Edsel's and Buicks, late in the afternoon, that we have just coaxed into barely running late that morning down Ute Pass is a pretty iffy proposition. Those old heavy cars with their sloppy steering are hard to keep on the road much less in your lane. Charlie buys a Spitfire GT hardtop a cool car that looks like a miniature XKE and Mary buys a Road Runner Super Bird, the one with the big spoiler. Not to be out done John sells his 48 Willis and buys a Spitfire that we go roaring around town in. John schools me on how to light a cigarette in a convertible and pull in front of oncoming traffic when making a left. You can cover 15 feet doing twenty before the oncoming traffic can go 100 feet doing fifty. Yikes!

I'm in John's backyard playing with Duke, John's one year old German Shepard. John is in the house on the phone. John popped out and asked if he can take my bike for a spin. Sure I tell him. When he comes back he has Sandy on the back; I haven't seen her since our "trip" to Denver. I get a big hug and a kiss, I thought that I'd never see her again. I've been rather dour lately but she cheers me right up. Sandy is bubbly and all smiles; she is wearing shorts, a tank top and tennis shoes, God she looks great. Sandy has snuck away and has to get back to the house before she is missed. We make plans to meet at Giuseppe's, our high school hangout pizza joint later on. I am over the moon. John and Sandy hop back on my bike and head for her house. Unfortunately a woman in her station wagon on a residential street backs out of her drive in front of them and John has to lay down my bike to keep from hitting her broadside.

Sandy gets a road rash on her pretty legs and shoulders. The pipes on my bike get scuffed and the levers are bent John rolls down the street and gets a little road rash. The Cops are called and John gets a ticket. Sandy's Dad comes and picks her up. that's the last straw for me. I'm thinking restraining order for John and I. John fight's the ticket; his Dad told him to always do that the cop who wrote the ticket may not show up. He has already beat several tickets. I go with John to traffic court. The woman driving the station wagon testifies that "she couldn't see the bike coming because John didn't have the headlight on." How did you know that asks the judge? Case dismissed but that doesn't help my case with Sandy's Dad. I'm the jerk with the motorcycle who kidnaped his daughter.

John's older brother Bob gets engaged. He is going to marry a nice girl from La Junta. John and I jump in the spitfire and head South East. We have a great time I can't remember much other than drinking beer, decorating the grooms room and car with condoms and cruising around sugar beet fields with John's new in laws in a cool hopped up 52 Chevy Coupe.

John and I have a beer with Charlie and some of his friends back from Nam on leave. They went to Palmer too. They are Huey door gunners in the first Cav. They are on leave because they re-upped for the money and will have to return. They are gun-ho, it's not too bad, things are winding down, a big adventure. Lots of stories. They can't see what they are shooting at, they just hose down the tree line. Killing jungle, then you fly back to camp for a beer. Another day at the office. I'm tempted.

Charlie gets me a part time job for when I 'm not in class, on his framing crew, where I learn a lot of useful practical stuff like how to frame a stick built house from plans. Charlie is the lead framer and layout man. Mostly the job is humping lumber and sheathing for the real carpenters. Work slows down. I get laid off by the foreman with the words "you're too smart to be doing this sort of work". I don't know what he means by that, what I should do for work or why I'm still hanging around imposing on my friend. There is nothing to do but throw in the towel and head to Texas, that or join up; I'll try Texas first and see how much grief I get. Maybe Dad has cooled off and Mom has been giving him a guilt trip for running me off.

I stop and get gas in Trinidad, my last stop in Colorado. It's late summer but early in the morning, I am chilled already, especially my hands, just hoping to get over Raton pass and into some warmer weather. I rig a string to the throttle, winding it around the handle and tying it to the mirror stem, so I can occasionally put my hands in my pockets. Its 600 miles to where I'm going. The long stretch to Texline is a dangerous two lane road with high crosswinds from the south that has me swerving and leaning over to one side a lot. Halfway there just south of Amarillo, it's warm enough, must be over 70, but I am hypothermic, my usual layers of cotton and leather jacket do not hack it, the wind strips the heat right out of you. My heavy split hide jean jacket is not very warm. I pull over at a rest area, park next to a concrete picnic table, climb up on it and lay flat on my back in the Texas sun to bake. I fall asleep, for I don't know how long, without taking any layers, my helmet, or bug splattered face shield off. I wake up to a blue bottle fly buzzing under my face shield, attracted no doubt by the dead bugs on it and the sweat dripping from my nose now. Warm enough, that's the Texas I know. Time to go. I shed a couple of layers and I am on my way again.

I roll into Abilene late in the afternoon I haven't been pushing my bike very hard for the last few hours, I'm worried about it and glad that we both made it here at the same time. I am exhausted hot, hungry and anxious. I don't know how I will be received.

# 1971 Dyess AFB, Abilene, Texas

**JJS.** I finally return to Texas, where I started. Coleman is just a short hop SE of here. The area is dotted with little airfields everywhere where a pilot in trouble can let down. I am assigned to Dyess AFB as Chief of Operations and Training Division concurrent with the responsibility for the Operations of the 96th Bomb Wing. Out of TAC back in SAC. I fly the Convair T-29 for my remaining four years of service; then deliver that airframe, the last radial engine aircraft using 100-octane gas in the inventory, to the boneyard. Dyess has a mix of aircraft. There are: B-52s, KC-135s, C-130s, C-123s and of course my T-29.

**RES.** So I roll up into the driveway my megaphones making a racket. Motorcycle riders can't help revving their bikes. A lot of them want to quit running or don't idle well. A habit. To my relief everyone is glad to see me, even Dad, who wants me to go with him to get burgers. Seems he has discovered a great burger joint called

## 1971 DYESS AFB, ABILENE, TEXAS

Whataburger, he tells me "it's the tallest building in Abilene", ha ha. I am sure that he wants to talk too. On the way there and back I tell him that my plans to stay in Colorado and go to Craigmore had failed and told him about getting laid off. I also tell him that I don't think it's a good idea to go into the military during a draw down. When we get back to the house and are standing in the garage about to go in I tell him that I want to be an engineer. This elicits a big laugh, which doesn't help. He was thinking Culinary School which I had not even considered. I like to cook, but I cannot see doing that for a living. Dad has a lot of respect for engineers, they are the smart guys like his brother, which doesn't translate to me. Engineers are the people who came up with airplanes in the first place and are the ones who now make them better and keep them in the air. He has worked with some very good ones. He says he will try to help, but he won't pay for a degree in flower arraigning or naval gazing.

The parents have been thinking of me and tell me that if I want to go to school I can live at the house and commute to a Junior college just down the road, take a few courses and see how I do. Bets are, I will fail miserably and losses will be minimized. The school is Cisco Junior College, basically an Ag school 45 miles east of Abilene. I take up their offer.

It's been a month or more since the move. Jim had smashed Tim's head against the mantel and cauliflowered his ear not long after they got here. Tim had to be stopped from taking him out with a claw hammer. Jim has also beat up Steve and Bill. School can't start soon enough. I have to have a word or two with Jim. We are going to be roommates again, he won't like that, I'm sure that he thought he was rid of me.

Having little to do while waiting to start school I burn out a dirt track for my bike in the adjacent five empty lots on our block.

The terrain has a natural uphill rise and is pretty lumpy making some tight turns with natural jumps, and dips that I dig deeper spinning the rear wheel of my bike. You can get some air in one spot if you're aggressive. The run toward the house is a series of woop-dee-doos. I am having too much fun giving my brothers a show. Jim wants to try. He is clumsy and gets off to a slow shaky start, he makes the first turn and almost kills it ramping up to the jump as he comes around and hits the heavy wash board his feet come off the pegs. They come off because he is sitting on the seat and not standing on the pegs like you should. He slips backwards on the seat twisting the throttle at the same time because his grip on the handle bars is the only thing keeping him on the bike. This causes him to miss the last turn. That launches him and the bike onto the 40 foot long front porch of the house. The bike bounces the length of the porch off the walls windows and the wrought iron columns coming to a stop when it, with Jim still aboard, hits the brick wall at the end of the pouch. Taking out a window, plus my handlebars and mirrors slowed him down enough so when he hit the wall he was not going too fast. Jim skinned up his left arm, hand, and knee on the brickwork, he jammed some fingers on his right hand and banged his foot on a column hard enough to knock his shoe off. Jim is real lucky, good thing he had my helmet on when he hit the wall. It has some deep redbrick cuts and dust on the left temple. I am pissed. I want to beat him up but I couldn't do a better job than he just did to himself. Jim is going to have to pay or beg for the money to fix my bike. Dad is at work. Mom is concerned for Jim, she's mad at me and my bike. She doesn't like it that I have been giving Maureen and Gary rides around town either. I wheel the wreck around to the garage and get to work. Jim goes inside with Mom to clean up. I haven't even been here a week on good graces, I don't need this. The bike has to be fixed if I am going to ride it to class next week.

I spend what money I have left fixing my bike, bending most of the parts like levers peddles and shafts back into shape where I can, and hand Jim a bill he never repays. School is good. Cisco is a complete campus in the middle of nowhere with dorms bands and athletic teams. They have a girls drill team that does high kicks in little cowgirl outfits with high white boots, the Cisco Bells. I end up dating Lisa, one of the Bells, she is the cute little one on the end, she is pretty, petite and chatty, reminds me of Mom. The Bells and the band are a big deal, they get invited to march in Macy's Thanksgiving Parade, in New York City. Campus is pumped up about it, that's forty Bells and seventy-five Band members about ten percent of the student body and twenty percent of the residents. Almost everyone on campus is here on some kind of scholarship or another, AG, band, Bell or athletic. I am the only one on campus with a bike,

Classes aren't too bad. I am taking Analytic Geometry, English Literature and American History. I get a nickname, Colorado, I must be whining a lot to my classmates. In October I start a temporary job at Sears in men's wear for the gift giving season. Mom got me the job, she is friends with the wife of the general manager, who she met at church.

Commuting is ok until it starts getting cold. A ninety mile round trip on a bike in bad weather is pretty demoralizing and not conducive to study. More conductive to napping when you get somewhere warm. Plus some cowboys in a pickup ran me off the road once just for fun. I haven't had a haircut for a year and a half. Dad wasn't around to cut it or say anything about it so I look like a Hippy. I occasionally borrow Mom's car when it's raining or real cold, but that can't continue. I also have to wear nice clothes and a tie at work. To look nice, I have to change at work when I ride my bike helmet hair and all.

I find an ad in the paper for a 67 MG-B at a great price. I call the guy. He has a workers comp disability claim in for a back injury he suffered in a fall at work and can't be seen climbing in and out of and driving a sports car. He offers me a good deal because he needs the money and is motivated. The car also needs new batteries. I tell Mom about it, she is onboard, she never liked me riding around on a motorcycle. Jim's foray down the porch and seeing me giving Maureen and Gary rides around town only intensifies her concern. Dad is of the same opinion although he questions the choice of vehicle which puzzles me since he had one himself. He says that they are unreliable and require a lot of maintenance. I counter that it will get good gas mileage and it is pretty cheap. I sell my bike to an airman and Dad chips in some additional funds for the car and money for new batteries and cables. He handles the registration and insurance. The car takes two 6 volt batteries hooked up in series behind the seats all in all a poor DC electric setup where you want your cables as short as possible but it makes for good weight distribution.

I love this car. I am always tinkering with it and soon get it running like a top. It came with a shop manual in the trunk which helps a lot. I should have taken the hint then as I had to refer to the book quite often. MG's need constant attention. Dad sees me tinkering and confesses that he couldn't keep his TC running and expresses some doubt that I can do any better. Besides he says "you can kill yourself lighting a cigarette in that thing". Sounds like hard won experience; but I don't smoke. One thing I have learned about Dad is, unlike Grandpa, he and tools don't mix; Pop can just look at a pair of pliers and he'll start bleeding. Most of his fixes involve tennis balls. Dad has solved his A/C problem in his truck by hanging a broom stick behind the front seat, draping some clear plastic over the broom stick, then securing the plastic sheet in place with jumbo binder clips he got at the office. He at least, stays real cool.

Before you know it its Christmas break and I lose my temporary job of course. Mom has come through for me and landed a job on base working for the BX. She must have leveraged my bag boy experience to get it. My job is to clean up after a rodent infestation in the building and the subsequent poisoning of the invading vermin. After the critters were poisoned they crawled under the display shelves and expired. The whole place stinks of death. My job, working alone, is to one row at a time, pull all the merchandise off the shelves, completely teardown the shelves, collect the dead rodents, disinfect the shelves and floor then mop and wax the floor, clean and reassemble the shelves, then restock them. Great job. I need the money. From ties to flies.

## Spring 1972

I have registered for Physics, Calculus, and English Composition for the spring semester. That's ten hours because there is a physics lab. The plan is to keep working and commuting knocking off some basic requisite courses. Courses are thirty bucks an hour, I'm working and buying my own gas, so this is below the parent's pain level. I have an application in with the BX should a position open up and there is a good chance of that happening. Things are going well until one day in February when I catch a chunk of torch cut steel in my radiator. It bounced off some welder's truck I was behind, just short of Cisco one morning. I limp on stopping where I can and adding water that just leaks out. I have to keep the radiator cap off.

The car overheats and warps the head as I climb the hill to campus and park it. Crap that hole in the radiator knocked a hole in my long term plans and let all hope leak out.

I missed my first class but catch the other two. It's hard to concentrate. I don't want to think of tomorrow and I am dreading the Mayday call I have to make. I've dug a huge hole of promises that I can never fill. I want to go to school and was going to pay for it by working, Dad said he didn't have the funds to send me to school but if I paid for my books and tuition, he would provide room and board at his house.

I make the call and talk to Mom. She calms me down and tells me to call back in two hours. When I call back Mom tells me to stay put and not worry, she is sending Jim with all my stuff. Mom has checked me into a dorm, Cluck Hall, the men's Academic Dorm. Mom has worked a deal; she has talked them way down since the semester is well under way. I hope that Mom hasn't stuck her neck out too far for me; there will be fallout and a reckoning the next time I see Dad to be sure. I walk myself over to the dorm and introduce myself to the resident dorm mother who is there to keep a lid on the academic dorm. The other men's dorm is the athletic dorm. There is only one dorm for girls and it is full of Belles. Lisa is glad that I will be on campus more; we can have meals together in the cafeteria.

Jim shows up and I move into my assigned room. My roommate Don is there, he says he is a Marine, always wearing USMC tee shirts but I find out that he just attended boot camp for high school Jr. ROTC students; he talks a big game. Don is actually a band nerd. Marine band maybe. I'm not impressed. I have a major calculus test a week later right in the middle of all this crap and have to pull a cigarette smoking No-Doze chewing all-nighter. I should have been studying but I've been spending my spare time working on my car. I have pulled the radiator and the head, they are in a couple of local shops for repair. It's a good thing that all my tools were in the trunk and you can take a MG apart with three wrenches and a screwdriver. The repair will cost me fifty bucks, everything I have. Anyway I'm beat and looking forward to crashing in my bunk. I open the door to my room to see that someone has taken everything off our common table and floor, including the ashtray with food sweepings and dumped it on my bed across the room. I spin around to see my roommate siting on his bunk with a smug look on his face. No words, I charge him, he jumps up and heads for the door, I catch him before he clears the door and slam him into the wall across the hall. He is down, stunned, I have my hiking boots on and he is fixing to get the shit kicked out of him when the dorm mother two doors down pops out and stops me. Lucky for Don. She hardly ever comes out, mostly ignoring our goings on, guys also walk around in their underwear or towels, we have a common bathroom, but she surfaces for this. Don is a band nerd turd that is why he had no roommate, no one likes him. The house mother acknowledges her mistake and suggests a change. I agree.

I move in with one of the Hibbits twins, guess they are tired of being roomies. My new roomie is cool. He and his brother were small town high school jocks who are always ready to throw the ball around so we get along fine, and they have a S10 Chevy pickup for hunting. My room turns out to be a favorite haunt for basketball players, they have an understanding with Hibbits. My roomie has refrigerator. The jocks keep it stocked with beer and a carton of cigarettes on top, things that they can't have at their dorm. Hibbits gets a share of the stash. There is a running card or domino game most of the time with lots of smoking and drinking. I start smoking more to be social and to tolerate the atmosphere in my room, Dad used to smoke so it's not a stretch for me as long as it's menthol. I'm not that fond of beer and can't afford it but I'll bum a cigarette. I like the jocks some of them pay me to help with their math homework while they play cards. I'm not in the room too much I am usually on a study date with Lisa, she has a job on campus plus all the Bell activity so the library is as good as it gets until I get the MG fixed.

Living in the dorm I meet a representative of every small town in West Texas from the valley to the panhandle that's a lot of goat ropers. They teach me useful things like how to spotlight coon hunt and rope trash cans standing in the bed of a moving pickup. Besides me there are only three other non-Texans in the dorm, one Vietnamese kid and two Middle Eastern guys. The Vietnamese kid has a room to himself, I have no idea what he is doing here. He and I rebuild a Volkswagen engine in his room for another kid because he wants to see how engines go together and work. I am the local mechanic. One of the Middle Eastern guys is a Jew from Israel and the other is a Muslim Palestinian. The dorm mother in her wisdom put them together since they are from the same place. They are more comfortable with each other than with the Goat Ropers. Walid had been shot in the gut; he has huge scars where they opened him up to save him, medical care being what it was for him. His plan is to marry a large breasted blond American woman and never go back. Walid cracks me up, he could be the poster child for the PLO with his heavy beard, goatee, muscular build and classic Arabic profile.

I go coon hunting with some of the Criminal Justice guys. They seem to know the local area well and where to hunt. Real cowboys. They get to do ride-alongs with the local cops. Some of the guys are missing things, possessions, a few at a time, so I ask them about it. They say they will ask around. Some of my stuff turns up missing too, my class ring, a nice baseball mitt, a bone handled Japanese hunting knife with a dragon on the blade that they saw me skinning coons with. I come back from class to find Deputies inside the dorm. The CJ guys have been busted. The Sherriff finds stolen items stashed in the air-conditioner register in two of their rooms. I do not recover my stuff. Turns out they were running a big theft ring; stealing cattle sprayers and fertilizer tanks to make drugs along with anything else not tied down. They learned the sheriff's routine on the ride-alongs and stole stuff where they knew they would not be that night. They got greedy. They got caught when they robbed the nest and their brothers.

Another cowboy is always bragging about shooting at wetbacks on his Daddy's ranch in Del Rio. A Texas Ranger comes and escorts him away too.

## Paint

Mom has a friend with a problem. Her husband, an Air force Col, Died and their only son, a chemical engineer with a PHD in chemical engineering, committed suicide. He killed himself after sampling his own wares, homemade LSD, but not before driving home from Dallas at a high rate of speed in his split window 63 Vette. The poor widow lives in a large house and has little rental properties all over town. I go to work for her all summer mowing lawns, painting houses, doing light carpentry and whatever she asks me to do including firing up and maintaining the Vette. She can't drive the car but she will never part with it. I paint houses mostly, hoping to earn school money. Pretty mind-numbing all that scraping and painting in the heat; the shadow of the house dictating where I can work, I wonder who has all those jobs now.

Dad gets me a job with Carol C Carter Concrete to show me what he means by digging ditches. I go to work with a bunch of guys that don't speak English. We ae digging foundations for an apartment complex. We dig and clear the trenches for the foundation's beams, set the forms, lay the moisture barrier, set and tie the steel. My fellow workers invite me over for a beer after work. They must have a dozen people in a single motel room. They tell me "no mas rapido tiene streche da job". They say this while placing the fingers of their hands together and pulling them apart. I stay on the site long enough to learn about masonry work, which I am interested in, before I am drafted for the survey crew; we lay out streets, curbs, and sidewalks in Coleman. Before you know it the summer is over.

Jim has landed a part-time job at Sears, his first. I can't believe it, he has never wanted to work before. His love of money has finally put some starch in his lazy bones. He works with this guy, Kevin. Jim tells me that Kevin is coming by and asks if I want to go riding around town with his buds. Keven rolls up to the house in his maroon Aggie colored four door 67 Chevy Impala and honks. We hop in. There's another guy in the car, Stormy. I shake hands with his friends and we are off. Stormy is a tall 6' 3" muscular eighteen year old, with long, frizzy, blond hair that hangs in his eyes and to his shoulders, he has light blue eyes, behind wire rimmed glasses, he's dressed like a rocker. Kevin is about 6'1 lanky with blond hair and bangs, he is the older of the two, about my age, dressed in boots, jeans and tee shirts. He has a couple of semesters at UT Arlington to his credit. Kevin, after about ten minutes of conversation, turns to Stormy and says "see, I told you there had to be a smart one" this pisses Jim off. Stormy laughs, pulls out a joint and light's it up, he offers it to me but I decline, doesn't matter, I get quite a buzz sitting in the car with the windows rolled up. I know because I get real paranoid. Don't know what to think about these boys. Pretend hoodlums, they talk about breaking into drug stores, robbing them and selling the drugs. They are as full of shit as Jim is.

Next weekend I am up in the bedroom, Jim hisses at me from the bathroom and motions to come here. I go in the bathroom; he closes and locks the door. He says "I want to show you something". Jim pulls out a bottle of Darvon capsules. He pulls some capsules apart spilling their contents onto a small plate, he separates out the tiny red spheres from the yellow and white ones with a pocket knife, explaining to me as he does it that the red ones are Codeine. Jim next pulls out his kit, a needle and a spoon. He demonstrates his proficiency with his new found vice. He has done this before. Jim melts the Codeine in tap water heats it in the spoon and draws it into his syringe through a piece of cotton ball he rolled tight between his fingers. Jim ties off with a belt and injects himself. I am amazed. Why did Jim want me to see this? I certainly don't approve, he must be dammed sure I won't rat him out. This is Jim performing ritual suicide; he always said we would be sorry when he was dead and gone, here he is playing with that notion. I cannot see why anyone would enjoy what he is doing. I hate needles after all the poking that I've received, courtesy of the Air Force. I ask him why he doesn't just eat the pills. He says he likes the rush, the real rush he's getting is being a junkie four paces from the Col's bedroom.

I take another ride with the boys just to get out of the house after dinner. On this ride they waste no time in lighting up. We are headed out Buffalo Gap Road where there's a maze of red dirt roads to get lost on. On the edge of town, Kevin pulls off the road and around the back of a dilapidated, deserted, two story frame house. We park and get out, somethings up. We go inside where Stormy lights a candle that's poked in a wine bottle sitting on a rickety, water damaged table. They have scored some coke. We are in a crack house. They all get out their rigs and get after it. Now I'm really paranoid. I'm offered the wonderful opportunity to participate but turn it down telling them someone has to drive. I drive my ass home and get out to leave them to carry on. Maybe they aren't full of shit. I have other plans. I'm going to school in a few days, back to Cisco.

## Fall 1972

It's a Friday early September, the weekend, so I drive my MG home from Cisco early to be with my family. The phone rings at the house, It's Timmy, he is somewhat incoherent, he wants someone to pick him up at Abilene High, not more than 5 minutes away. I go and get my brother. Tim is standing on the curb in front of the main entrance. He is wearing jeans, cowboy boots, a blue chambray button down and Dad's old silk beige and tan rough weave sport coat; the one Dad wore when he was in high school. The coat is draped over Tim's shoulders and he is smoking a cigarette he holds in his left hand. Something is definitely wrong, Timmy is right handed and students are not allowed to smoke on campus. I see that Tim's right arm is in an OD sling When I pull up my top is up and I have to help my one armed little brother roll into my low slung car. We head for the base. Tim tells me what happened to him on the way. One of the Coaches was out sick so upper and lower class kids were combined to make supervision easier. Phys Ed was held indoors at the basketball court. The activity is something called murder ball, indoor smear the queer, rugby with no scrum, constant action. The coach pairs off the seniors against the sophomores. Tim gets the ball and heads down court. An opposing player grabs Timmy's arm trying to rip the ball away snapping his arm above his elbow. Tim is in tremendous pain, he has only had an aspirin from the school nurse and a menthol cigarette to ease his agony. I am very concerned for my now swooning brother. I check him into the base hospital. The primal end of his ulna is totally separated. He spends one week in traction to reduce the swelling then he had orthopedic surgery to reconnect the end of the ulna with an approximately 4.5cm long stainless steel screw. This incident reduced Tim's elbow extension and rotation also affecting the main nerve in his right elbow causing permanent numbness in his arm and hand. It is the same arm that he broke in NY.

## Spring 1973

So I head off to Cisco for school. Maureen's birthday is on the 16th of January in less than a week. I want to get her something nice, something that I know she wants. She wants a pocket watch. I have been looking. One of the guys in the dorm said that he saw some pocket watches in Breckenridge about 25 miles north of here. I jump in my car and head north. Sure enough I find one and the price is not too bad. Unfortunately while I am out Dad calls the dorm asking for me. Some wag exactly who I never find out tells him "He took Lisa to Dallas to catch a concert". The first time I hear that is when I return his call. I can't get a word in edgewise and am ordered to bring the car home. As usual Dad is short with his phone calls and hangs up before I can tell him where I really was. How in the hell am I able to call him back so soon if I'm supposed to be at a concert in Dallas? I make the long drive home and get a lecture that I do not deserve but I bite my tongue. Dad says that the car is a major distraction for me, I am to park the car and Jim will take me back to Cisco. He has me. He knows that I want to go to school, I am enrolled and he's paying; he is blatantly leveraging that to clip my wings. He doesn't like Lisa either, another distraction. Happiness and contentment for me equals distraction. He can't stand it. It is quite unfair to make me choose when there is no sense to it. I just turned nineteen. I mistakenly think that he will cool down and I am just parking the MG at home for a while.

I take one of the Carin Terrier pups back with me to ease the pain. I name him Mackie Macbeth of Cluck Hall. One of the pilots a Col. was transferred from England to Guam and he couldn't take his pair of AKC Cairns with him. Mom and Dad took Tolly Toes of England and Toby Checkers of Essex, Mackie is one of their pups, just weaned. Mackie is a great hit at Cluck Hall all the guys spoil him rotten. There are no complaints; hell some of the guys are raising Chinchillas in their rooms. After a week I get a call, Jim is coming to get the pup. Dad has sold it for $500.00 to some lady, a young officer's wife, who just had to have it. Can't protest, that would more than cover my tuition. I never want or have another pet. I'm cured…

I am wrong. Dad sells the MG for a lot more than I paid for it. My mistake was trusting him, letting him register the car for me. He put the car in his name, the bike was in my name. Dad gets a great price because of all the work and money I put into it the past year. He takes the money and buys himself a new white Toyota station wagon. I find all of this out myself when I bum a ride home for spring break. No one had the nerve to tell me, not even Mom. I am livid but say nothing. I really am stupid, I forgot the end of the refrigerator game, not to trust anyone, especially him. My bad. I'll not make that mistake again as long as I live. I have Jim immediately take me back to Cisco. Dad writes me a letter saying that he actually did me a big favor, removing that distraction and complaining that "your mother thinks that I have wronged you somehow". He also whines "where it is written that I have to buy you a bike and a car". Dad always was a self-described "Indian giver". The Indians were wise and proud. How could anyone own something that was here before they were born and would still be here when they had long turned to dust? He asks for a letter back. I don't write, that would be a bad idea right now.

I get word that Grandpa Nuno passed away. He had a priest at his side to give him last rites. My problems seem small after hearing that. I say a prayer for him and my Mom who could not be there for him and is very sad. I am sad too that I did not get to know him better. Mom says that I remind her of her Dad.

# 1973 Austin

As soon as school is out I head for Austin with Kevin and Stormy, I don't even bother to go to Abilene, nothing for me there. Lisa and I broke up a month before the semester ended. Her plan was to get married, get a job, and start having babies. The Wrangler Belle plan. A lot of Belles have June wedding plans; we have been going on couples dates and it's contagious. Some couples already have apartments and there are some pregnancies of course. Lisa wants some of that and now. I am too blunt and tell Lisa that my plan is to get an engineering degree and eventually practice, not have babies and work in a factory to feed them. I am looking at three more years of school before I can be serious.

I feel that I sound like the Colonel after I say it out loud, blunt, rude and unfeeling; I hate myself for saying it. I cannot lie and I know it. I have been very bitter the second half of the semester. Lisa is not onboard with my plan. She is mad at me for wasting her time. She is majoring in MRS. and she is about to flunk the final.

## 1973 AUSTIN

I must admit that I have become a little irritated with her since she started working for La Raza, had to look that up, the race. Sounded pretty racist to me, plus she had started calling me white boy, which I intensely dislike. There are a few weeks left in the semester so she dumps me and gets right to work. She lands this dumpy looking guy within a week, showing plenty of PSA for the effect it has on him, campus gossip and myself. I had purchased a hammered Navajo silver bracelet for her that I was going to give her for her birthday. My Mom sold it to me to give to her when she was working at Burro Alley for Christmas. I soaped it up deformed it and managed to force it over my left hand with soap. That thing is not coming off. It will be there riding on the wrist my Father maimed to remind me to never trust anyone, anyone at all, for any reason. I lost: my mount, my dog, and my girl, but I still have my gun; sounds like one of those sad country and western songs I hear in the hall. Instead of Amarillo by morning its south to Austin.

Stormy, Kevin, and I, head out in Kevin's 4 door 68 Impala. They get a place at the Racquet Club apartments on Lake Austin. Tennis courts, and right on the lake, sweet. I register for summer semester in the School of Mechanical Engineering at UT. The requirement and the process is easy. They let me in with my in hand carried report cards and SAT scores, if I pass, I can stay. I sign up for Calculus 2. One course three hours for the last fifty bucks to my name. Probably should have taken something easier, like Texas History, but I am inclined to bite off too much, I'm too cocky for my own good but don't know that yet. I am going to do this on my own. The Colonel be damned.

Shortly after getting settled we get a come get me call from Jim. He has graduated but is not sticking around for pomp and circumstance. Jim has already quit his Job in the paint department at Sears. Kevin lets me borrow his car to pick him up. When I get there the first thing he says is "what kept you" he's mad as hell. He is already packed up, sitting in a lawn chair in the garage surrounded by all his stuff. We load up and split without a word. Jim had his plate breaking day and it was not pretty. Jim doesn't even want to talk about it. Dad will forever say that Jim made a big mistake the day he quit the paint department at Sears; it was steady and he is a natural salesman.

The Racquet Club has more students than tennis players and those students sure like to party. Jim likes to party too and is in full circulation. I spend most of my time at school. Summer school Calculus for engineers is intense, I have a class every day and lots of homework. You ride a shuttle bus across the river to campus and once there you are not inclined to leave. My roommates like to party and they are carrying me, I'm broke, so I can't say anything. I have to find a part time job so I can carry my weight.

I come home one evening and Jim has scored. He met the Walker boys, Mark and Mike and through them Jeff DeMuynck, all Army brats, Colonels sons from Fort Hood. In hunting around for some dope to deal he finds pot and a job for all of us as carpenters in one deal. I am the only carpenter among the four of us and I didn't think anyone would hire a part time student carpenter. I was thinking of something on campus.

The boys come and get me after class then we go to the jobsite to check it out. Turns out it's the projects out East 51$^{st}$ st, public housing in the way of huge blocks of three story apartments. The site looks like it is on a flood plain, there is mud everywhere with deep tire tracks and I don't remember it raining lately. I walk by a trailer sized sump pump pumping out a three way twenty four inch drain junction box, the water level never drops while I am there.

We walk farther onto the site to the Tool Van where the partners are, Richard, Dave and Fat Cat. There is only one overworked carpenter, Aubrey Mayes. They want us to start right away. There is a lot of wood to haul by hand since they can't run a forklift on the soggy ground. They will pay us in cash that evening if we will jump right in and start throwing 2 x 4s up to the second floor of the back building. It sounds good so we turn to. At the end of the day the partners yell "roll em up" meaning get up all the tools and chain up the air hoses and extension cords then put them in the truck. It also means that Dave, the source of the pot Jim wants, is rolling a joint and braking out the beer from the water cooler because it's "beer thirty". The boys get to cop a buzz on the job and the boss don't care. He joins in.

We take our cash, take a shower, go out and have a huge Mexican dinner, the most food that I have eaten in months. Feels good to have some money in my pocket again. The sun and fresh air plus the exercise and food make me feel great. I decide to skip class tomorrow and make some money. If I work, that makes: two carpenters, two helpers and a wood carrier. We could really get some work done manned like that for the partners.

The next day we all show up for work. One of the partners, Richard is not on site so we ask about him. He got arrested for passing counterfeit bills making me want to know more about our employers. Dave Osser is missing a thumb. An Airman, he lost it when he got mortared out on the apron at Ton Su Nut Bien Hoi. The Air Force made a thumb out of his index finger. Dave is also a pool hustler and is quite good at it, maybe too good. He was dominating a table and winning a lot of money when a sore looser started a fight. Dave was throwing punches at the guy's head and thinking that the fool was wasting energy hitting him in the body when he took a knife blade to the heart and dropped. The Air Force MD's cracked his chest and fixed him. His chest has two horizontal and one long vertical suture scars where he was put back together with staples. Fat Cat, I never really learn his real name, is a very large overweight menacing looking Cajon, he shows us shackle scars on his wrists and ankles where he was chained to a wall in solitary. He offers to show us the fly tattoo on the end of his dick but we decline. Cat tells me that if I want to get rich I should start up a church or a NGO, a tax-free charity. That may have led to his imprisonment. I am listening for sirens and wondering if they stole the truck and tools.

We go by Sears after work to buy tool bels, and some personal tools. I prefer the Rig Axe I can drive a 16 Penny nail in one shot with all my tennis playing, I also get a speed square, a framing square and a Skill saw. This is fun and the partners make sure they keep us happy and busy. We are their Hippy crew, we move faster with framing hammers that any other crew can with a nail gun. I end up being a knockout carpenter, going back and fixing stuff that the code inspectors found fault with. There are buildings not square, some without sway braces, just waiting to fall down with the first strong wind, and widows out of line, doors wrong size, etc.

I have been neglecting my studies. It's hard to concentrate on a Calculus problem when the boys are playing Thin Lizzie full blast and partying in the next room along with all that second hand pot smoke. I am tired from the heavy manual labor I've been doing. Stormy and I had a contest of who could carry the most studs and run 50 yards, as if normal work wasn't enough. I admit to myself that I just can't do it, something has to give. I am not about to go hat in hand back to the well, humiliate myself again and beg for help. I decide to sell my youth and strength. I drop out; my current

## 1973 AUSTIN

grade status makes me eligible for readmission to the school of Engineering. I feel a great relief, it's been years, I'm out from under the yoke. All I have to worry about is working and eating. Dad finds out what I've done two weeks after the fact and writes me a "we could have worked something out" letter right after he wrote me a "I thought you were going to work" letter. I don't write him back, he'll figure it out. Didn't even know he was aware, or even cared that I had enrolled; it was my money. Don't believe him anymore, he's just writing a letter after the fact to make himself feel righteous. He doesn't mean it, he is sending Kathleen and Steve to Boulder this fall; I'm an afterthought, his conscience pinging.

We finish the projects doing the framing decking and siding for three buildings and we are done with our part. But not without incident. I knock out my front tooth, the stainless steel one, with my own 16 oz. claw hammer. I was hanging by my knees, upside down trying to pull a nail out when the hammer slipped. Never pull your hammer toward your face. I exhaust whatever I have saved as a down payment for a new tooth. I am wearing Chukka boots because that is what I have. I like them because of the crepe soles. They do not slip when you are walking top plate like a thirteen-year-old girl gymnast walks a balance beam. Unfortunately they will not stop a puncture. I jump off a wall and land on a one by four with a 16ga. nail sticking up. It goes right through my shoe and foot, I can see the point sticking through the top of my shoe. Lucky for me the nail is in the end of the board. That allows me to knock it off with my rig axe with one shot. Back to work. I don't take my shoe off till I get home; I keep working, it's only three hours to go to quitting time. Clean wound but what a mess. My sock is soaked with blood. I go to the dispensary at Bergstrom and get a tetanus shot and limp around for the next couple of days. I can't get Kevin to walk top plate, he is unsure of himself so he hangs a leg either side of the plate and scoots along on his butt. We laugh at him and kid him but he doesn't care, he is safe. Because of this he puts a nail in his inner left thigh with a nail gun while trying to toe nail in a header. The nail glances off the poor, sap filled, grade C lumber we are using on the projects. It only goes in half way, about an inch and a half, so I pull it out for him, I pour peroxide on it and apply a compress, Kevin is laid up for a few days.

Its late summer, the weekend, I'm watching the boat races on lake Austin right by our apartments after playing a little tennis with my cute neighbor Susan. My hair is real long now I wear it in a ponytail and I have a big bushy mustache. My body is a dark tan with hard defined muscles from all the manual labor, I'm pushing 195 pounds. I have my shirt off to cool down, enjoying the moment. A shadow looms beside me and I look around, it's the old man. I introduce him to Susan. He sits down and watches the race. I look at him and make sure he sees my new look; he says nothing. Mom is not with him. Dad has purchased three lots in Lago Vista on Lake Travis upstream of where we are now. He is down to close the deal. Dad tells me about it and says that I should go and check it out. He also tells me that he has flown Steve who graduated with the same class as Jim, but at the head of it, and Gary out to California to spend the summer with his Mother. She's been all alone since Grandpa died a little after we hit Abilene, a year and a half now, when I was having my own troubles, he sort of let up on me then. The boys are going to paint the house and fix up the place and keep Grandma company. Good for you. Why should I give a shit? He has given Jim and me the boot, and now we don't need him. He has six other kids he likes better. He gets up and leaves without saying anything more, inspection tour over.

## 1973 AUSTIN

Jim stumbles on to a job for us off Riverside just up the block from where we live. There is a lot of construction in our neighborhood, I like the commute. It's the same kind of stuff, framing and decking apartment blocks but these units have better material for them and are of a superior and more complex design, pretty actually. I bid on the trim and soffit and get it. I do a good job. I do all the measuring and cutting the boys do the nailing. It looks so good, we are asked to do another project. We roll from one site to another. I will always remember decking the roof of a three story apartment, when I stand up and I'm in the midst of a stream of Monarch Butterflies 100 yards wide by 1000 yards long and 30 yards deep, simply amazing. They are swirling and flitting around my waist. I watch them until they are out of sight, headed north across the river. A long way for a bug to fly for milkweed; just drifting with the prevailing warm gulf winds from the South. I like those winds too. Austin is the civilized part of Texas. First place that I have ever lived where I didn't feel like there was a target painted on my back.

A carpenter's life is good; Jesus was a carpenter. When it rains, carpenters have the day off! I love to sit on our covered veranda and watch the rain come down. Life is good. We went to work one day with the promise of drizzle and rain in the sky, not checking the weather or watching the local news. We want to deck and dry out a three story building and beat any potential rain, which would waterlog swell and ruin our two weeks work below. We sling our power cables through the framing and the trusses for our Skill saws and hand up half inch plywood 4 x 8 sheets for the roofing. It starts drizzling. We work on. The rain picks up. I am on the roof. My saw won't work. I poke my head under the deck and see insulation flaming and dripping off my power cord. I drop my saw and swing like a gibbon through the rafters and framing until I get to the pole.

We are wired in direct to the line without a breaker or a fuse. I pull my rig axe and chop the cord where it lies against the pole, sparks. We are lucky. The rain suddenly comes pouring down, a deluge typical to Austin, and the fire in the rafter's goes out, only one pitch of the roof has been decked, a fortunate decision. We need to buy another 300 feet of extension cord.

Dad forwards me a letter authored by Bill, endorsed amended and signed by him, telling me to make Jim pay the $3.00 he owes Bill. Seems Jim borrowed $20.00 from Bill and shorted him $3.00 on the repayment of said loan. Dad puts his chop on the letter and amends it to include the price of two pair of knee socks plus interest for the loan to the tune of $4.26. I am ordered to collect. I bring it up with Jim as ordered; he goes high order. Jim is in my face ranting about how I owe him, Dad owes him, everyone owes, him; he is spewing all sort of offence done to him real and imagined. Kevin has been pumping him up, massaging his ego, his weak spot, calling him Jeffe. Jim is continually hustling people and drugs. We have been letting him be our front man, collecting the checks and talking to people, because that is all he is good for. Jim is still lazy as hell. He ducks the hard work on the job; a malingerer. He stands on the top deck gazing at the horizon, like Cortez claiming all that he gazes upon for Spain and his personal aggrandizement instead of packing lumber and pounding nails. Jim is intent on assaulting the messenger and throws a roundhouse right at me. I counter with a left jab that he wades into as he is charging me. Jim's mouth is open cuz he is yelling invectives, Jim's upper and lower right canines punch through his lips with my stiff armed closed fist blow. It's all over as Jim rolls onto his back onto the veranda through the open sliding glass doors, blood all over the place. I get him a towel with ice. I load Jim in the Impala and take him to Bergstrom AFB for four stiches and a tetanus shot. Jim's debt to Bill is never paid.

We land a six week long job putting up a three story apartment complex, the exterior walls are prefab with windows and siding, they are real heavy. We get the trusses, the decking, sheathing, soffit cornish and trim too. A real muscular fellow, Billy who says he is from Nicaragua, walks on site and asks if we can use him in Spanish. Jim is off somewhere so we have only three people trying to set crane lifted thousand pound walls. I tell him we will pay him cash to help us. We work on. After setting the whole quadrant, squaring, tying, bracing and nailing it in, Jim shows up. Billy sees Jim. He starts pointing and laughing. Then he says Rasssssss-Pooooooo-Teeeeeeennnn! Yeah Jim does look like Rasputin by now with his hair down to his arm pits along with the Foo Manchu mustache. This cracks all of us up; Billy is going to fit right in. We are making good money, getting paid every two weeks. Jim, jeffe, usually talks to the people in the trailer. Near the end of the job Jim manages to let them put him off and get three weeks behind when they skip town over the weekend. I mean the office, the crane, the forklift, all the trucks, and heavy equipment is gone; only an old POS station wagon with flat tires is left. Collectively we are out fifteen man weeks of pay and the rent is due. We are pissed. Jim want's to borrow my double barreled 12 gage shotgun so he can shoot up the station wagon. Yeah that will help. It will take some time to track down these guys, if we ever manage to do that, besides Jim failed to get anything at all in writing. They are from Houston and that's a big town. In the mean time we start scrambling for work and try to stretch out what we have.

Our old friend Aubrey Mays offers us a job doing composition and hot tar roofs, nasty backbreaking work moping tar and shoveling pea gravel….After doing several apartments I am ready to throw in the towel. Between buying an expensive Gibson SG electric guitar, all my carpentry tools plus paying for a new tooth and missing nearly a month's pay I am broke. No school for me my plans are torpedoed. I ask Mom to come home. Tim says he can get me and Stormy a job at the Abilene Reporter News where he has a job cutting in adds to the copy. I need to reset and try again. Jim stays on in Austin. He is never coming back to Abilene.

After I am "home" I go for a ride with Steve and his girlfriend just to get out of the house. We are driving around down Grape Street when a car with four guys in it pulls alongside of our car. The occupants start whistling and hooting at Steve's girlfriend, who is driving, and pointing at their crotch. This goes on for a while. I am slouched down in the back seat of the two door Maverick coupe. I finally roll up and flip them off. Big mistake, they start chasing us. We head down First Street. The theater has just let out and there is a glass fronted Duncan Doughnuts right across the parking lot from it. I tell Steve and his date to park and to go in the store thinking that they would not start anything in front of a crowd of witnesses. Cars line the wide side walk in front of the store; a sidewalk wide enough for only two people side by side. We start down one end of the walk, the door is at the other end; I am bringing up the rear. The four guys are right behind me. One of the guys starts yelling at me, telling me that he is going to kick my ass. I face him and tell him that he is obviously drunk and in no condition to fight. I ask what the hell is he doing frightening and chasing a poor girl? I tell him to leave and sleep it off. I then turn to go in the store. He runs and jumps me from behind and breaks my glasses. I spin around in his arms breaking his hold, then its two lefts and a right hook. He goes down like a box of rocks. Six weeks of shoveling pea gravel has just hit him square in the face. I am mad as hell, this jerk just broke my glasses, now I have to beg my parents for a new pair. I grab him by the hair and start banging his face against a car bumper. Then I turn to his friends and ask them if they want some of this; they do not. I tell them to pick up their friend and get the hell out of here before the cops come. They pick him up off the ground in front of me and haul him off. I go sit in the car. Some kids come over and tell me that that was the coolest thing that they had ever seen. I am not at all proud. Later I find out from Stormy that my assailant was one of his cousins just out of prison. He is arrested for armed robbery and assault the following week. Never flip off anyone. Welcome back to Abilene…

## Fall 1973

Jim is entrusted to go pick up Steve and Kathleen in Bolder then bring them home for Christmas break. I am working at the Abilene Distorter so I am not available. Mom want's Jim home for Christmas too. Time to make peace. Mom loves Christmas, me not so much. Dad gives Jim a couple of credit cards, a hundred bucks cash and the War Wagon; then he's off. Two days later Dad gets a call from Kathleen. Where the hell is Jim? He's way overdue. The dorm and cafeteria are closed. It is cold and snowing. She and Steve are standing on the curb waiting for Col. Yancheck to come rescue them. No one has heard a thing from Jim, nothing. Dad buys them both a bus ticket and tells them to forget Jim and head home. Steve and Kathleen arrive in Abilene 30 plus hours later. There is a blue corona around Kathleen like a natural gas flame, she is livid and lets everyone know the depth of her distain without end.

Well after Christmas Jim shows up, drops off the truck and the credit cards, when no one is around. He leaves town with no explanation. Jim is toast; he knows it; he doesn't want to hear it. Jim takes Dad's 20ga, Ithaca pump, the one he bought when he was seven, Dads German stag horn folding knife that he has skinned all the game he's ever taken, my duffle with all my hockey gear and Kathleen's bow, as a Christmas present to himself. Much later I find out what happened. Jim took a day for himself to tour the Springs instead of staying on task, ending up in Manitou Springs. He had been drinking and was arrested and charged with being drunk in public after curfew. He didn't know that the city had decided to crack down on vagrants and had established an early curfew. He fits the description of a hippie to a T with his long hair and Foo Man Choo. By the time Jim got out of the pokey the next day, the truck had been towed. After getting the truck out of impound he drove up to Boulder and found the campus closed down. Kathleen was not standing on the curb. Knowing that he has screwed the pooch Jim gets some hooch and heads for the hills for a bender that will help him contemplate his immediate future; evidently this takes him awhile. He finally runs out of pot and booze, gets tired of sleeping in the truck and crapping in the woods so he heads south, and straight back to Austin.

## 1974 The Oilfield

Tim and I are driving around in the War Wagon getting gas, taking the long way to waste time, joy riding, and milling around, it's raining hard, a Texas frog choker. There are no real proper storm sewers in Abilene, just ditches, so every intersection has a four way dip filled with water. We are going down 10$^{th}$ when a Rambler full of local Hispanics locks brakes, hydroplanes through the intersection on the ditch water and T bones the truck on the starboard side, right into the three door pillar post. They ran the stop sign but are inclined to argue. The front of the Rambler is caved in and the side of the truck is caved in. It's still raining buckets, we trade phone numbers and drive off; both cars are still mobile enough to do that.

Dad is understandably pissed when we arrive, we were not where we were supposed to be, we are late, off task, on an unauthorized pursuit, in an off mission area; it's our fault, that is undeniable. Dad calls the owner of the Rambler and offers him $200.00 cash which he charged to my spiral notebook account of what I owe him, the

Book of Life, and which the Rambler owner gladly accepts. I fix the truck by first pulling the truck into the alley behind the house. Using a tow chain, I throw a loggers choke round the base of a telephone pole then connecting it to a strap around the midpoint of the truck's pillar post with a come-along and with the doors open and the wheels blocked, I jack the truck on its side with the come along till the driver's side wheels are in the air. Tim and I jump up and down on the up side of the truck repeatedly until the pillar is pulled and the doors close and seal properly. This makes an impression on Tim, a repair method that he will unsuccessfully attempt to repeat in the future when he uses the retaining wall along the driveway for a dead-man, pulling it over.

With no chance of attending any school at all, I get a job as an auto mechanic apprentice with Dan Hixson Chevrolet. Once again Mom lands me a job, she is friends with Mrs. Hixson from church. She has cornered her and obtained a concession. Lord love her. They make me an apprentice truck mechanic and whatever they ask chassis monkey. We start out on South First Street now the Arrow Ford Dealership, then we move to a new building off the loop east of Dyess. I learn all about: warranty work, chasing squeaks and rattles, new car make ready, oil changes, tire jerk and anything else that won't make a mechanic any money. There is a book and code for each task that says how long the task/repair should take and that is what it pays, no matter how long it actually takes you to do the job. I borrow Dad's Craftsman tool box tools and ratchet set, you have to provide your own tools. I take them home every day, the other mechanics have locking roll arounds that no one can just pick up and carry off. I have to clean up and mop the floor with Naphtha after we close and before I leave. There is a uniform fee too. Working with a lift is nice, better that laying on your back

## 1974 THE OILFIELD

to work overhead. Truck mechanics have to crawl over under and around the truck. I change bearings and a clutch in a dump truck. Crawling under the hood of a truck or swinging under the frame is no problem for me so that is what they make me do.

If I am working for a Chevy dealership they should make me a good deal on a car. I am tired of borrowing the truck while Dad drives his new Toyota. I have been keeping my eye on the used car lot. I spy a sweet 70 Mercury Cougar it is loaded, with leather seats so I talk to the salesmen, they don't want to lower their margin on this car so they lie to me and steer me to another. Never trust anyone in a polyester suit with wide lapels and a wide tie. They show me a green 69 Mustang fastback 6cyl with a four speed. I bite. I go the base Credit Union where I have an account and see about a loan, which they approve. When I return to the dealership I find that they have sold the Mustang out from under me for cash. I settle for a copper colored 73 Vega with beige interior. Looking back I probably should not have settled, a boy and his wheels you know. It looked sleek, a sporty hatchback with good mileage. Dad is against this. He says "never a borrower or a lender be". I cannot see living at home for two years to save enough money to buy a car. To me it is monthly cash flow not a sum total anyway you look at it. Dad has never been in my position; he doesn't understand.

The young long haired mechanic in the bay next to me owns a new hopped up Chevell, he goes on and on about the music scene in Austin, the Armadillo, progressive country. He drives there every weekend. We are both working to pay for our cars, he is factory trained and makes a lot more than me.

I am a disillusioned working stiff mechanic who needs to make and save more money than I am currently making so I can go my own way as soon as possible. I manage to land a job in the oilfield through one of Tim's friends. The price of oil has gone up and there is a lot of activity. Abilene is an oil town. In the oilfield it is either bust or boom and it is boom time. I go to work for a fellow named John Patrick Goodfellow. He is an independent Oilman and he needs a mechanic for his lease. JP has a lot of old equipment to maintain and bring back into service. He likes that I am and engineering student and thinks that I can really help him out. He is going to pay me twice what I am making at Hixson's. JP is a tall barrel chested dark haired man wearing thick black rimmed glasses. He is all smiles and handshakes which puts me at ease.

My first job is to drive a trailer load of PVC pipe to his lease in Ranger. He gives me a dirty OD 70 Plymouth Duster with matt black racing stripes and a 198cu in slant six in it that gives the lie to the racing stripes, and no A/C. This is the tug for the 16 foot trailer? I thought that we would at least have a pickup truck. It would have been better if the Duster had a 318 or a 340 in it. This will be a long trip. The trailer is a light dual axel rental rig, it's already loaded and ready to go. I'm off.. I am afraid to go over 50, assuming the Duster could do 50, I am to act dumb on site but observe what's going on at the lease. I look pretty dumb towing a trailer with an underpowered Duster so it won't take much convincing. JP is in a limited partnership with a fellow named Johnson, who is on site, an investor, and presently pushing the tools.

I drive down some dirt roads past a tank battery to the rigs and pull up to the larger one. Johnson walks out to me and waves where he wants me to go. He is wearing an aluminum doughboy hardhat and coveralls that used to be white, now yellow from oil, dirt and many washings. Johnson ignores me, he is a busy man and I am just some flunky JP hired. He waves over some roustabouts and they turn to the load in the trailer. I walk to a little 24 foot camper trailer about 500 yards away near the fence line that serves as the office. It has a phone, water and an air conditioner. I grab a glass of water from the water cooler and sit down at the dinette I am starting to have fun; it's been a long time since I had any fun. I call JP and tell him I made it, he wants us to hang around and watch what happens. I notice a drawing taped up in the trailer wall that shows the lease and what's going on. Prior to this all I know about oil is that the price is going up from $13.00/bbl to over $30.00/bbl. The existing wells couldn't pump enough oil to pay for the electricity to operate the well at the lower price. They were allowed to be left idle for years without plugging them. The effort is to rehabilitate/service these existing wells, and drill intermediate saltwater injection wells. The injection wells will chase the oil to the existing wells for a more efficient secondary recovery effort. JP has a contract to dispose of some neighboring leases saltwater that he will use for the injection wells along with what he produces. You always get saltwater with your oil, the first thing it hits, out of the well is a separator. There is a lot of saltwater to dispose of and this is the first injection well JP is drilling. Sounds good. Steel pipe is dear during the boom. Theft is common. There is an industrywide shortage. JP is trying to use PVC in his shallow saltwater injection wells. He has tested the pipe enough to satisfy himself that it will hold the pressure required to complete the well and pump cement down to seal the formations drilled through. PVC will also stand up to the highly corrosive saltwater that will be injected down the well.

 I walk over to the "big" rig, they are still coming out of the hole, it's a single; the tower is only tall enough to hoist up a single pipe joint at a time. Instead of stacking the drill stem in the tower, each 30 foot stick has to be laid down on a float or tractor trailer flatbed. You swing the lower end over and set it on a roller skate and slide it on to the trailer as the elevator lower the upper end, then roll it off onto the stack wrench off the adaptor and make it up to the end of the drill string sticking out of the slips in the rotary table, lower the elevators latch on and do it again. Slow going and a lot of work compared to a triple where you can stack a 90 foot stick on end and leave them in the tower. Good thing that these wells are shallow, only 2000 feet deep, 67 sticks of drill stem and collars per trip up or down the hole. The last sticks to come out of the hole are the thick walled drill collars, used to put weight on the bit, the others are thinner walled with heavy tool joints. The other rig is a truck mounted water well drilling rig; it's a single too.

While the roughnecks are pulling the drill string the roustabouts are gluing the 2.5 inch PVC pipe I brought together they are going to snake it in the hole by hand once the drill stem is out. I am dubious. I'm not a big fan of glue, JP also cheeped out and bought short couplings instead of the long heavy duty ones with more adhesion surface. Longer collars will also resist bending failure at the joint. When they are finally out they start putting in the PVC, it's pretty clumsy the PVC breaks, pops out, at a collar dropping about 300 feet down the hole. Instead of waiting and "fishing" for it with a wire line and a tool Johnson decides to drill it out. That's when I leave, I've seen enough.

I drive the Duster to the Abilene Country Club as requested, to meet with JP, I'll find that AP takes most of his meals at the Country Club, he puts it on his Tab. Its late, there is only buffet leftovers. Shrimp cocktail is in abundance. I pig out; I love shrimp but don't get it very often. I am debriefed. Dad has raised me to be the perfect Lieutenant, JP recognizes that in me and intends to take full advantage of it. JP is interesting to me. I think that I can learn a lot from JP. He doesn't order me to do stuff, we talk about problems and come to a logical conclusion and a plan of action that we both agree to and execute. Very different from my relationship with the Col. JP has respect for me and my capabilities. I think that he genuinely likes me.

The first thing I do is get the International Harvester semi-tractor running, without it we can't move anything, the rig, the doghouse or the mud pump. I find that the clutch is shot after driving it around. I test the PTO and it is nonfunctional I need to pull it. Johnson is still trying to drill out the PVC but he lends me the hand of a roughneck. We block and jack up the motor and transmission then disconnect the driveshaft. While pulling the transfer case the jack slips dropping it onto the leg of my helper breaking his tibia. I am under the truck with him when this happens. I roll the housing off of him and drag him out from under the truck. His leg is obviously broken so I fireman carry him in to the dog house and splint his leg. My helper is a big Cherokee from Oklahoma. I ask what I can do for him. He tells me to call his old lady to come and get him. He has a pretty white F100 step side SNB pickup with three on the tree. He cannot possibly drive it and he will not leave it behind. I call his wife, she says that she will be there in about four hours, maybe. When I get back to the dog house I find that my injured helper has slammed down about four or more beers that he fished out of the water cooler while I was in the trailer calling his old lady. Four hours is too long, lots could happen in four hours. I put my large drunk broken Indian friend and helper in the bed of his truck and drive him to the clinic in Ranger. Johnson allows a chase car to bring me back.

When I crack the PTO gear box I find that the gears are shot, teeth are sheared off. I call JP with the news. He tells me to contact a machine shop he deals with in Fort Worth. I load the gear box in the trunk of the Duster and I am off. The shop is not much to look at when I get there. I talk to the shop foreman he doesn't know if he can help, he needs to ask the Engineer. I meet with the Engineer. He is an old unassuming black man with a shock of white hair. He is dressed in a tee shirt jeans and a greasy stained denim apron, the smartest man in the building. The machinists' show him a lot of respect. He counts the teeth and makes some measurements with a mic, pulls out his slide rule, and makes some calculations on a note pad. He tells me that he can't cut hypoid spiral bevel gears but he can make me a set of hardened straight tooth gears that will work. I agree with his proposal and cut him a check. I call JP and tell him what's up, he tells me that he will make a deposit in my bank account to cover the cost. They will ship the rebuilt gearbox to Ranger on a bus when they are finished.

When I get back to Ranger I find that everything has gone to hell. Johnson has twisted off and lost the drill string. The mud pump motor has quit. Johnson has left the job site. I call JP; he says that he will run the operation through me for the present. People are standing around. I ask the driller to show me the break, it's on the float. Johnson has left all the drill collars in the hole about 300 feet of hardened steel. The Driller and I rummage through the dog house for fishing tools. We find an inside and outside tool, a cone with threads in the inside and a spear with toggle teeth along its axis. Both are clearly worn. I call in.to JP; he tells me to bring in the tools and send the hands home. I take the tools to a machine shop in Abilene on my way in. They will rethread the tools sharpen the toggles heat treat and harden them. I tell the crew to go home, we will reconvene in two days. It's the weekend, they don't mind. Looking at the twisted off end I believe that the overshoot tool has the best chance of recovery. I call JP, he agrees. I'm nervous I have encumbered about 3k in debt on JP's behalf and have no way to pay it back. That happened extremely quick; faster than I thought possible. I am uncomfortable, I have just ordered a CNG ready 390 V-8 ford crate motor to power the mud pump, encumbering an additional personal commitment of 1K. I call JP and ask him to cover it, he says he will. I am really sticking my neck out, I don't have money like that. This is very challenging work, intoxicating really, it's getting complicated. I need to separate my personal bank account and establish a business account I haven't been keeping proper track. I realize that I am spending my own money on the lease. I start keeping a ledger.

I pick up and install the new mud-pump motor over the weekend, it cranks right up. I run the engine intending to break it in for a while before I put a load on it but it quits within 15minuits. We are out of fuel. I call the phone number on the tank and ask for service. They tell me that they are owed the last two loads. I tell them that I don't know anything about that but I will give them a check if they will come out and fill me up; which they agree to do. This episode gives me pause. I call JP, he says he will take care of it. I believe him. It's over an hour one way from the lease. When I get home all I want is a shower, a hot meal and to hit the sack.

The fuel truck is there bright and early and so are we. I picked up our tools on the way in. I have to work the deck as a roughneck, since we are down a man and may have to make a lot of trips. I know nothing and I am fixing to get some fast schooling with no mercy. The first thing I learn is, the mud around the rig will tear the soles of your boot off. Bentonite, clay added to drill fluid/water, to make mud viscous enough to float rocks, looks like gray pudding. I climb into the dog house, pitch my ruined boots in the trash and borrow my Indian friend's gum boots, literally filling his shoes. I grab a hard hat and go on the floor. The driller has gotten on me about my bracelet, he tells me that I'm going to catch it on something while throwing chain and end up ripping my hand off. I can't easily take it off so I push it up my arm, pull my sleeve over it then wrap duct tape over my sleeve.

We are going to jet our way down. There is still a column of fluid in the hole. We manage to get a stick of drill stem started down with the overshoot tool and an adaptor on the end without caving in the hole, after that we can get the Kelly on and start pumping. The first 30 sticks go in pretty quick then it slows down, but not much. The driller is complaining about the mud pump, he says it's shot, the output pressure is not high enough for him or physics, we need a minimum 1000psi for a two thousand foot well. I think that it may have sucked up some PVC cuttings, I don't think that they settle out in the pits like rocks and sand. And we didn't have a good screen on the suction. Every time we make or break connection the deck is hosed with drilling mud that you skate around on while working with counterweighed tongs and tools, we have to manually drag and place slips in the rotary table then seat them with a sledge hammer before breaking connection, the driller pushes us hard till the weight comes off the elevators, he thinks that we are on, he torques the drill string until he thinks he has a good bite, then announces that we are going to circulate for at least twelve hours to washout the hole around the bit and drill stem, he thinks that the PVC is barber-poled around the drill collars. It is after sundown. I shed my muddy clothes thinking that I might just crash in the trailer. I'm dead tired after eating everything I've brought with me, and can find in the trailer, plus chugging half the water in the cooler. I almost forget to call JP to tell him the good and bad news. He thanks me and tells me that I need to come in. I need to swing by his place in the morning. He will let me know what's up. I put my muddy clothes back on, ooze onto the duster's vinyl seats and head west.

I run by JP's place the morning. He lends me his car a beautiful brand new black on black 74 ford Thunderbird. It looks just like a Lincoln MK IV. JP is worried about the pump and had the driller shut it down when he called in his hours and his own report; he wants a second opinion. I am to meet a fellow named Shorty Ice at the Derrick Café, where else. I'm to have breakfast with him, pick up the tab and take him out to the lease to look at the pump. Shorty is short all right, about five foot two in cowboy boots with bull dogger heels and about eighty years old. While we are eating he tells me that he invented the first movable drilling rig in Texas. He shows me an old photo he pulls from his wallet it's a picture of a wood derrick with wagon wheels being pulled by a mule team, outriders on more mules are holding guy ropes. I hope he can help, we don't have a mule problem.

We head on out to the lease. Shorty is riding shotgun. Shorty drones on while he puffs on his pipe. I like Shorty, he probably mentored JP at one time, there's some gold in there, some in everyone if you care to filter. When we get there we are the only ones on site. The driller didn't turn off the pump motor all the way so the battery is flat. I hate to do it but the only vehicle available to jump the pump motor is the T-bird. I pull up to the skid mounted pump and get cables out of the dog house. I pop the hood and wind down the window because it's hot while I rev the motor up before I start the pump motor; it cranks right up. I take off the cables and shut the hood leaving it running so I can back away. Meanwhile Shorty has been wandering around the pump, a Gardner Denver Triplex, listening to it, looking at the valve manifold. The flow is currently set to the Kelly, he looks at the manifold gage; then shuts the valve to the Kelly off, looks at the gage, then opens the mixer flow to the pit and jets the pit opening and closing the valve while looking at the gages. I'm watching what he's doing so I don't miss anything. Shorty then throws a third valve which directs a high pressure three inch stream of drilling mud right through the window of the T-bird. In less than a second the inside of the car is coated with gray drilling mud. My first instinct is to throttle Shorty but he looks like he's about to stroke out. His face is drained of color, his mouth is open and his pipe is lying in the mud at his feet. I scoop all the mud I can out of the car and wipe it down as best I can with red rags. JP is going to kill me.

Shorty is very apologetic, I can't be any harder on him than he is with himself right now. I ask him his opinion. He says that we should pull a head to make sure, but he thinks that it needs new piston swabs and cylinder liners. I pull one for him and sure enough there's is a lip at the top of the cylinder; worn out, Shorty is right. We need to call JP. He tells us to come on home. He will get the car detailed. Time is working against us; we have to get that drill string out fast before it's stuck there forever. It's a long soggy ride back to town. I drop Shorty off first. I'm whipped.

The next day JP has me check out the equipment laydown yards around town. Piles and piles of rusty junk I don't find much but I get to see a large variety of equipment. I do come across three loose rusty sleeves that we could bore and trim, machine them to work, so I buy them as a plan B for $20.00, cheap enough. I come home and call JP; he has found what is needed in Midland so I unload what I have and head out to pick up the parts before they close for the day. Midland Odessa is a continuous equipment yard. I somehow find the place, they're waiting for me to show up so they can go home. Everything has been prearranged so they help me load up; one sleeve in the trunk and two in the back seat foot wells, the swabs and rings go in the passenger side. The Duster doesn't like all this weight. It's another long slow drive home and an equally slow drive out to the lease the next morning.

The whole crew is present the next morning with a welder. The Kelly needs patching and he can help thermal cycle the sleeves to break them loose. I use a torch to heat the case then I shrink the sleeve with some propane. A few smacks with a sledge hammer breaks them loose. We do it again to install the new sleeves, the swabs are much easier. The Kelly is a thirty foot long square tube driven by a square hole in the rigs rotary table. The Kelly also has a rotary fitting on the top to pass the drilling fluid down the drill stem and attachment points for the elevators. It sprung a leak, splitting at a corner, when Johnson twisted off; the roughnecks say the deck jumped, with them on it, when it happened, like a loaded spring breaking, it was so wound up. It was spraying us the whole time we were working our way down. Pretty soon we are back circulating.

We need more water in the pits to make up for absorption and evaporation. We get our water from a big tank. The lease, the Wilson Ranch is huge, both sides of the road and the highway. Must be over 2000 acres. Wilson has cattle, fine looking red Brangus. He also has sheep, goats, and fields planted in oats and sunflower for bird hunters, cultivated fields for fodder and timber breaks for deer. The tank is too far away to walk so I get directions and drive. There's a gas AG pump at the tank, a six cylinder gas motor with a pump on the bell housing. The whole thing is mounted on a two wheel axel with a battery and gas tank. I get it going and drive back to the rig. There's no water coming out the other end of the pipe. Chances are that I might of started up the wrong pump motor this place is so huge, so I walk the line. Its dammed hot and dusty today and coming on late afternoon, Siesta time around here, the hottest part of the day. I hike over two thousand yards till I find the break, should have started from the other end of course. The flow line is two inch white PVC. The line has been laid out unburied without any expansion joints so it has been crawling around like a sidewinder on the ground expanding and contracting until it separated at a joint. Not a big fan of glue, just plain unreliable. I hike back and drive to the tank and shut down the pump. I ask for some help I want to make some big Us in the line with some elbows and extra lengths of PVC pipe, this is when I am introduced to the Shovel Gang, the Roustabouts, Billy and Hector. Billy and Hector are Mexican Nationals on the run, they live on the ranch in a wall tent and cook over and open fire. The driller told me that they were Johnson's boys and now they're all mine. Neither of my charges speaks English and my High School Spanish is severely taxed; I should not have been reading all those Hardy Boy books during class. My boys are all smiles and a big help now that they know someone will take care of them. We throw in about four expansion joints and get the water flowing. Then I call JP; he confirms that the boys are my responsibility. They have already told me that they are hungry and need food. All they want is bacon, beans, eggs, and tortillas. I make a run to the grocery store for them before I head home, I also buy them some extra stuff that I think they may need; they smile at that. Billy, my age or younger, is a five foot three bundle of energy. Hector, barley 120 pounds, could be Billy's Grandpa. I say goodnight, I'm catching dinner in Ranger then coming on back. The driller and I are going to stand four hour watches while we circulate all night. I'm sleeping in the trailer. That sucker's coming up tomorrow.

I get a call from AJ the next morning; he wants me to get Hector and Billy started working on some leaky tanks all in the same battery, they are empty now. JP wants to grout or pour cement in the bottoms to patch them; I'm dubious but we head on out. I open the top and can see gas escaping as it bends light and also get a good whiff. A flashlight shows me it's empty, the ground must be soaked underneath there's also a containment berm around the battery. We pull off the lower manholes. This thing will have to air out for a long, long time before anyone can go in without alternate breathing equipment. That's what I tell JP. I'm not going to make those fellows go in there. He has bigger fish to fry. The Railroad Commission is going to shut us down if we don't fix a leak in a pipeline. There's oil pooling on the ground. He tells me to get the boys started trenching the oil away from the leak and digging up the pipe line. While they are doing that I'm to drive to a pipe supply and get some clamps, big split rubber grommets with a stainless steel metal clamp you bolt on. When I get to the store they ask me who's it for? Maybe he has an account so I say JP Goodfellow. They give me a dirty look and say he has no credit here. I pay with a check. I need to be careful with my name dropping, it will be, the Wilson lease, if anyone asks. I come back with the parts and we get the clamp on and the leak stopped. Now we have a small pond of salt water with oil floating on top of it. Turns out this line is a feeder for the salt water injection pump. All this water is going to a separator on another tank battery. The salt water is rotting the pipe because the pump is not working. I trace the line to the battery and find a lake of salt water. First things first. It's very hard to light crude on fire, I use railroad flares with little luck, a brush pile in the pond plus some gas soaked rags finally gets it going. The saltwater pump is shot, it needs new ceramic pistons and cylinder liners bearings and rods, a complete rebuild. I go back to the supply house once again, they have all the parts, I write them a check. I call JP on the way back and tell him how much it costs, about $1500.00, ouch. It takes me the rest to the day to get it going but it had to be done. Hector and Billy have channeled the spill away and ponded it like the other spill. They are going to burn the oil while I check on the rig.

The crew is mad at me. I missed all the fun, they are out of the hole, hooray, now the discussion is what to do next. They are all for moving 20 feet and starting over except for the driller who wants to fish for the PVC with a wire line. The water well rig is usually used to start a hole then the rotary table rig is moved over the hole once the hole is deep enough to get some weight on the bit, at least 300 feet. Nothing is moving until I get the IH tractor back together. I have to check on my gears tomorrow. We all head home to sleep on it. I need a meal, a shower, and a good night's sleep. A day off would be nice. I grab a double chicken fried steak with fries at the diner in Cisco. I can't depend on anyone leaving any scraps for me at the house, bunch of alligators.

The next day I call on my parts; they tell me that they will be on a Greyhound Bus tomorrow afternoon, hold for pick up in Ranger. I have time to kill so Tim and I load the junk liners in the duster and head for the scrap yard. We get two bucks for them. Then I head on home. JP calls. I am to meet him at a steakhouse outside of San Angelo for supper. I don't ask questions. I take the Duster on the scenic route northwest through the hills, a nice drive. Soon I'm there. It's a long low building one of those places just the locals know about, only open a couple days a week. The aroma wafting toward me s incredible, fatty smelling mesquite smoke. I go inside and ask for the Goodfellow party. I'm escorted to a back room where JP is holding court at the head of a long table with about twenty people, middle aged and older couples mostly. They are just finishing up their meals, I'm late. JP stands up beaming as he introduces me. JP comes over and shakes my hand. Everyone else gets up to meet me and shake hands. I'm embarrassed, wasn't expecting this. I've been sandbagged. Everyone is smiling and real friendly though, they have heard about my adventures in the oilfield I am the young engineering student that is making things happen for him. That puts me at ease and I make friends. The table is cleared while we are mixing. JP excuses me, he tells me to go get supper in the dining room since I must be hungry; they are going to talk business. I'm glad to oblige. I am starving. I need at least six thousand calories a day so I'm all about eating and don't give what just transpired a second thought, I sign for my meal and have a nice drive home. I respect JP, he is very good to me. He treats me better than anyone I have worked for ever has.

    I get up early and head for the lease in the duster to help breakdown the rig and get it ready to move to the starter hole. We are going to abandon this hole and deal with it later. The water well rig has to be moved too, it hasn't run for a week or so. Promises to be a long hard day. I had grabbed a couple of muddy packs of Kool's out of the T-bird when it got hosed. JP always kept a carton on the dash. I've taken up the habit again for those long drives. Cigarettes, fried pies and a Dr Pepper for breakfast. DQ for lunch when I go to the bus station. It's getting dark before I have the tractor going, too late to start anything and complete it before sundown. We will all be here bright and early for the move. When I get home I have to make a double take, the pump liners I just sold for scrap are back in the garage right where they were. I ask about that over a chunk of chuck steak and green beans. Tim is the only one at the table with me at this hour. He tells me that Mom asked him what happened to them. He told her. She thinks that we made a bad deal and buys them back for five bucks driving a hard bargain. I tell her that she will have to sell them for twenty five to break even and they are really scrap. She disagrees. She paints them red fills them with dirt and flowers making tall planters out of them; you can't win, she never made a mistake in her life.

More cigarettes, fried pies and Dr Peppers. I have a pack of bologna and a half a loaf of Mrs. Baird's white bread for lunch. We are making hole by sunset. A good day. My charges have been helping with the move. Billy has learned what I did on my first day. The mud can rip the soles of you boots off. He has lost the heel of his right boot and the sole of the left one, he has repaired them with duct tape wound around his boots. They are also low on food. I take Billy into town and buy him a nice pair of boots at Gibson's before going by the market. After I take him back to the ranch its dark thirty. I catch my usual dinner at the Cisco Diner before I get home and pass out.

The next day, Billy is gone. I guess I shouldn't have bought him that sack of socks too. Hector gave him all their provisions so I have to tend to that again after work. Hector is too old to be doing what he is doing. I tell him that his job is be the gate man, a Shepard. He is to keep the bulls, cows, steers, sheep and goats in their proper pastures, out of the oats, hay, sunflowers, the pump jacks and the tank batteries. Hector just has to walk some rounds in the morning then keep in the shade by the main gate, where I put a lawn chair, so I know where to find him. I have Hector tag along with me when I drive around the lease so he won't be lonely, don't know what he does at night now that Billy is gone. I have lots of stuff to do. I only work the deck when we are making a trip. That all ends when I site the little water well rig.

A dozer is there in the morning to cut the pits. The operator will cut them just like the other ones. The location has already been staked out. It doesn't take long. I back the rig in, extend the out riggers and raise the mast. Everything is mounted to the truck chassis, the mast the draw works and the mud-pump. Everything is hydraulic, running off the truck engine PTO; the mud pump has its own motor. Everything runs on diesel. The unit clamps onto the drill stem and pulls it down with a six foot stroke. It uses the weight of the truck to put pressure on the drill bit. Because of this the rig needs constant attention. The operator needs to watch the gages and constantly adjust the pull down feature. The big rig is easy, once you are turning you sit back and log how many feet per hour you are making. The little rig doesn't have a platform, just some railroad ties to keep you out of the mud but they don't. The crew doesn't like it because they end up working both rigs at the same time and the little one is more work. Roughnecks always know better than the boss. You have to use a couple of thirty-six inch pipe wrenches and a sledge hammer to make connection on the little rig. It has s swing out color to catch the tool joints while you do it. Hector and I end up running the water well rig. The truck has all the pipe thread tooling on it too. I get to be the driller and fiddle with the levers and valves. I operated it some before the move so I know how it works. First we have to fill the pits with water and mix some mud. After a few days of going thru soft stuff we are down to making about five to six feet an hour forty to fifty feet a day, half the rate of the big rig It should take a month for a two thousand foot hole. It's a shame we lost the first one, a month of work, five man months factoring in the crew, almost a half year effort.

    We are running low on fuel for the mud pump servicing the big rig again, every other piece of equipment runs on diesel. JP wants me to try to use wellhead gas. I'm to tap into the well casings and use that to run the mud pump. I give it a go. I know where everything is on the lease by now and we have plenty of PVC. There's a six inch bore

V twin compressor by a cracking stack that Shorty pointed out to me. I can use it to move the gas. Shorty said that they used to make drip-gas in the stack. I plumb eight wells into a manifold, pull from that manifold, blow it into another well casing close to the rig and use the engine vacuum to pull from the casing. To run the compressor I have to buy a 21hp Koehler V twin industrial air cooled motor and attach a gravity gas tank, spending 2K in the process making me nervous. There is no pole power near the compressor and even if there were the power draw would be too high. I hope that JP knows what he is doing and this will be a net energy win and a cost savings. The money I am spending would have bought eight tanks of propane. After a couple days effort it works, but not for long. Paraffin builds up on the compressor pistons which knocks the heads right off. All is not lost. It's a tough piece of equipment with a great failure mode. The answer is to preheat the input gas to the compressor. The paraffin is easily cleaned from the heads, pistons and reed valves. It is easily reassembled, it has solid copper rings as head gaskets. I plumb the manifold to the cracking stack bleeding off enough gas from the compressor outlet to fire the stack. It takes a lot of fiddling but I finally get it to work. I keep my fingers crossed that the heads on the mud pump motor don't come flying off from the same thing.

Its 1000 hours, morning break. I'm in the trailer making my morning report to JP when he interrupts me and says "Turn everything on. Make sure both rigs are operating and people are working them. Turn on all the pump jacks too. He will call back in two hours to explain. No problem with running and manning the rigs I am puzzled by the pump jacks. I have checked out every pump on the lease about sixteen in number. All of the wells are on a timer but turned off. When I activated them most of pumps worked, the motors turned and the donkey head nodded up and down, but I don't know if they were making any oil. After sitting idle so long they probably all need servicing. In fact JP had me look at an old twin boom well workover rig he is buying for cheap. I got it running and he is working on a road permit to move it the two or so miles on the county road that it needs to travel on to get to the lease. I tell the crew to split up and look busy, can't tell them why, just that's what JP wants. After turning everything on and waiting a reasonable amount of time to make sure none of the old gear blew up or caught fire I make my way back to the trailer to wait for my call. JP is flying some people up to Possum Kingdom for lunch, I am to drive up and meet him there; he's going to buzz the lease on the way. All I have to say is "You have a plane?" My Dad is a pilot and he doesn't have a plane. I am a mess. I tell him that I can't show up like that but he tells me that is exactly what he wants. Not long afterwards here comes JP down on the deck like Sky King flying a Cessna 310 twin, he circles the rigs a couple of times beats up the lease and then climbs up and to the north. I drive around in the Dodge turning everything off and then I'm off. Life is good, hell it's great! Time passes fast. I'm there way too soon.

Possum Kingdom is a repeat of the San Angelo steakhouse experience. JP has flown up five people, more folks have driven up, must be about twenty people in the room. The flyers are excitedly telling the drivers about their flight. Everyone grins at me, the young engineering student/tool pusher that's making it all happen. Again I am excused after making my manners. They are ready to get down to business. I am again uninterested in what is transpiring. I pig out and have a nice ride back to the lease just in time to shut down operations for the day.

Monday morning I get a call from JP. One of Wilson's prize bulls got tangled in a pump jack and had to be put down. I have three pumps working one hour every six; as luck would have it he has to be by one that was running. The bull should have been secure in his own pasture. Either the fence has a break in it or someone left the gate open. Hector should have caught all that. I need to get out there and talk to Wilson.

Wilson is a huge guy he looks like John Wayne's bigger brother. He tells me that he has about a thousand pounds of barbeque now as he pushes his straw hat back on his head. I like the man and I really hate that I've disappointed him. I tell him that I'll find out how it happened. I go looking for Hector. I jump a herd of goats that are where they shouldn't be. It takes me quite a while to get them back in their pasture. Goats are disgusting, they always turn, crap, and flee, disgusting to chase after. I also find some sheep in the oats, two of them are already belly up from eating too much. They are easier to herd. Goats are always looking for somewhere to escape and sheep are always looking for someplace to die. I finally find Hector at his tent. He is not looking so good. He tells me that he is sick. He shows me his bloody stool. The hell with the critters. I immediately load Hector in the Duster and take him to the doctor In Ranger where I'm a regular now. The doctor checks him out, his Spanish is better than mine. He tells me that the best thing that I can do is send him home to his family. I go to pay the bill but the doctor says forget it. He gives Hector something to sooth his stomach ache and something for the pain. I take poor Hector to the bus station and buy him a ticket to Matamoras. I give my poor sick friend a hug goodbye and slip him all the cash I have on hand, about sixty bucks, he looks happy to be headed home to folks that care about him. I never asked him how long he had been away.

We are still making hole. Mid-week I get a call from the driller first thing in the morning. He says that a trucker has arrived on site with a bill of sale for our drill string and is there to take possession. I tell the driller not to do anything and wait for me to arrive. I call JP. He tells me to get right out there. He would drive out himself but he has too many speeding tickets. That explains why he never drives out to the lease. I head east as fast as the Duster will go. When I get on location the crew is just standing around, the drill string is gone, they helped the trucker load it, could not have done it without their help. The driller tells me that the trucker had a bill of lading and a bill of sale. I ask if they even got a receipt for the load. No they had not. I tell them "congratulations you have just put yourself out of work, you are all fired. Gather all your PSE and get off the lease". I call JP, he tells me to call the Stephens County Sheriff who he knows. JP says "he has the fastest car in the county". I call the Sheriff a large imposing Black man. He asks me to jump in and we go in pursuit down the only likely road in the area that a truck might take. JP is right the Sherriff has a fast car; he's a good driver and he knows the road. The truck has an hour at most head start on us. We burn a lot of gas with no luck. The Sherriff and I are buds by now. He drives me back to the lease. The Sherriff takes my number and promises to keep on it. I thank him for his help. I am pretty down. Just when I had things sort of working. After all that work fishing it out of the hole just to have the boys give it away. I grab dinner in Cisco. I am in no hurry to go home.

The next morning the Sherriff calls the house Mom answers the kitchen phone, she's immediately alarmed that the Sheriff is calling the house asking for me. She has questions. I haven't told her anything about what happened. Good news. He has located the tractor trailer with the drill string still loaded on the float. It's in Breckenridge, in front of the trucker's house. I ask him if I can drive up there and "watch the consignment" the Sherriff allows that

I can. Mom has been eavesdropping of course. I fill her in. Mom is adamant that I not stake out the truck. She gets more insistent as I grab a pack of lunch meat and a half loaf of bread. She follows me upstairs as I pack a bag and into the bathroom as I go for my tooth brush. Dad has overheard everything. He tells me to take a pistol. I don't think that is a good idea; that might put me on the wrong side of my friend the Sherriff. I am going to be all smiles. I head for Breckinridge in the duster with a carton of Kools on the dash. It's a Wednesday morning and its already hot, DP and fried pie time.

The truck is easy to find. Its parked in front of the trucker's house in an old tree lined residential area just off the main drag. I park in front of a vacant lot a few houses down. I immediately regret not having the foresight to bring something to read. I'm able to get a good AM station KROX out of Fort Worth though. Mainly I'm obsessed. I worked so hard to get things going. The big rig is useless without a drill string. I am still mad at the crew. They could have left half the string in the hole and laughed at the trucker, but no, they actually pulled the string and loaded it on the truck for him. I don't get out of the car or approach the trucker's house. I do drive by real slow, that's my drill string alright; I recognize it by the tool joints and drill collars. I really don't know what to do or what I'm doing but I'm doing it. I called JP before I left he had nothing to offer except concern for me wishing me good luck. There is only a dozen houses on the street. 50's style postwar clapboard, real quiet except for the constant buzzing of the cicadas. By night fall, everyone on the block must have called the Sherriff or one another about me, he's already talked to the trucker. The Sherriff is just going to let this play out. I sleep in the car, not letting it out of my sight.

I have breakfast lunch, dinner and morning absolutions at Dairy Queen; there seems to be a DQ in every small Texas town. The air conditioning is a welcome respite from sitting in a hot car. The only break in my routine is meeting a gaggle of bikers up from Houston, they are headed for Anchorage, and they are taking backroad America to get there. Their south sides are much sunburned. I think that not a few will ever get out of Texas but they are a happy bunch with good attitudes, all decked out in red white and blue.

Truckers do not make money unless they are rolling. From the looks of his property and rig this trucker could use some money. That old cab-over Freightliner looks pretty sad. Motivated, no bonded commercial driver would be trucking a hot load. I have no idea who is behind this or what is going to happen but I am determined to know. I'm like Mom, a dog with a bone now. Don't know who is going to blink first, can't even think of eating another DQ Dude and I am down to my last pack of cigs. Its late Friday afternoon. I really don't relish spending the weekend here. Mom is probably going nuts. No news is good new right? Not for her.

Its late afternoon. There's some movement at the truckers house. The trucker's portly wife and skinny son come out of the house. She is wearing a loud flower print Mumu and flip flops. He is sporting Jeff Davis facial hair; he's wearing jeans and a wife-beater, a cigarette is hanging from his lips. They fire up the truck and idle it for a while warming it up. I keep waiting for the old man to make a showing but no, the truck eases off the curb and starts rolling. What a chump, sending his old lady and kid off with a hot load to do his dirty work for him. Maybe I unnerved him sitting there for three days, don't know. I start the Duster and head out behind them. Finally something is going to happen.

They head south; I stay on their six. They are headed for Abilene. I am extremely excited; time passes quickly. The truck pulls into an equipment yard on the northeast part of town. The yard has a big sign up that says Auction Saturday. They are going to try and sell it; that's clear to me now. I follow them into the yard and beat them to the yard manager who has come out of the building. There's a thirtyish tall blond woman behind him. I tell the manager that what's on that truck is stolen property. I tell him that it was stolen right off my job site. At this the woman behind him explodes. She is Johnson's daughter. He is sending his kid to do his dirty work too. Perfect. A couple of spineless old men cowed by a nineteen year old kid. She screams scornfully at the manager spittle flying "do you know who he is? He's JP Goodfellow's errand boy". The manager wants none of this, he turns to the truckers, their mouths are hanging open, they evidently know nothing; he says "get that shit off my lot. If it's found here they'll shut me down." Ms. Johnson is furious, she curses all of us. The truckers pull out, I follow. I guess they are going home. Ms. Johnson jumps in her Cadillac and is right behind me.

It's dusk I have my lights on, We are not thirty minutes on the road when the truck slows on an uphill grade, pulls over on the shoulder and stops; we all exit our vehicles. The air smells like something is burning. Joe Dirt tells me that the clutch is all burned up. Ms Johnson has been tailgating me, weaving and flashing her high beams. She reeks of alcohol; she is not pleased that the truck is stuck. She curses all of us roundly and heads for her car screaming that she is going to go get her gun and shoot me. She makes a sloppy U turn and speeds off weaving. I think that she partook of some liquid courage before we even showed up at the auction yard.

I talk to the Mom and the Son. I sympathize with their plight, stuck on the road and all. The son and I talk mechanics and how to fix it. We tip the cab up get out flash lights and put out reflectors. The Boy and I are friends now. He reckons that he can get a part and be out in the morning with some tools and his Dad to fix it. Just then a Taylor county Deputy Sherriff pulls up and puts on his flashers. Nice fellow. The truckers tell him their sad story; they are reluctant to leave their truck and the load sitting on the road. They can't afford a tow to Breckenridge. I tell everyone that I will watch the truck, heck I have been watching it for days and I add "I'm the only one getting paid to be here", not sure of that, but it gets a laugh. The truckers are concerned about me, they tell the Deputy about the threat on my life. He wants to wait and see if she make good on her threat. We wait about ninety minutes then he runs the truckers home. I immediately head south and call JP from a phone booth. I tell him what happened. We are in Taylor County. He and the Lawyer get a judge to grant an impound order that night. He has a tow truck and another tractor out there within an hour. I follow it home to Abilene. Don't know if I did any good here. Still don't have the drill string. It probably will be tied up in legal proceedings, but that is between JP and Johnson. I take a long hot shower in the wee hours; I feel real dirty in more ways than one.

I have breakfast with JP' the next morning. Impounded drill strings make no holes so I still have a problem. JP thinks that the little rig can make it to two thousand feet. The drill string is about two thirds the diameter of the large rig's. It can only turn a six and a half diameter bit max. I tell Bob that I will give it a shot if I can hire a couple of hands to help. I will maintain my expense and hour log that I had established for the old job. This is the beginning of a long grueling effort.

Kevin and Stormy are kicking around town doing nothing at the time mostly playing foosball at the Arcade, so I ask them if they want to do some cash work, they do. We rent the penthouse suite, essentially the whole top floor, in the Old Historic Eastland Hotel to eliminate the commute. The Penthouse has a much appreciated window A/C unit, a kitchenette and an adjoining room. We eat mass quantities of everything and try to hit the bats that fly up and down the hall with tennis racquets in the evenings; we are too slow but we get some laughs. We work seven days a week marking time and the ROP rate of penetration. We complete two holes with no incidents. I go to the tank to get more water flowing into the pits one afternoon. On my way back I see the derrick waving and wobbling on the horizon. I tear down the dirt road and on to the lease as fast as I can. Kevin and Stormy had been "smoking" while I was away. There they are, napping on the pipe rack while the ass end of the truck is jacked about four feet in the air totally off the out riggers supported only by the turning drill string. They could have been killed. Like I said this rig needs constant attention. As we complete each hole we string galvanized pipe that we thread and make up with centralizers on the collars down the hole. Centralizers are birdcage clamps that keep the pipe in the center of the bore hole when you go to complete or cement in the well. I throw a big end bearing in the Duster going to Midland to get the centralizers. The Duster hits the dust. JP gives me a pink Cadillac Coupe De Ville with a white vinyl Landau roof to drive that would have done a Mary Kay rep justice. The windows do not work but the A/C does. One of JP's girlfriends drove it into a lake when she got mad at him. Water must have got into the motor too because it seizes up after a few days, then I have the T-bird. I lose it when the lease is up. I am on my own driving my Vega before the summer is over. Haliburton comes out to log, cement, perforate and frack as I complete each well. We use what we have in a tank battery for fracking fluid. We plumb the newly completed wells to the salt water injection pump manifold. That is quite an operation and quite expensive I presume. I never see the bill. Operations sort of come to a halt when fall comes.

I apply to UT Arlington, its close enough to the lease where I can go out there and work some. There are more wells to drill but I never drill them. I am the only one on the lease now. I try to do what I can by myself. I really hurt my back fixing a leaky flow line by myself. I lay immobile on the floor of my apartment for a couple of days helpless and mad at myself. Gives me time to assess. I worked like a dog and have little to show for it. I did manage to fund some of my school for this semester but it's not enough, I need to keep working. Arlington doesn't offer much in the way of engineering. I am taking English history and government which I enjoy and Latin which I don't, I'm tinkering with the Idea of being a Lawyer who specializes in technical contractual law thinking that there would be big bucks in it, but now I want to just get an engineering degree and simply practice.

I know the lease like the back of my hand now. I tell JP that there are four tanks full of oil and ask about selling it. The tanks are hooked up to a pipeline owned and operated by Permian. JP says that I only have to raise a flag on the County road, and identify the tanks I want to sell with a note in the jelly jar below the flag. I do it. I get a call from the Permian guy to meet up. He says that he can't take my oil. He tells me that he hasn't bought oil from the lease in years. He tells me that the oil in the tank has been in there for so long it has stratified. All the tars have settled to the bottom of the tank and all the light oil is floating on the top, a gravity crack. I must have looked real sad because he says: "tell you what. You get a hose and hook it to your car's exhaust; put a length of pipe with a tee on the end and attach it to the hose. Then you heat and stir the oil pushing it back and forth across the bottom until it is homogenous again", pasteurized crude. He says you can tell its mixed when you drop a depth tape in the tank and the weight on the end comes up clean". Sounds pretty dammed dangerous like the spark of your ignition could cause a huge fuel air explosion on top of a crude oil cocktail. They won't even find enough left of me to put in a match box. I do it; he buys it. I'm driving JP's new lease car when I do this. I find some of his presentation material in it which I read at a safe distance. They are Irrevocable Bonds. Instruments he is selling in his Ranger operations. To be realized at some future date. I know that this place is all out go, and no income, past the salt water injector system and the four tanks of old crude that I sold. I have been fiddling with the timers the best I can do by my estimation is fifteen dollars a day from each well and he only gets a percentage of that. Can't be more than 3K a month. Everything JP has, except his plane, is rented, borrowed, or leased. He likes to live like a rich guy, a high roller. I must admit that I like his lifestyle but there is no depth. I keep thinking about Johnson's daughter the absolute hate on her face and the bile she spat at me. I truly believe she would have shot me if she had been sober enough to drive back. JP must have done Johnson dirty, dirty like these Bonds. Can't help but think that I'm next. I was working for cash. No withholding. Treating me like a contractor. I have no health insurance. I think about all that money funneled through me. I think that JP is going to eat my soul or do me dirty too, neither is a good thing. Can't trust him. No integrity.

I go home to visit for the weekend. It's Friday night Mom will be expecting me home. She will be pacing around drinking coffee until I arrive. She wants to talk. Mom has come through for me. Neither she nor Dad like JP but they. Mom thinks that JP is ruining my credit and putting me at risk. Neither one of them liked me chasing drill joints, the dangerous work and me living in a dive in Eastland that has bats in the halls. The parents have just come back from a USAA trip to Guadalajara. Mom was very popular as the group's interpreter. She managed to make friends with the Whites'. Mr. White is an engineer with LTV Aerospace who is building the F-16, they live in Arlington. He is going to get me a part time job at DFW airport, a public relations job, I just have to wear slacks, a coat and a tie; that's different.

JP wants to talk to me but that bridge is burned, there's no going back. I like JP; I like my brother Jim, but I wouldn't work for or trust either one of them. JP is just a bachelor with baggage and an eighth grade education, doing the best he can. Toiling in the oilfields since he was thirteen, doing whatever it takes. Makes you ruthless. I once asked Jim about his long range plans he said that he would hire competent people to work for him. Yeah, you see that all that time, competent people working for incompetents I can't deal with people who have no integrity, their word means nothing; you can't trust what they say, half-truths; some say it keeps you on your toes. The Chinese would call it Monkey Clever. I call it lying. A couple of semesters for me at Arlington then next fall I will go to UT Austin. I don't know what my new work situation will be like.

## DFW

My new job opens my eyes to large scale human factors. DFW, from the air and on paper the layout concept is great. You have a central artery for road traffic parallel to the runways, linear instead of a hub and a perimeter road. There are two runways on either side with the terminals, semi circles in-between. The airport layout lends itself to future expansion. There are parking lots on either end of the artery, you don't have to drive through the airport to park. The main drawback to the layout is when you want to go from one side of the artery to the other. A plane landing on the opposite side of the facility from his terminal must taxi down to the end and all the way around the terminals to get to the other side; the same with people who need to make a connection with a different airline that has their terminal on the other side, common with international flights routing through customs, and then there is the baggage and freight. To knit all this together and keep the people moving LTV has designed and built the Air-Trans system. Air-Trans is the first unmanned computer controlled robotic airport train system. A lot of assumptions were made in the design the impact of which are only realized after it was built. Everything is made of concrete so is the train track; a cast U form it runs around like a roller coaster on the ground, under roads up and over itself. Since it is on the ground like a big gutter the three high voltage rails, to power the electric train cars, are run on the side of the U to keep them from shorting out if the gutter gets full of water. DFW was farm land. The airport disturbed and displaced many critters. The track is a better mouse trap. Anything in the track that wants out will eventually try to get out by climbing on the power rails with the inevitable consequence. The thing is full of dead animals great and small.

There are individual cars with Styrofoam filled rubber tires that run in the channel. They are remotely controlled by a master computer. The software is still in development. The cars have been running empty while they debug the software. Now it's time to put it on line. We are required to be there for any disruptions in service. Some of us ride the cars while others man the stations. The people on the cars are there because there is no driver and this may upset some people. We are also there to calm the riders if the car should stop, which happens quite a bit, we are to prevent them from trying to exit the car. Folks get excited when they think they are going to miss their connection sitting in a stalled driverless contraption with no phone service. The cars also inch worm, surge and stop, they are trying to maintain a separation between cars and there is resonance in the software as programmers try to optimize its flow real time using primitive processors.

The people in the stations have another job. The system is late and over budget. All airlines are required to pay into and use the system. Some are resisting, insisting on using buses. Someone had the bright idea of charging passengers to ride. They have installed a turnstile that takes quarters. There is also a change machine for those who have no quarters. Passengers do not like going through turnstiles with luggage. They do not like fishing for quarters when they are in a hurry and their arms are full. The novel, at the time, change machine will actually give change for a paper dollar; but it only gives seventy-five cents to the dollar and you still have to put a quarter in the turnstile. Once you are past the turnstile the passengers are lost and angry. The signs are mounted too high and are very confusing, they are alphabetical lookup correlated to gates, unrelated to an airline, instead of color-coded routing. There is no disembodied voice announcing the arrival and destination of the trains in the stations or on the cars. Some trains will eventually get you to your gate but after a complete tour of the airport if you get on the wrong one.

I tell flustered agitated people to just jump the turnstile and help them with their luggage. Most people pop in a quarter and grumble. I point up to the map when they look at me and answer any questions they have. No one ever gets mad at me, I smile, agree with their comments, note them and assure them that changes are underway, in fact they are. I have this job for about nine months before my services are no longer required. The job is ok, I don't get dirty. I dress like a gentleman and I get to work on my people skills. I can read my textbooks in the stations when it's slow and when I ride the cars all the way to the parking lots. I do my homework at the airport hotel over my dinner/lunch break. I work every peak holiday period when everyone is traveling and everyone wants off to try to put back some money for Austin, hard to do working part time paying car payments and rent. Lots of lessons learned. American Airlines headquartered at DFW will eventually put in their own excellent elevated Sky Trans system that address all these issues. I manage to keep my job until all the change machines and turnstiles are removed, the signs are all changed at the same time. It's the end of summer.

My pre-law efforts at Arlington are not enjoyable. Latin is a bust, I hate it, it's not like the Spanish I took for four years, a grievous miscalculation, I thought I needed Latin for all the law terms for my engineering law degree. I am disillusioned, I don't like the folks vying to be lawyers. I love English History and the origin of Common Law. My English History Professor is a man who mesmerizes me with his lectures. But I soon hunger for the black and white of Physics and Math instead of the gray area of law, it all seems a matter of pettifogging or obfuscations and the sowing of doubt …now called spin. I want motors moving parts predictable repetitive behavior, Mechanical Engineering, it's something physical and predictable a tangible thing that you can see, feel, and hear; either it works/runs/does the job or it doesn't. I am more than ever determined to go back to Austin, go to school at UT and be a Mechanical Engineer.

# 1975 Retirement

**JJS.** 1 April 1975, after thirty two years seven months and two days of service, I retire from the Air Force with over 12,000 Hours to my credit. I fly my T-29, the last 100 octane bird in the inventory, to the boneyard; it retires with me.

**RES.** The Spring of 1975 is a time for endings. Poor Grandpa Nuno dies In March of this year. He passes in his own bed in his own house with a Priest at his side. Grandpa suffered from congestive heart disease among other ailments. When he passes mom makes no claim on her inheritance. She signs over the stores to Sal and his boys who worked hard for it.

Uncle Bill contracted an undiagnosed wasting disease where he was unable to take nourishment that turned out to be stomach cancer, a terrible thing to happen to a strong physical specimen like Bill or anyone, God rest his soul. Bill was an Army quartermaster during the occupation of Japan, making trips to feed the victims of the Atom bombs may have been exposed to and ingested radioactive material about which not much was not known at the time. Bill had moved from Albuquerque back to the LA area as the regional manager for Winston Salem. Grandma Nuno had become less able and had moved in with Bill before she passed making her final years a pleasure for the both of them.

When Dad retires he first tries selling insurance. Dad knows a lot about insurance. But probably as a customer. I have a list, a spread sheet he made of all the term life policies he has, over a dozen plus, one on each of us under two organizations. Here is a guy that thinks a lot about dying.

You are what you do. It must have been hard on Dad to retire, like a divorce. You lose your identity of thirty two years. No one has to salute anymore. You are no longer responsible for a thousand people and ten thousand things. Then there is what to wear? The Air Force has been telling him that every year and every season of his working life. You can always spot a recently retired lifer, nothing matches, socks with sandals, striped or checked shorts with cowboy belt, topped with a short sleeved button down and bolo tie. Selling Insurance doesn't last long; Dad is no salesman.

Dad starts at Abilene Areo. He teaches ground school, instrument rating, single and multi-engine, and navigation; he also flies charter. They really like and value him. We all get to fly with Dad. Gary takes ground school from Dad. Dad flies pipeline inspection with Gary and Tim for the Railroad Commission, down on the deck looking for leaks and dodging buzzards. Gary is still in high school, he has good eyes and is very coordinated. He would make a good pilot. I take a couple of trips to California with Dad, just the two of us, one to my Cousin Dan Scanlon's wedding who I haven't seen for eleven years. Pat is there too. I get to meet his lovely wife Robin and Dan's bride. Dan is a Professor of mathematics at UC Irvine, Pat has a sweet job With Disney. Happy people. Happy times. Dad gives me a course in aero map reading and radio navigation, operating the autopilot, he keeps a running dialog of what he is doing asking me questions to see if I am following him. I get some stick time in too. We head for the Hoover Dam to check it out. Dad's head is always on the swivel. He is spotting things in the air and on the ground that I simply cannot see, eagle eyes. As we approach the dam he spots a sight-seeing plane miles way just a spec, he peels off and we are on its six in an instant. Dad skids the plane, banks and peels away while telling me that you have to kick the rudder to rake em. They never knew we were there. This is Spring break for me. I attend spring, fall, and summer sessions w/o a break. We stop in Las Vegas and stay with Bob White, Dad's old High school and Army Air Corps Buddy. Bob White helped design the airport there. They are nice people. They take us to the strip and show us around. Next it's off to LA, Orange County, and John Wayne Airport. Coming over the mountains the brown orange smog is so thick Dad heads out to sea so we can drop down and come in under it. Then some amateur tries to land on the runway that we were directed to so we have to go around. When we land Dad wants to look at planes like we do everywhere we land. Dad rents a car and we head over to see Lloyd and Madeline Clark in Laguna Beach. The Clarks have built their dream home on Crescent Cove, property that they bought in the 50's. It looks like a Japanese umbrella with walls of glass under pinning it on the cliff's edge overlooking the sea. They have a pool separating the house from a pool house apartment that I have to myself. I have been working like a suck egg plow mule for four years running. I don't know how to relax. I am very anxious and don't know what to do with myself. I go swimming and spearfishing in the mornings with Lloyd in some too big borrowed trunks. We have Chiviche for breakfast most mornings if we can spear something. Crescent Cove is an artist community, a very wealthy artist community. Lloyd is an artist too. It's an artist fair every day. I take long walks past miles of uninterrupted wealth. I am quite out of place here in in my Ropers and jeans. I've brought some study material but I can't bring myself to look at it even though midterms are coming up. Maybe one of these days I can do whatever I feel like in my dream house. Flying home we stop at the boneyard, after overflying it, to look at planes what else. We fly back to Abilene. This was a good trip, just the two of us. Dad also flies Bill up to Canada in a Mooney to go Guide fishing. A good trip for the two of them.

**Fall 1975.** Steve and I are enrolling for the fall semester at UT Austin. I drive to Bolder in the War Wagon and help Steve move out of his place in Boulder Canyon, since he can't attend CU anymore. The state will no longer grant instate tuition for Military Families, even if they own property and have paid taxes for many years. There is a long continuous physical residency requirement now, hard for military families constantly on the move to meet. Anti-military backlash by the state, stemming from the Vietnam War, curious because of the many military establishments in Colorado and the money that they bring to the economy. Dad sold the house in the Springs the summer of 1975. Steve has been attending summer semester. Steve and I drop off the War Wagon in Abilene, load up the Vega and head south.

We meet Kyle, Kevin's brother, at the Armadillo Beer Garden. He is going to let us crash at his place till we get settled. You can just go through a hole in the wood privacy fence around the beer garden, left unrepaired for the underage. We are happy to be in Austin. We have lots of fun before school starts, hitting the Back Room and the Pub with other students. It's nice but I need to find work.

Steve and I have spent everything we have on tuition and books, then there is my car payment. We have also run out of food. Kyle is in Abilene visiting and getting more of his stuff. All we have to eat is a sack of frozen green beans, some flour and some margarine. We have fried breaded green beans for several days before we land a job as pizza makers for Mr. Gatties. We ae hired to man the new flagship store on the corner of 19$^{th}$, (now MLK) and Guadalupe where the old Waffle House used to be, right next to campus. I remember seeing Johnny Winter there eating breakfast like me. The Waffle House has been torn down and a new building constructed on the site. They have hired a flagship crew for the flagship store everyone is young and fairly attractive. We have to attend training to learn the company standards and procedures required for a pizza chef. We learn all about pizzas and how to make them the Mr. Gatties way. Steve and I are so starved that we both pop toppings in our mouths whenever we can. We both get food poisoning from that, right after eating greasy green beans for a couple of days with predictable results. We still have to go to training, so we soldier up and only eat Provolone to put a cork in our gut.

Kyle comes back and feeds us starving boys. We start work a week before classes begin, when students start moving in, then its class in the morning and work afternoons and evenings. We get our first check, it's pretty meager but enough to keep rolling. We get to do some fun stuff, see midnight movies at Dobie mall; go places on the drag and Armadillo World Headquarters.

Work is pretty slow at first, there is usually only two of us working at any time. If you don't have a rag in your hand the manager sends you home. Our manager is one of the people who trained us.

He looks like Chuck Norris. He likes to wear a brown polyester suit and a wide brown tie. We have polyester uniforms too, brown, yellow, and red, vertical striped polyester smocks that are hot as hell to wear in front of the oven. They get gross real quick. Business picks up after the first home game when the frat/sorority community discovers us. We have a large open restaurant. Besides making some great raised yeast dough pizzas, we also serve draft beer by the pitcher, spaghetti with garlic bread, and fresh green salads; we even have an Ice-cream bar. The girls like salads and the guys like pizza and beer. Pretty soon there is enough traffic to keep a crew of five busy all night 1700 to 2300. My shift settles down to Steve, Cowboy Jon, Diane and Debbie. I do most of the cooking, I never burn a pie. We are a happy lot, always joking and laughing.

Steve and I find a cheap apartment at the corner of HW 183 and Lamar way on the north side of the river, not the best part of town. We drive to intermural fields, halfway, to catch a bus to campus. We drive my car to work since we get out so late. The girls at work are funny they like to sneak a spoon of Blue Bell while we are cleaning up after closing, thinking no one sees them but their giggles give them away. We chow down on burnt pies and beer pitcher dregs. The restaurant has a great sound system but it only plays Janise Joplin with Big Brother and the Holding Company on an endless loop player. Diane, a pretty petite brown eyed girl who wears her mass of brunette hair in a long thick braid down her back has been flirting with me. I have been flirting with Debbie; I can't help it, she reminds me of my high school sweetheart. Nineteen year old Debbie exudes femininity, she looks like a little Dutch girl with her blond hair, blue eyes and dimples. I tell Diane that. Diane gets mad; she tells me that Debbie is married. I am floored. I haven't seen a gold ring on her finger so I ask Debbie; she tells me yep, she's married. She has a stone ring on her ring finger like the ones you see at Oat Willie's, a head shop. In a dream like voice she says she and her husband just walked out on the beach at sunset, pledged their love and bonded themselves together for eternity. Great. I have pissed off Diane and Debbie is married, no harm done, we are still friends. Steve and I give Debbie lifts; she lives on the north end of town too, in Hyde Park, it's on the way.

I am working the oven when Debbie comes up to me and pulls me by the arm around the corner to the washing station. Debbie tells me that "Polyester Chuck" the manager, had asked her to come out back where he hit on her, offered her cocaine and then pinned her against the wall and started slobbering on her and groping away. Of course I'm pissed and want to confront him. Diane sees us and asks what's up. We tell her. Diane has the cooler head; she tells both of us to act normal. She is going to report "Chuck" to management that night. We have a new manager, Mike, the next day. Never trust a man in a polyester suit. Mike is a tall jovial thirty-year-old with a beer gut and a bushy Mario Bro mustache; he is one of us. He wears a smock like us and pitches in when things get busy. Mike is a good manager, he makes sure that we are happy in our work.

Debbie is mopey the next night. She is not her usual ethereal self. I think that she could walk in the woods and with just her presence, all the woodland creatures would attend her like she was Snow White. She tells me that her husband has turned into an abusive alcoholic. She has been trying to deal with it but now she is afraid; she is leaving him. Diane has offered to let Debbie stay at her place. Debbie asks me to help her get some things from her house after work. I pull up to their house and let her out. The house is dark. He hasn't left a light on for her. She asks me to wait. After ten minutes or so Debbie emerges with a small bag. Her husband is dead drunk, passed out. I drive her to Diane's place. I don't go in. She gives me a hug because that is who she is. It's very sad.

Trying to cheer Debbie up takes my mind off my own troubles. I keep telling her to stand up straight, stop hanging your head and sighing; I try anything I can to get her to smile again. Steve and I have after work parties at our place. Manager Mike lets us close up. He tallies up way before we finish cleaning up. We have glass pitchers and mugs that must be washed. The mugs and pitchers are placed in the freezer for the next day. We have aluminum pizza plates that have to be washed. We also have metal cutlery. The dining room has to be cleaned, it's carpeted. The kitchen needs to be wiped down and everything mopped with bleach before we leave. Since we are serving only draft beer it is easy to grab a couple of pitchers and some "burned" pies for the after midnight dinner party at my place. Debbie is the first one there. When I get to the apartment she is already inside. I wonder if I left the door unlocked and if my guitar is still there. Debbie says with a smile and a laugh that some friendly bikers broke in for her. Debbie has tamed the Banditos with a smile and earned our place a modicum of protection. Steve and New Age Cowboy Jon are real late; they got pulled over. New Age Cowboy Jon's 1950 Chevrolet pickup has a tail light out. When the officer walks up, he finds Steve with a pitcher of beer in each hand and boxes of pizza on the floor between his feet. They tell him their sad story: how this is their dinner, they are poor students, who just got off work... The police had mercy on students back then. He lets them go. Why ruin their life for that. We have our party. Debbie spends the night.

Debbie tells me that she loves me right off. I fall hard for her too I can't hide it. I tell her that she is on the rebound, vulnerable and still married! Debbie says she doesn't care what I say, she loves me. I tell her that I love her too, she convinces me. Debbie is an irresistible force of nature, we snap together like magnets. She's right, we were meant to be together. Debbie is the most sensitive, romantic, creature that I have ever met. I am smitten and entranced. She doesn't seem to have a mean bone in her body, just like my Mom. Her husband is a fool. How could he possibly even think to abuse her?

Diane helps Debbie move out the next day, dog and all. Debbie has a collie pup named Zeus. I wish that I could help her more but I can barely help myself. I just make enough to pay my bills and I'm not paying enough attention to my course work. The pizza shifts are taking up too much of my time. I finally lose it. It is a big at home game day; the place is packed. I am making pizzas till we are out of thawed out raised dough and cheese. I have put two more kegs on line. Mike the Manager calls central distributing for more stock and tells me to try to nuke some frozen dough, it's packaged in aluminum foil so I need to break it out and separate it with wax paper. We are out of pizza plates and glassware too so Mike the manger tells me to go buss tables and wash some dishes. I go into the dining room to find that the frat boys have decided to make a sculpture out of all our crockery cemented with half eaten food and beer soaked napkins. They have stacked everything three to four foot high on a table and the four chairs around it with mugs pitchers and pizza plates. I start trying to clean it up putting stuff in the buss tray I'm holding when a sorority girl comes up with a stack of dishes expecting me to take them from her. I look at her, drop what I am holding on the floor and take what she is holding, then drop that too. I announce that I quit as I march through the kitchen pulling off my smock and tossing it on the counter on the way out. I need to find something else. I can no longer engage in servitude to these morons, who are majoring in partying, Mrs., liberal arts general studies and making social contacts, coasting on their daddy's money. It's like high school kids but with money and no supervision. Their conversation is nonsensical narcissistic blather. There is a comic strip running in The Daily Texan that captures this exactly. The Academic Waltz by Berke Bretherd. The characters are all Izod, Kaki, Button downs and Aviators. Steve and Debbie are not too far behind. I don't eat pizza for years. When I hear Janis Joplin, it makes me want to puke. I swear then if I ever have kids, they are never going to work in food service.

Kyle, gets us a construction job for the short term after finals early December. We are hoping to work through the Christmas break. An ex-Navy Chief is building a controlled climate metal building storage facility this is going to be his business to support his retirement. Some of the slabs have been poured and now it's time to erect the beams, the Chief wants to set up the beams with a small tractor with a bucket and back hoe that he is operating. The chief is working this out as he goes along. We bolt the beam together with the legs on blocks. He has me throw a chain around the beam and hook the ends of the chain on the tractors bucket. He picks up the apex of the beam high enough to get his bucket under it then he had me block up under it and he lowers it down. It looks like it will slide and fall and I point this out. The Chief makes two men go buck the legs with their boots. He tells me to re-rig the chain as he releases the load and repositions the bucket under the beam. After it is chained with a loggers choke to the bucket he tells me to jump in the bucket and ride up with the load. I refuse. I tell him that it is too dangerous because he will be moving and will not have the out riggers deployed. I tell him I will climb up and unhook him after the beam is bolted down. The Chief lifts the load as we get clear. The machine is clearly undersized for what he is trying to do. The tractor is tipping and swaying on its tires under the load as he gets underway. His control is so bad that he is never able to set the beam on the lugs sticking out of the concrete. I tell him he is going to kill someone, that he should rent a crane. I add the job will go quicker be safer and you will save money. The chief gives up and sets the beam down. Steve and I are paid for the day then let go. Kyle tells me that I embarrassed the Chief; his advice for the future is to play dumb like he does. I was dumb alright, now I have no job. I still have money in the bank to pay my bills. I send in my car payment last month's rent is coming out of our deposit. I start looking for a cheaper place to live. Steve and I will have to split up since we don't have enough deposit money for a new place between us.

I see Debbie every day. We are madly in love with each other always hanging on each other, holding hands. Debbie buys a book of poetry with pictures making notes in the margin about us. We are getting around in my Vega applying for work. Debbie has a gas guzzling Jeep Cherokee Chief with an International Harvester V-8. The starter goes out when she drives over to see me. She and I spend some of our last funds to buy a new one. I install it; her parents thank me. We live right behind Dales Auto. I don't need a jack to do the work because her truck sits so high. The truck is great for driving around with her dog.

I come out of the apartment late one night after showering and shaving, I'm off to see Debbie at her place. My car isn't where I left it. I frantically run around the small parking lot when I can see every parking place from where I stood. I call the police; they inform me that my car has been repossessed. I can't believe it I have had that care for over a year and have more than $1200.0 in it; I just sent in my payment for Christ sakes. I was headed over to Debbie's because she had left her Mother's heirloom gold bracelet on my stick shift.

There is nothing I can do till the morning. I call Debbie, I'm falling apart. I tell her what happened. I walk down Lamar in the dark from 183 to Enfield Road so I have some time to think and some exercise to take the edge off my emotions. I calm down somewhat, When I get to Debbie's it's after midnight, she lets me in and loves me. She holds my head and tells me that everything is going to be alright. She is there for me because I was there for her, that's the way it works.

I call the bank the next day. Seems that I hadn't added in several bank monthly $3.50 service charges and so was over drawn about $3.00, even though I still had $75.00 on account. The credit Union rules insist that a minimum Balance of $75.00 be maintained to qualify for continued Membership. So they bounced my check charged me $20.00 for the favor and will sell my car to settle the debt. I am livid. I have no car money or food. Any business I have with them must be settled in person in Abilene; I don't even have bus fare. Proud me I go without eating for a couple of days, Debbie notices and starts feeding me. She takes me to her Mom's and robs the fringe. We go to her Grandma's house by the DPS headquarters off Koenig where that sweet lady stuffs me with baked chicken and cookies. Debbie even begs BBQ sandwiches for me from her ex father-in-law's restaurant, it's fatty and greasy, I wolf it down. My dear sweet friend helps me get by. Debbie is there for me at the lowest point of my life, she keeps me from despairing. She is the best friend I have ever had. She tells me that we are soulmates; I believe her to my bones she doesn't need to hang with a guy down on his luck but she does. She keeps me going to class and not giving up.

Steve gets a job as a bank teller and starts rooming with a coworker. I get a call back on one of my job applications it's a janitorial position with Montgomery Ward in Capitol Plaza about four miles away from where I live and no way to get there except hike. I can borrow a 10 speed from another student, so I can work. I can't keep the bike too long because he needs it back

I start my job riding through all kind of weather and traffic. I have to be there by 0430 to meet the crew for breakfast and assignments so the traffic is not too bad, I have no lights; dogs chase me when I cut through the neighborhoods to doge traffic. I get off at 1000 when the store opens for business. I can ride the shuttle bus to class from the Plaza and leave my bike secured at the store. Debbie helps me through until I get my settlement. The bank finally sells my car and distributes the proceeds. My share of the sale is $380.00, less the bounced check fee, plus my deposit, when I close my account; it nets me $455.00, wholly unnecessary. The people who repossessed my car stole Debbie's bracelet.

One of the guys in the warehouse loading dock crew posts a bill. He's selling his wife's 1967 MG Midget for $250.00; I buy it. The car needs tires, has no top, and runs like crap; it needs a valve job. The body is good even if the paint is not, no rust, no wrecks no dents, all the chrome is there. I'm very happy. I pull the head and give it to Dales to refurbish. It takes a week to get the head back, it has new guides, valves, springs and seals. I install it with a new head gasket set and new ignition parts; I have wheels. I throw some blankets over the shabby seats and take my girl for a ride cuz she has been neglected while all my attention has been on old Annabelle, Debbie's name for the car. Annabelle made her jealous, we had our first fight over her. Now we have made up. Debbie and I are very passionate, our feelings for each other run deep and strong I love her passion. Debbie wears her heart on her sleeve. We drive all over Austin in Annabelle.

I take the bike back. I have to apologize for the shape of the bike. I had a big wreck when a pit bull ran me down one morning.

I move to a place off 48th in the Avenues off Airport. The guy who holds the lease has sublet me a room. I found it through an ad posted at the Commons, it is closer to work and the shuttle bus stop. I wish Debbie and I could get a place together but she is over most of the time anyway. Her place has two other girls living with her and she doesn't want to share. I love this girl, I follow her around like a pup. She treats me like I'm her husband; I especially love that. I meet her brother David and his very pregnant wife Marcella. We help them clean and paint a basement apartment that they rented near campus, to make it baby friendly. He's going for his Masters at UT. I help Debbie find a lawyer for her divorce. I am there for her through all that. It's over for $250.00 a simple dissolution. She is mine.

My new roommates are both weird. The guy who holds the lease is a Naval Academy dropout. His dad was with the AVG, now he's a male stripper. He says the Academy did it to him. His girlfriend is over a lot but they are not there most evenings until very late. She has to keep an eye on him when he dances. Crazy guy, he has a parachute rigged over his king sized waterbed. My other roommate is a mediocre art major who pins up mediocre nudes all over the place. He has no girlfriend.

# 1976 Austin Brentwood

## Shade Tree Mechanic

I fix my friends cars for free. It never occurs to me to charge them. I just want to help in the naive notion that they will help me at some future time. The ROI is .1 percent but I do it anyway. I want people to think kindly of me. I am in the nice guy business. I pull a Chevy van engine, a 327 and install it in Kyle's Camaro. I kinda owe him. I also have a soft spot for females with car problems, I think that this gives me points but I am mistaken. They only want their turnkey freedom back. Girls get mad at me when their cars won't work, like I'm responsible for that. I do brake jobs on the War Wagon and my MG which I'm always tinkering with, I convert it from ball to tapered roller bearings in the front end. It still doesn't have a roof, the seats are crap but have some nice wool blankets, and the trips are usually short.

## The Humane Society

Jim has been living with Buzzy Brown, an old Cisco chum of mine, off Hearn Street near Town Lake, downtown Austin. Then with Mark Walker and Jeff DeMuynck since I bailed. Whenever I introduce Buzzy I introduce him as Buzzy Brown and his Band of Renown. Buzzy, a base player, is down here with McEwen, another Cisco alum, to hit the music scene. Buzzy is a lanky long haired towhead with bangs and thick black Buddy Holly glasses. He is from Eastland, McEwen looks like Ozzy with a scraggly Fu Manchu. Jim is still hustling. When he is waiting on deals he starts hanging out at the nearby Humane Society, first playing with the dogs, then volunteering to clean cages and walking dogs. They hire him for the night shift. Someone has to be there at all times for the animals. The night shift thins the pack. Jim has been working there since he failed to pick up Steve and Kathleen that one Christmas. Jim managed to get Stormy hired for the night shift to. I ask Jim about work for Steve and myself when we were out of work. He told us that he would check into it but he never did. He didn't want his brothers to work at the same place where they would expose his bullshit for what it was or show him up.

Jim rents a house off Brentwood with Stormy next to Muller Municipal Airport. To get back in Dad's good graces he asks me and Steve to move in with him. Trouble with that plan is, there are not enough bedrooms in this two bedroom house. Jim has the Master Stormy has the backroom. The plan is to stick me in a mud room leading to the back yard and put Steve in the FAN trailer till Jim can turn the garage into a room for him. Dad hauls the trailer down and sets it up, he leaves the War Wagon with the trailer, it needs something to move it if need be. Jim inherits the

War Wagon. He has no car; he and Stormy have been commuting on Stormy's Honda SL 350. That's a good enough reason for Dad to give Jim the truck.

This is the Brentwood house the scene of much goings on. Our friend Mary Davis rents a nearby house while Jeff DeMuynck and Lyn rent another, both within a block of us. Mark Walker is a permeant fixture on our living room couch along with a guy named Lester The Lizard. Mark is learning to play his Les Paul Junior on my Little Devil Amp. It sounds better than his Pig Nose. Between Stormy's stereo playing full blast and our jam sessions it is Loud, Loud, Loud, but then again we live next to the runway. Who could complain? Years later I go to a party at a nearby house. The new owner said that the "hippie house" was still a legend in the neighborhood.

Debbie is none too fond of the Hippie House, neither am I, it's dirty and we had more privacy at the last place, there's also a lot more traffic over here that just gets worse as time goes on. At least Jim and Stormy work at night and that gives us most of the evenings alone together.

Jim takes me to work one night to show me what "Doc Scanlon DVM" self-proclaimed doctor of Veterinarian Medicine, does all night. We drive on down in the War Wagon. I am going to help him hose out kennels, he lends me some rubber boots. We are doing the rounds, shifting hounds from kennel to kennel as we clean them. Jim has kept up a running monolog about this dog or that that, I later learn the relevance of. Jim is making his selection to meet a quota. It's spring and the place is over full with critters. After we clean the cat house he goes to work. Jim makes a mental note of any poodle mix or small yip yappy dogs, vicious dogs, dogs with medical problems etc; same with the cats.

Jim has two chambers. He can get several small dogs and cats in the perforated stainless steel sheet metal basket or one big dog thus his preference for small dogs. It frees up more kennels quicker, cats go even faster; sometimes he has to use a choke stick. He places the animals in the wheeled basket, closes and secures the lid. The basket is wheeled into a cylindrical vacuum chamber laying on its side, he closes and secures the lid then the animals are taken up to 70,000 feet, left there for 15 minutes or so till they go to "sleep". When the lid is opened and the basket is wheeled out, it is anything but sleep; most of the animals are turned inside out, there is a lot of cleanup involved. The animal carcass are piled on the floor in front of the roll up door in the back of the building. Jim injects pentobarbital into the hearts of the kittens and puppies. He says if you put them in the chamber they will often resuscitate in the dead pile. About 0300 a truck backs up to the rollup door outside. Jim rolls up the door. It's Mr. Nordyke. It says Dead Animal Pickup on his truck's door, he's the guy that gets the road kill; he looks like Quasimodo. Jim asks him "how's business Mr. Nordyke". Quasimodo says "picking up" they both laugh at this old saw as he gets after it with a pitch fork and a flat bladed long handled shovel. I help Jim hose down after Q leaves. There are floor drains all over this place for that purpose. There is no way to hose away what I just witnessed. Debbie would not like to know that I was even here to see it.

This is a happy time for me. Debbie makes life good, doesn't matter what we do she makes it nice. Like rides in a convertible on a warm Texas night, simple, uncomplicated. There are no mental games or agenda's, just love. My job is menial. I was hired by Mr. White the building engineer at Montgomery Wards. The other guys are Clint, Big Head Bobby and George, three Black guys that drive all the way in from Elgin and Giddings for this job. George, the wise guy, is always talking and teasing Bobby, a large good natured fellow, and Clint an older meticulous fellow who drives an immaculate green Karman Giha with a white roof. I take breakfast and breaks with them. George and Clint aspire to Mr. White's job, Bobby talks a lot about joining the Army; I'm the only student. Mr. White briefs me on the chiller, the furnace and the electrical distribution panel, just in case…most of the time I empty all the trash and vacuum all the rugs in the building. Sometimes Debbie will call, just to say she is thinking of me and that she loves me and when she will see me again. I run to catch her call in men's wear when I hear the phone ring. Who else would call that number at 0600? A call like that means a great deal to me; it gets me through the day with a light heart; no matter what I'm doing, swabbing toilets, burning trash or struggling with my assignments. I have something to look forward too when my day is done.

Debbie and I hit the Armadillo for entertainment or go to the 501 club on Red River to catch Plum Nelly, one of Debbie's favorite bands. Having a beer and dancing with my girl, life is good. On weekends we usually take a drive or do something with Zeus. We mainly go to Decker Lake up where her parents live to go swimming with the dog. Zeus looks like a mat of seaweed with a stick, his pointy nose, poking out of the mass at one end. Just the sight of him makes Debbie squeal and laugh. Debbie is the best friend I have ever had, she saved me from despair, she shared what she had when she had nothing herself. Riding back in the car I look at her and say to myself," if this is what it means to be married to Debbie Schmidt, that's what I want".

I am doing well at school. My summer semester grades were excellent, that's what being happy and having a schedule will do for you. Summer ends and the fall semester starts, I attend school year round.

In August Timmy moves in to the garage, he is going to commute to San Marcos and attend Southwest Texas State. He is down before the semester starts. Jim takes Steve and Tim out to hit the town. They go to Mother Earth off Lamar and 10th to see Nitzinger. He has a song that is called "you can tell this band is from Texas by the shit on their boots". I don't club because I have to be at work at 0430. Mary Davis assures me that my name and number is in the girl's bathroom rendered in 3 inch black magic marker, so I am there in name only. The boys close down the place and head home. Jim had probably done his usual rails followed by not a few highballs I do not know what he "treated "the boys to. Jim manages to miss the turn off the interstate access road to our neighborhood hitting a steel guywire supporting a power pole taking it out. Lots of sparks, transformer explosion etc. Jim backs off, drives down the block and pulls into the drive. The boys pile in. Jim calls the power company to complain about the outage. The lights and the air conditioner do not work. I am almost late to work, no alarm. Steve tells me about it later. Jim fixes the truck by replacing the fender with heavy 8 inch C channel. That with the holes in the front from that 64 Cadillac that Tim rear-ended, holes that look like bullet holes, give the War Wagon a real menacing look.

## Wisdom Teeth Fall 1976

I've been having recurrent pain in my jaw, my teeth are moving around; I grit them at night and torque my jaw. I also have a constant headache. I have been told that I need my wisdom teeth pulled but I have blown it off for a couple of years. I no longer have privileges at the Air Force clinic and I can't afford it myself. Word gets back to Mom and the word comes back that that Dad will take care of it. Doc, one of the old breed that plays doubles with Dad and who

I have also played at Rose park is willing to do us a favor, a cash deal, on the weekend, when his office would be normally closed. It's Halloween and a Saturday. Dad is able to pull it off, the planets and the stars have aligned. He is flying a corpse from San Antonio for burial in Oklahoma City in a Turbo Twin Beech Barron, 58 the one with a double cargo door, a pretty fast bird. He also has a return trip lined up, another corpse he will pick up in Brownwood that afternoon for deposit in San Antonio, I'm the dead weight supercargo. Austin and Abilene will be intermediate stops in his flight plan. I do my regular shift at Montgomery Ward and I don't have class. Debbie spent the night but left when I went to work at four in the morning, she's going to a party tonight with a bunch of her friends from High School. Jim gives me a lift to Austin Aero in the War Wagon after Dad calls. Dad is all smiles; he's decked out in bolo tie, gray Stenson, and boots. He's wearing his tooled belt with his name on the back, with the silver turquoise inlaid buckle that we all chipped in for and gave him for his birthday this June. Our passenger is strapped down securely in the back. Dad says that he would take us to lunch but he can't leave his passenger sitting in the plane and he's not eating. Our "passenger" is a large Hispanic gentleman about 6'3", weighing over 300lbs, he's secured with cargo straps over blankets. I'm not hungry.

Dad drops me off in Abilene Mom is there to pick me up; she drives me straight to Doc's office from Abilene Aero. We are let in by a gentlemen who looks like a tall, overweight, eighty five-year-old Captain Kangaroo. Doc is a spare lanky man of average height with a big nose, a short aged Mr. Green Jeans. CK tells me that Doc is indisposed, he will be doing the extractions. Great, subbed out. CK gives me two pills and a cup of water then sets me in a chair for X-rays. After he develops the film, CK tells me he will have to break up the teeth because the roots are either splayed out or curled under themselves or adjacent molars. Lots of nova-cane is injected. CK splits my teeth with something like a nut splitter. I can feel him wrenching it down, sense the vibration and hear it when the tooth cracks. I am wide awake when he twists the pieces out with a grind against adjacent teeth and my jaw bone. Captain Kangaroo is done; its prepaid. We leave after he instructs mom what to do with a woozy boy. I get no prescription or pain pills. We head to the house where I get to sleep it off a little in a recliner.

In what seems to me a very short time, Dad calls; Mom drives me to the airport while he is refueling. We are off as soon as I get there. It's a short hop to Brownwood, we don't gain too much altitude. We land and taxi in front of the hanger on the apron and shut down. A hearse pulls up with our "passenger" he is a tall lanky man who died on his back in a bed that elevates the torso and legs; rigor has set in, he's all bent, like a chair. The undertakers have to work to fish him in and get him secured. We stay in the plane for this 30 min. evolution while the plane rocks and sways on its chocked tires. We are off in short order, there isn't anything in the way of traffic at this airport. As we climb Dad nods to the back and says "this is the best kind of passenger, they never complain".

There's is a geologic feature in Texas called the Balcones Escarpment. The Escarpment separates the black dirt riverine country to the east from the limestone hill country, the Llano Escondido separates the high plains and high desert to the west. The escarpment runs right through Austin, it is the site of heavy weather and flash floods. Warm humid air, carried by the prevailing wind from the south, hits the escarpment and trains thunder storm cells north and south when it hits the dry lines coming in from the west off the high plains. As we climb Dad is setting the radio and bringing up the radar, he says "there's a quarter a million in the dash of this bird"; I don't need a radar to see that there is a line of weather to the south east as we hit 6000 feet. Dad tells me that he doesn't like to fly in clouds especially clouds like those ahead. He climbs over 13000 and starts weaving around under and over the base of the anvil topped clouds. It's like this the whole way, a little over an hour in this 200kt plane. The cabin is un-pressurized. The passenger isn't complaining. As for myself, I can feel my skin tightening as my face blows up like a balloon, my lips don't work from all that Novocain and I think I taste blood. The letdown doesn't shrink my hamster face.

When we land Dad doesn't linger, he has several hours of flying left. I call the house. No one picks up the phone so I walk. The place is dark when I get there. Everyone is off to one party or another, that's fine with me. I'm just glad that the party isn't here and for once there's no one sitting on the couch waiting on a Jim deal. I crash and pass out in my room. The next thing I know Debbie is there jumping on top of me. She is wearing her Regan Raider Cheerleader outfit, having just come from her party. The only light in the room is moonlight coming thru the screen door to the backyard. Debbie might have been drinking, she may have head butted me in the face trying to kiss me. I may have been unkind. I just cannot remember. Did I just dream it? The next day is Sunday. When I wake up there is blood on my pillow and sheets. I stagger into the bathroom to look in the mirror my face is all swollen and smeared with blood. I sleep the day away and study all night for a major test the next day after work.

Monday I call Debbie's place but no one picks up. After my test I try to get ahold of her again; I leave a message with her roommate. I never get a call back so I try again; her roommate says "she doesn't want to talk to you". Later I do a drive by but her car isn't there. Does she just need some time? She never said anything about that, quite the opposite. She is always coming by to see me, Debbie usually leaves a love note if I'm not there. Maybe after a few days she will tell me what's up.

A week goes by; I drive over to her place to see if I can catch her. It's evening. There are a lot of cars parked at the house. There seems to be a party going on in the backyard. I walk down the driveway. There is a chain link fence around the back yard to keep the dogs in, I see Debbie through the fence. She's sitting on the back porch steps next to a lanky guy with long black hair. She is holding her face in her hands. Is she upset? Is she crying? He has his arm around her. I go to open the gate when Diane comes running up with a broom. She tells me to "leave, just leave" running me off like some cur, she doesn't want trouble. I feel like I have just been pole-axed, I'm stunned. Am I dreaming this too? I leave. Don't know what else to do. I just lost the best friend I have ever had, a girl I wanted to marry and have no idea why.

I drive by Debbie's parents' house, both of them are teachers, her Dad is outside watering some plants. I pull over and get out of the MG. After some small talk, I unfairly ask him if he knows what is up with his daughter, of course he knows nothing. I breakdown and start bawling. He barely knows me, but he is kind to me, a good man I would have liked to know better. I never see Debbie again I am dead to her. Did I say something unforgivable or did she cheat and can't face me; I'll never know. I morn her loss for seven years, losing her colored all my future relations with everyone I meet. People really do get puppies then take them on one-way trips down a country dirt road. Debbie teaches me that a friend is just someone who will eventually let you down. Like the Col said "don't trust anyone". I'm still wearing my bracelet; she made me forget what was so painfully learned.

# CHAPTER 21

# 1977 Austin IRS

It's spring. I see an ad on TV; the IRS is hiring part time workers. The IRS has a regional service center in south Austin, they are a major employer in the area. I apply. The IRS has benefits, Ward's has none. I can type over 40wpm but I don't want to do transcription, all returns are manually transcribed on to a data base matrix for the IBM mainframe running Cobal to consider. I get a position as file clerk. My day is turned around instead of working 0500 to 1000 now it is 1700 to 2200 with a break for dinner. Clerks pull physical paper returns for audit, error correction and or examination. We get a stack of IBM punch cards with Julian date numbers printed on them. All documents are given a unique Julian date number in the order that they are received and entered into the system. The original paper documents are stacked on shelves until the end of the tax year in the order they are received. It's in the stacks that I meet my good friend and future roommate Tom Luckenbach. Tom is walking on his hands between the files showing off for the girls. Tom is a six foot four 230 pound blond haired blue eyed Adonis, Tomas mas guapo. Tom is very popular, I meet a lot of folks through Tom. Tom is a student at UT too, his roommate, Zach Anderson, is working here as well but as an Examiner. The Luckenbachs hail from Alice Texas but Tom spent most of his youth in Egypt and Saudi Arabia; his dad, Luke was a Tool Pusher for AARAMCO; we are kindred spirits. Zach calls him Killer. Tom was caught in a riot in Cairo. He ran his motorcycle through a crowd of rioters maybe hitting a few of them trying to escape. Tom is very bright and athletic, the middle child of five boys, he's studying mathematics and accounting, we hit it off right away. Luke is retired now, living his dream; he has built a new house and a detached garage/boat barn/apartment on his Granite Shoals lakefront property on Lake LBJ. Tom likes to get together a gang, mostly couples, to go out there on the weekends to party. Tom's Mom and Dad seem to enjoy it, the fun is contagious. We go swimming, boating, water skiing, have cookouts, drink beer, and camp under the stars or in the boat barn. Sometimes the gang goes country and western dancing at the HW 281 club. I meet a better class of people. I'm tired of Jim's posse, he likes to keep low company; Tom is paranoid when he comes by the house, I am too and I live there. I spend as much time as I can away from that hell hole. I'm no longer part of a couple so I don't go to the lake very often; feel kind of left out. Tom is dating Doreen Schneider a gorgeous young woman, of course. I like Doreen, she is an Army brat. Her dad was stationed at Tinian in the southern end of Taiwan about the same time I was there. Doreen is a hoot, always has a new joke to tell, she is the dinner manager at the Magic Time Machine, a theme restaurant just off Riverside Drive. She has a brother at UT studying Mechanical Engineering. She treats me like a brother.

 When the weather turns hot Tom teaches me that paid sick leave is for water skiing, in that it is just the thing to make you feel better.

Samantha goes with us, Sam is another UT student working the files; she's studying Computer Science. Sam is a, slim, clean cut, auburn haired beauty who never wears makeup, an athletic girl, with piecing dark brown eyes. She likes to run and play softball, she can also waterski. Sam's favorite tee shirt is from the Capitol 10,000 marathon. Tom is dating Sam now, a relationship change that I'm not privy to. After a long hot day on the lake we are headed back to town, there are four of us riding in Tom's Dad's old pickup truck. Sam elects to ride in the bed with me leaving Tom and Zach the cab. We have a long conversation during a beautiful warm Texas sunset heading East down HW1431. At some point during that ride Sam decides that she likes me. I'm not hitting on her or anything, after all she's dating my friend, in fact she tells me that I 'm too serious, Debbie has left me none too happy. No joy in Bob land. A few days later, back at work, Sam asks me to go to Eeyore's Birthday Party with her. Eeyore is Winnie the Poo's sad donkey friend who always loses his tail. Eeyore's birthday is celebrated every spring the last Saturday of April at Pees Park along Shoal Creek; costumes are encouraged. I don't think anything of it we are friends. I meet Sam, we don't know each other very well but talk is easy. Soon we are holding hands. The park is full of revelers but we find a quiet spot under a tree and sit on the grass. We ignore the crowd and start making out. It has been a while so I am lost to everything but the girl in front of me, I can't believe it. Over my happiness a shadow creeps, a real shadow. I look up and see Tom in a toga with a beer in his hand and disappointment clear on his face; he turns and walks away. We talk about what just happened; we resume our activity but that is cut short again when some drunk decides to pee on our tree. An omen? We move on.

Sam has me over to her place where I meet her roommates, cousins and sisters Frankie and Mattie, Czech Chicks. Frankie is the eldest, it's her house; she has a steady State job and a steady boyfriend, Sid. Sam and Mattie think that Frankie can do better but Frankie always defends him. Frankie is a thin happy soul with brown hair that she wears in a bob and has a big unreserved laugh. Mattie is a slim beautiful blue eyed blond who could be the poster girl for Sweden, Mattie has a lot of guys after her. Frankie and Mattie are from Smithville. Sam and her cousins are very tight, they have been hitting every dance hall between Austin and Flatonia since they were old enough to sneak out. These girls are funny and a lot of fun, I drop by after work occasionally to visit if I have time. We sit around the kitchen table nursing warm beers, smoking cigarettes talking about nothing and everything. I meet Sam's Mom and Dad and some of her sisters, they have a family of eleven. Sam looks like her Mom, she's holding a baby girl on her lap when I meet her, a beautiful woman. I go to church with Sam, we attend mass at St Ignatius where she went to school; she laughs at me because I turn serious. It has been a long time since I have been in a church, there was every prospect of being struck by lightning. I go jogging and throw around a ball with Sam, even play in a couple of softball games when her team's short a player. She would make a good tennis player but we never get around to playing. Sam gets a golden retriever pup that we name Zloty, after a gold Czech coin. He is great fun to play with in the evenings.

    The 1977 spring semester is ending. The IRS is reducing its workforce now that the April tax return deadline has passed. The IRS has many part time workers who are content to only work the tax season to make a little extra money. Sam quits for the summer to concentrate on school. Tom takes a position as a math TA at UT. I need this job and they keep me on. I see Sam when I can, maybe once or twice a week or during the weekend. We go dancing, Sam teachers me how to CW dance. We catch concerts and go camping during the summer. On one camping trip Mattie has her Dad's boat that we tow with Sam's car. We meet up with her date at Cypress Creek boat launch Lake Travis, launch and we are off. I don't know what Mattie is doing with this guy, Mark, who I have never met, he's a large, pudgy, unattractive fellow very unlike the guys Mattie usually dates. Mark tells me he is a butcher at HEB. We head to little Devil's Hollow skiing on the way. After we set up camp, no tent we are sleeping under the stars tonight, I corner Mattie and ask he about Mark. She tells me that he was just some guy who happened to ask her out at a time when she looking for someone to go camping with so she wouldn't be a third wheel. I feel sorry for Mark he is over the moon and he has no chance.

Its late summer 1977 exams have just finished up. I haven't seen Sam for a week, we have both been, understandably, very busy. It's the weekend, Saturday, so I go for a morning drive in my convertible to see my girl and my good friends, a single source of happiness for me. I hate to spend time at the Brentwood house. I am there to eat sleep and shower. I usually stay on campus or in my room/hallway/mudroom where I try to read and play my guitar. It's good exercise for my dodgy wrist. I can't even play my guitar now. I always keep the SG in its case, because I don't have a stand for it. Last week when I went to pull it out for a little stress relief, guitar players know what I mean, I find the neck broken. The headpiece is snapped off, there's a diagonal break below and across the back of the neck below the nut to the key for the high E string. There are also dents and scrapes on the back from a belt buckle; I don't wear a belt when I play it. I'm pissed. Of course no one in the house knows a thing about it. What can you do? I take my guitar to Waterloo and ask them to see what they can do, haven't heard back if it's kindling or not. Makes me sad; that was the nicest thing I had. Doesn't pay to have nice things. Some person, unknown to me, has destroyed something I really cared for. Doesn't pay to care either.

When I get to Frankie's house they are busy with weekly chores, laundry and cleaning, so different from my roomies and the dump I live in. Usually the cousins are happy to see me but today there are long faces, I feel like I am intruding on something. I make a crack like "who died?" I ask where Sam is; they tell me that she is in the back sewing. Sam is sewing a light skimpy looking cotton print summer jumper. I ask her what's up. Samantha is happy, unlike her cousins, she tells me that she is making something for a date she has that night. She matter-of-factly says "I met this guy, we are going to a ballgame, I like him and I want to see where it goes." I don't linger, don't know how to respond to that, and don't want her to see my unhappiness. Well at least she told me, even if I had to drive over here to find out, Tom had to find out by walking up on us while we were rolling around on the ground. I wouldn't have been able to handle finding out that way. With that I am dumped. I was sure that Sam was "the one". We never argue. I could find no-fault in her till now, this is a big one. I get it in the neck too. My world turns very dark. Doesn't pay to care for people either. I will miss my "friends". Guess the kitchen table caucus got together, found me wanting and agreed that Sam could do better. I am just a poor boy with no immediate prospects or money. Kind of wish I had not gone down this road. Now I'm twice as sad.

Well It must have gone somewhere for Sam and her ballplayer. I don't call her, and she doesn't call me. I am over at Steve's place. He has escaped Brentwood and now lives in a big house with some other students off avenue H in Hyde Park. I hang out with Steve and his roommates: Debbie a Nursing student, Gary, Debbie's boyfriend, a

History major and shuttle bus driver, Craig a Psychology major and Cecilia who is in Fashion Design, then there is Steve, who is majoring in Government, a very eclectic group. Gary is one of the Hippy shuttle bus drivers. He is on strike. The Hippy part time shuttle bus drivers are striking for higher wages and benefits. He is very vocal about it. I don't argue but I can't see why they are striking for a temporary job. Isn't he in college to get a good job after graduating or is he going to make a career of driving a bus. I predict that they are just going to ruin this gig. Professional drivers will be hired and they will soon be out of a job when the current service contract expires. We have dinner parties, lively conversation and good music. At Brentwood there is: smoking, drinking, loud heavy metal, no talking, just Jim holding court for a bunch of bozos who are waiting to score. We are all out in the intersection playing Frisbee when a car pulls up. Its Mattie's car. Mattie and Sam get out and walk over. Steve introduces everyone. They were just driving by and recognized us. It's pretty awkward, I wish that they hadn't stopped or I was somewhere else, they probably can see that in my face. They leave without saying much of anything. I leave too; let Steve fill the curious roomies in. Seeing Samantha just reminds me that I don't want anyone else, she was perfect, except for the fact that she doesn't feel the same way about me, she's looking for a TBD better man.

In the late fall of 1977 I apply for an Examiner position, there is paid training, the pay is better but the position is only open for 8 hour shifts. I sign up for the 1700 to 0200 shift, which has a lot of openings. I get off work when the bars close. I have no life beyond work, class, and study, no room for friends, fun or feelings. Engineering requires four hours of study each night. I keep my head down and keep working. Time flies, two semesters go by in a flash. I make the best grades I ever have made. Working so far south presents occasional transportation difficulties when I have car troubles. I have to hike from Riverside to Ben White a seven mile round trip at the height of summer. I get in trouble for wearing shorts when I have to do this. I don't take any shit from my supervisor because I am looking at a three and half mile hike at two in the morning when work lets out, followed by a bus ride home and another hike. My supervisor overlooks this and suggests that I put a pair of jeans in my back pack and change when I get in next time.

I do drop in on Doreen and her roommate Evelyn sometimes on my way to work, if I have an hour or so to kill before work. They live off of Riverside, halfway home. Doreen and I go out on dates. She still burns a candle for Tom and I am burning one for Sam. Doreen understandably dislikes Sam; she calls her "that girl whose last name sounds like a female part". We go swimming in Lake Travis, tubing down the Comal River, and Barton Creek. Doreen takes me to Port Aransas with Evelyn. We are purely platonic I'm just there to keep jerks away. Doreen takes me out to dinner and clubbing during SIN (Service Industry Nights), she gets lots of discounts.

Sometime in late spring 1977 I get a call from Samantha. I am surprised, thought she dumped me for some ballplayer over six months ago. She says that she landed a programming job at Applied Research Laboratories, a UT satellite campus in north Austin. She says I should apply too, "because there are a bunch of guys here like you", that's about as deep as it got. I drop by and fill out an application on a long shot tip, from some girl who no longer cares for me, if she ever did. They tell me that there are no current openings, they have hired everyone they need for the summer, when students are generally hired and work. Translation: you are late, you missed the boat, no chance till next year fellow, go away. I don't hear back so I don't think anything of it, bad tip, from a sorry ex, don't know why she even called me. But I do get a call, about three weeks later, requesting an interview. When I show up for the interview the person who was supposed to interview me isn't there so another Engineer talks to me. I take this as a bad sign. His name is Jimmy Byers, I tell him about my oilfield and auto mechanic experience. I show him a design portfolio I ginned up in an elective drafting class I took at Austin Community College, they don't teach drafting at UT. I have about twenty hand drawn, hand lettered, toleranced mechanical detail machine and assembly drawings. I get no feedback from Byers, a man of few words, he thanks me for coming by and I leave. I never hear back from anyone.

## Grandma Scanlon

Grandma had been steadily going downhill since Grampa died. Dad always thought that his Dad wore himself out taking care of Viola when it should have been the other way around after his heart attack and bypass surgery. Steve and Gary went out there in the summer of 73 and it wasn't pretty. Grandma was drinking her lunch. The Heinz boys were ripping her off and encouraging her to drink. Well she did what most old folks finally do, she fell and broke her hip. Dad flew out in a twin Beech and grabbed her when she was released from the hospital. He flew her to Abilene and installed her in his house spring of 76. Mom takes care of Grandma and they become good friends. Tim, Bill, Maureen and Gary are home so she gets to spend time with them before they leave the nest. Tim shared a room with her when he was working in the oilfield. I am not there. I am busy with my own life. Gary is her good buddy, he tells me that Dad is kind of mean/smart assed to her. He has cut her off of course, scorn. I only see her once passing through town, we boys are going skiing. I ask "how are you Grandma" she answers "waiting to die"…short conversation. Grandma passes away Oct 1988 with Mom and Dad at her side, a ten year wait. Hope she was happy, enjoyed the wait.

    The summer after Dad grabbed Grandma Tim, Bill and I drive to LA to clear out her house. We are to bring back anything that we want, Dad is going to rent the place to the Orascos, daughter. The Orascos are old High school friends. Tim has been working in the oilfield and has bought a new 1977 F150 crew cab truck with an 8 foot bed. He and Milstead are planning to rent a trailer in San Marcos after living in the dorm their first year. They are going to Southwest Texas State, LBJ's alma mater. We take off down the now familiar southern rout to LA. Tim and I share driving duties so we roll on without stopping. The road, old Route 66 is now complete. We stay at Grandma Nuno's luxury apartment, it's very fancy. We head over to Grandma Scanlon's place early in the morning. The furnishings are already spare. Tim grabs the living room and bedroom furniture down to Grandpa's ashtray stand. We also clean out the kitchen. The only thing I want is Grandma's skillet. I like to think of her cooking for her boys when I use it. We grab her china and silverware boxing it up for later distribution. Grandma did beautiful fine needle, lace, and crochet work, so we were sure to gab that. Grandma knitted each one of us kids an afghan baby blanket when we were born. We work until it is dark cleaning up the house then we head back to Grandma Nuno's to eat and sleep, that's when we discover that we cannot get the truck in the parking garage under the apartments. Our load includes a refrigerator which is sticking up too high. It's late and we don't want to pull everything out in a forlorn hope to do a better a better job than we spent all afternoon doing so we head on in to eat and talk about it. Grandma had gone all out and we have a great feed.

# 1977 AUSTIN IRS

We decide to draw straws. One of us is going to sleep in the truck. Billy draws the short straw. We didn't bring a pistol. All we have is a large knife from Grandmas Kitchen. I almost trade places with Bill but I think nah, nothing's going to happen. I've been too long in Texas. Sure enough three black dudes start rifling the truck in the wee hours. Bill sits up waving his blade and yells, startling them. They take off. This could have turned out badly. They don't come back. Bill tells us this over breakfast. It is unlike him to be aggressive or fight, he's an inch or so taller than I am now. Grandma is furious, she hates what her neighborhood is becoming and what could have happened to Billy, and she knows the people to blame. We hit the road. Things are uneventful until just East of El Paso. I'm driving, doing a little over 70. It's after midnight following a big thunderstorm when I have a rear tire blow out. The rear-end tries to pass me as we are headed to the bar-ditch approaching an overpass. Our load is threating to roll us. I manage to get it under control, I don't hit the brakes, keep the truck on the road and roll to a stop. Another close one. We have a full sized spare that we mount after we calm down some. Other than all of the above it was a good trip. We stop in Abilene to see Mom and Dad then drop Bill at San Angelo State in San Angelo. Tim and I head to Austin.

March 1978 Tim's truck is totaled at night on a two lane black-top outside of San Marcos. Tim is not driving. He is snoozing on top of some gear in the crew cab. A friend fell asleep at the wheel putting truck in the bar ditch eventually rolling it. Timmy was thrown through the windshield when the truck first hit the far bank of the ditch after launching off the crowned road, tearing up his right shoulder elbow and arm without breaking any bones. Tim suffers a bad concussion and a clot on his right eye socket. He is lucky to be alive.

## Escaping Jim

It's January 1977. Things at the house are intolerable, it looks like a doctor's waiting room, people I don't know siting on the couch with Dr. Jim monitoring his pager. Dad wants to know why Jim needs a pager. Tim and I tell Mom that we need to get out before we all get busted. Tim had moved into the garage of the Brentwood house for the Fall semester when Milstead skipped it. Jim is gone god knows where for a week or so. When he comes back he is going through withdrawal. He starts screaming at me telling me that I have got into his stash that he hid in his air conditioner. I have no idea what he is talking about. As usual he starts swinging. I knock him through a wall where a vertical gas space heater is mounted taking the heater with him. Tim and I were in the kitchen just trying to get breakfast before class. We split, Tim is settled off Enfield I get a cheap efficiency off 45$^{th}$ and Speedway. Mom helps me covertly. She covers my deposit. Dad is not to know the where or why. Tim has his own money. About two weeks after we move out, Jim gets busted with a quarter ounce of speed. Dad makes good on his promise and disowns him; he sends us a circular letter telling us this and advising us not have anything to do with Jim and not to show the letter to Maureen. Mom goes to bat for him in court. She badgers the judge until he releases Jimmy to his "Mommy" it's his first offence, at least as far as the judge knows, more to follow.

Sometime in September, five months after applying, I get a call from the Lab they would like to offer me a part time student position. I wouldn't be making as much money. It will be worth it to do Engineering instead of accounting even if I just got a promotion to Examiner 3. I accept the offer. There is a hitch I have to have a SECRET security clearance to work on the contract that they have in mind for me. I have already done some paperwork for the IRS but now I have to go back 10 years and list everywhere I lived. That's a lot of places when you move every semester. It helped that I was a military brat and that Dad had a Clearance. This takes some time while the FBI does a background investigation. I just hope they do not link me to Jim. After a week or so I am deemed worthy of a clearance. You cannot work at the Lab without one.

After a few days at the Lab Samantha walks into my office and hops on a drafting stool, she is all smiles, don't know what she wants, and don't know what to say. I am very uncomfortable, didn't expect to see her, our schedules are at odds with that. I haven't seen her for over a year. Really don't want to see her, a source of unhappiness. She says that she's working for a cocky guy named Huckabay. I'm working for Dr. Tom Muir who I haven't met yet. Here she sits, like we had never been apart. She asks me to go to a pig roast thing with her Saturday night. Some of the guys that she works with are having the roast out by the quarry in Cedar Park. I am confused. Sam has been here long enough to have had everyone who cared to, hit on her so why me. Without thinking, I say ok. She takes off to catch the bus to campus.

It's Saturday. I follow Sam's OD Chevy Nova, Novocain, to the roast. We turn through a gate with balloons on it off FM 1431. After winding down a rutted dirt track that is real rough on the Midget, we get to the roast. There are no structures on the property but it has power and light. We are in a clearing in the middle of a live oak grove. There is a big live oak fire with a huge 200 plus pound pig over it on a spit that is turned with a homemade chain drive rig. Some of the guys have been at it all day roasting, drinking, and getting roasted themselves. There are about sixteen people around the fire. From the look of the pig, which is still dripping bloody fluid, they will be here all night and the next morning too. There are very few females at this gathering, mostly someone's old lady. Few females are engineers, programmers or scientists at this time. There are several crestfallen faces among the older programmer dudes to see me there, guess they were looking forward to flirting with Sam.

I grab a lawn chair. Sam grabs two beers from the keg. Sam then does something quite unexpected and uncharacteristic that she has never done before, she sits on my lap and puts her arm around me. Samantha is anything but clingy. Now, I'm kind of slow on the uptake, but I get it, I am being used. Sam has a few beers, then she and another older female programmer get a bright idea in the bright moonlight. The full harvest moon is up and you can see the big quarry tower crane silhouetted. The girls decide to climb it and howl at the full moon. They are off on foot in an instant; I follow. I worry. It's not far and there is no gate. The girls climb to the top of the tower and taunt me to follow. I decline telling them that I will stay down to catch them should they fall. It looks rickety, rusty and ramshackle like it hasn't been serviceable for quite a while. After howling as much as they cared to, the girls climb down; they sure come down slower than they went up. All that beer is working. We go back to the party. We have had enough, the pig has not. Everyone is glum because it's not done, but we are. Sam and I split and go our separate ways, not even a hearty handshake. I should be mad but I can see Sam not wanting some old creepy nerd creeping on her. She should have just told me what she wanted. Maybe she didn't know what she wanted. We are now unofficially Lab official in the rumor mill. I get knowing looks in the hall.

Sam asks me to go camping with her a few weeks later. She wants to go camping out at Reimers, Ranch, Hamilton pool. It's

# 1977 AUSTIN IRS

cold out. It's November. It's a swimming hole. No one will be there and we can have the place to ourselves. That's the pitch. I am going because her cousins are going. Frankie and Sid, Mattie and one of her "guys "are up for it, now Sam needs a date so not to be a fifth wheel on the tractor trailer of fun. She is probably asking me because her first second and maybe third choice, turned her down. I am Mike the butcher.

It's cold with a constant fine misty drizzle as we hike in down the muddy trail. We set up camp and start a fire. We don't have a tarp so we crawl into our tents and leave the fly open. Sam gets cold so we decide to take a hike, that's what were here for. We crawl out; Sam lets me zip her up and fix her hat like she's a little kid. Sam laughs off my obvious affection, cow eyes she calls it. She leads off because she knows where she is going, I am unsure about any of this, it's my first time here and I sure as hell don't know where any of this is going. The long single file hike kills conversation, probably a good thing. I am firmly in the friend zone, one of her "guys". I remember back to all those kitchen table conversations about the poor pitiful souls who were hopelessly in love with Mattie and had no chance. They tisk, purse their lips and shake their heads. The girls talk about calling ex,ex boyfriends when they need help with something, leveraging their sex. We all went out to dinner with one of these "Guys" on his "date" with Mattie, she made him take Frankie Sam and me on the "date" and pick up the tab for everyone! I don't want to be that pay to get played guy.

We have a few more faux "dates", mainly boot scooting with her coworkers, Friday nights at the Lumberyard, a dance hall down the street from the Lab at the corner of HY 183 and Burnet Road. Sam teaches me the Two Step, and how to line dance the Cotton Eyed Joe and the Shottische. I go to see her sister's high school dance team perform at the Super Drum, women's basketball games, stuff like that, just because she asked me to. I never say no.

Sam asks me to go skiing after school lets out for x-mass break. A bunch of students and staffers are going. It's unfair of her to ask. This thing is long planned, I don't have any money for fun. She just wants to use me to keep the jerks from hitting on her. I tell myself that I don't care, but it bothers me that she may be hooking up with some other guy, maybe the guy she agreed to go on the trip with months ago.

I am behind. My supervisor let me slide on my hours to really hit the books for finals now he is cross with me. I have to generate about forty full scale drawings for a transducer array with random generated placement. All I get is a printout. For the five hundred or so transducers. This sucker is going to be a real whale killer. I manage to finish. My drawings are checked by two engineers, there are no errors. Steve also has some issues. He has inherited Mom's Buick, he's headed downtown on IH 35. Steve takes the 6$^{th}$ street exit in front of the Cop shop when an old fellow in front of him slams on his brakes and Steve rear-ends him. He gets a ticket of course, the guy behind always gets the ticket but he fights it and wins. It's illegal to stop on an interstate off ramp, even if you need to make a four lane change in under five hundred feet of road. The front of the Buick is creamed. I find another left fender, hood, and bumper which I install over Christmas. Steve and I both miss Christmas and New Year for the first time.

One of the creepy nerd programmers has a party every year where he shows continuous porn, the F-Film Festival. There are flyers in the hall and taped to the bathroom doors. How he gets away with this I don't know. The host, who sports an appropriate waxed handlebar and a Van Dyke, has been bugging Sam to come because she ducked the festival last year. She gets lots of invites. Samantha enlists me to go with her on short notice without telling me what's up. When we show up there is mixed crowd of gawkers looking at a 1950's porno staring Mr. Fat Soo and Ms. Uno Tractive. The host notices that we are there and is a little miffed to see me sitting next to Sam on the hearth sipping a beer. We don't stay for the feature film, he's probably in it. Samantha asks me to come over to the house, it's still early. I don't know if the invite is planned or spontaneous I haven't been at her place for over a year, turns out Sam is horny, so am I. That must have been the first porn either one of us have ever seen. Who would have thought that her button was bad porn? Sam gets inventive we are very passionate. The cousins are gone, we have the house to ourselves. We jump in Frankie's King sized bed, both of our beds are miserable twins. We have the best longest most passionate sex that we have ever had; Sam says "that was nice". I think so too. We are together tonight, it is nice. We do things together again. I see her almost every day even if it is just to chat. I am very happy. Never stopped loving Sam.

I like my place it is quiet dark and cool, a bottom floor apartment; a good place to study and find peace. Sam comes over one night late in the semester. I am studying for a metallurgy test. I have a quiz first thing in the morning so I ask her to give me a few minutes; it turns into 30. I have been hitting this subject real hard it's worth 4 hours credit, it has a lab. I have spent all week pulling samples mounting and polishing samples for the microscope and researching crystallography at the Library. I am determined to crack an A.

I work at my little kitchen table. Samantha falls asleep on my bed; it's late, after 2300. When I finish putting my notes together, I leave the kitchen light on and join her on my tiny twin bed around the corner. We hug kiss and make love but something is different.

Sam leans on her elbow, and bluntly tells me "I don't want you to get your hopes up; we are not together, I want to date other people." the way she says it makes me think that she has already made that move and has uttered this practiced line before. I ask her why she just made love to me when it was her intent to dump me. She says "maybe I shouldn't have done that". I try to say something hurtful but it only comes out lame because what I say is not true. I ask her to set me up with some of her friends so I won't be lonely. Sam is nonplused, she gets up, gets dressed and leaves; I didn't even last a semester this time. Was she trying to get a different response from me? I would like to say that she hurt my feelings but I don't have feelings anymore; Debbie Schmidt took care of that. Whatever was left, Sam raked into a pile, poured gas on and burned, salting the ashes so nothing would ever grow again, when she dumped me the last time. Maybe that is my problem or one of them, no feelings, no money, no fun, a boring Engineer. Sam want's stuff; a nice house mostly. We'd drive by, where some of her friends live. She'd tell me about how nice they were inside, envy. I can't buy her a house, at least not for a while. I can't even buy her a wall tent. I just taste bitter disappointment, I did kind of have my hopes up. I do love her. But to quote Thucydides "Hope is the last bastion of the doomed" or Bob Scanlon "Life's a bitch and then you die". Entropy. The last thing Sam wants to hear is that I love her, she already knows it. I will miss Mattie, Frankie and the gang. She gets all the friends of course. Doreen is dating Tony, the incredible Hulk; now I don't even have that. No friends once more. I tell myself that I don't care.

    I don't want to live here anymore, makes me sad and it's expensive. They won't let me keep two cars and my bike in the parking lot, that's too many vehicles. I have to get rid of something or move. I had been passing this garage on my way to and from the bus stop that had a blue midget sticking partway out. One day I knocked on the door and asked about it. The girl who answered the door said that the car belonged to an ex roommate that had skipped owing her money and leaving the car. That was two years ago. She said I could have it for $50.00; she just wants it out of her garage. I give her the money with my name and address, pump up the tiers and haul it five blocks with my red MG to my place. Now I have too much rolling stock. I can't sell the blue car because I don't have a title.

I move in with three Math TA/Doctoral candidates who have leased a dumpy un-air-conditioned 1920s bungalow at the corner of Red River and $28^{th}$, four blocks from campus. It has a garage they will let me use. I answered an ad posted at the RLM (Robert Lee More Mathematics building). None of us are ever there. They have air-conditioned offices in the math department, I am a constant presence at the engineering library where they have individual study rooms. I have my own office too, at the Lab, when my office mate quits. Very conducive to study. There are only two ME students out of a staff of over 500 at the Lab. There is one PHD ME Dr. Dale Evertson. Dale gets no traction at the Lab so he accepts a teaching position at the University of Tennessee, another UT. Most employees are Physicists, Computer Science or Electrical Engineers a line that is becoming smeared. They all work in peer groups. I work alone, I am a blank piece of paper guy. I have to build it in my head before committing something to paper a redraw is even more tedious than drawing up something in the first place. Not many EE's chose to go into power, most chose microelectronic hardware and software. MEs do motors, generators, and transmission lines. I was lucky enough to get the department head for eight hours of EE classes, he made everything clear and simple for me; I made A's. MEs have to be familiar with most every other engineering discipline. I am specializing in Mechanical Systems Design. I especially enjoy Dean Rylander's class and Dr. Krisel's beam mechanics class, where it all came together for me. Instrumentation and Engineering Economic Analysis classes also prove to be very useful.

Working at the Lab was a boon for my programming courses. I take FORTRAN and a graduate level course, Computer Augmented Design, which is programming intensive across three systems, plus learning 8086 programing language and machine code. At this time all programming is done by IBM punch cards. Students wishing to run programs have to go to Taylor hall and stand in line to access the five card punch machines there, the size of a kitchen counter. You buy your own cards by the box which are two feet long, then you have to punch a card for each line of code in your program. You then stand in line and submit your stack of cards to the operator who puts them in the que then feeds them to a card reader when it's your turn. You get your cards back with a printout, either it ran or it failed to compile, this cycle may take four hours during prime time. If your program fails to compile, then it's find the mistake with very few clues, stand in line punch a card then run, and repeat till the program runs. The use of a semicolon instead of a colon will sink you, and is hard to spot. Woe to the student who spills their box of cards which must be kept in order. With all that standing in line, and sharing limited assets, Taylor Hall computing center is a Zoo 24/7. All assignments require that you turn in your cards with your print out. No way to share cards/cheat. The printout should have comments which are unique to the programmer. Comments help you and others navigate your coding. I will always be grateful to Robert Altenburg, he gets me hooked up at the Lab. I can enter my program using a keyboard and monitor in my local user space. We have the same mainframe that they have on campus only souped up. I can seamlessly run and modify my program till it works at one sitting. Afterwards I have cards automatically punched and my program printed. This access cuts days of work and time wasted just going to campus and waiting in line. I'm as permanent a fixture in my office as my six foot drafting board. I don't have a phone, so if anyone wants to get ahold of me they have to call my office and leave a message.

    One of my roommates, Mike Jordan, gets an upstairs apartment in the old 1920's stone apartments at the corner of 28th and Speedway and asks if I would room with him. I room with Mike spring and summer semester 1978. It's better than the place we were just in, has fewer water roaches and it is air conditioned with window units. The Speedway Shuttle bus, still driven by students, starts running at 0600; it stops and makes a left turn right in in front of my bedroom window which faces the street. I don't need an alarm clock. Another redeeming feature is that it has a carport in the back, Mike doesn't have a car so I have room for my two Midgets and my Yamaha 360 dirt bike. My cars won't become bath tubs when it rains. I like living close to campus, I hardly ever drive. I just ride the bus between campus and the Lab. Never go out, never do anything, never see anyone. I eat once a day at the lab, they have $1.00 sandwiches; the fixings set out on a trays. You can make a real Dagwood for a dollar, a life saver for me. I see Sam at work once in a while, she may happen to walk through the cafeteria while I'm there. All there is between us now is occasional side eye if that. She joins the dive team. I'm sure that she will be real popular. She looks great in a swimsuit. The Lab has a team of volunteer divers that provide services to the Lab's acoustic test facility close to Mansfield Dam on Lake Travis. I would like to do that too if I had the time and money. You have to buy your own gear. The dive locker is on the moored test facility.

## Dumped Again

Fall semester my roommate is entrusted with living at his professor's three bedroom house on 45th and Red River while she is on sabbatical; Steve and I are roommates again. The roomies at Avenue H have all graduated, spit and gone their separate ways. The house is a nice furnished white stucco with an attached garage in the back and refinished wood floors that the owner did herself. The yard is big, two lots or more, big enough for me to pop wheelies in the backyard on my bike for fun. The house backs up to an ally I can ride up and down on. Mike buys a Kawasaki Ninja he can't ride, I worry about him. That bike is a real killer, something about the rake of the front forks plus it is stupid fast. We are only here for one semester so we decide to have a Newyear/goodbye beautiful house party. I don't remember the party or who was there other than Susanne. I didn't invite her. Dumb me. Mattie is dating Tommy G; I am moving to his place in north east Austin, Springdale and HW 183 next week. It will be a quick move, the sad thing is I can still move everything I own in one trip in the Midget, everything except my blue Midget that I will tow, and my bike which I can ride. Everything I own is shabby and worn, haven't bought any cloths in three years. Mattie and Tommy are here so I guess Sam tagged along just to torture me. I haven't seen her since she dumped me for a, TBD alternative. Midnight strikes and I kiss Sam, we keep kissing; it has been a long time, over a year. Steve gives up the master bedroom and his king-sized bed for us. I had flipped him for it when we first moved in, he won. Again we have an extremely long passionate session, there is no sex like sex with an ex. To say I was pent up would be a huge understatement, I'm still mad for her. Again the immediate letdown; Sam says "this doesn't mean that we are back together". I am stunned once again. Gob-smacked. What just happened? She won't stay the night. She immediately gets up, gets dressed, and leaves to underline her point. I was right she came by to torture me.

The next morning I think of what happened and replay it, you know where you think of what you should have said in the endless loop playing in your head. I ask her just, what does it mean? Do I mean anything at all to you? Why are you so mean to me? Will we ever be together? Do I have a chance? You don't seem to want to let me go, don't know why, and you are ruining me for other women. I have a year to go before I graduate. Sam has been the only girl that I have been with now for two years, even though I had opportunities, I haven't been interested in anyone else.

After that I am on the do not call do not resuscitate list. I am lucky, the Black Widow eats her mate after sex, first decapitating him. My head is still attached but I am eaten up on the inside. Sam graduates in May, six months before I do. She takes another job doing systems programming instead of the application programming she was doing at the Lab. Sam meets another set of guys. She tells me about them, prospects. Like I want to know.

I see Sam every once in a while when she shows up at our place with Mattie. Sam doesn't want to hear about classes, quizzes or homework, she is done with that. She is done with the Lab. She's making good money now, while I am still a poor student who can only afford to eat once a day. She brings by her new Celica and shows it off, asks me what I think and has me test drive it. I am no fun, a boring engineer. I can't keep up; can't compete with whoever she may be interested in. I don't have the money or evenings free to go out dancing, or get dinner and see movies, not that she wants to do any of that stuff with me.

Austin used to have a festival on Town Lake called Aqua-Fest held the first week of August. Sam, Frankie and Mattie are down there working the beer counters for the commodore, they do it every year. Tommy goes down to meet up with them. I decline to go I don't have money for beer, I don't really like to drink, I have class the next day, I hate crowds, I don't want to run into Sam and whoever she is seeing these days. I hit the hay but wake up to a commotion when Tommy comes in real late. I can hear more partying in the living room going on. Tommy is putting the Doors on the turntable, his favorite artists, he and Mattie must be getting frisky. I roll over and go to sleep.

I wake up and get going before Tommy and hit campus then the Lab. When I get home after 2000 Tommy asks if I want to go to a friend's house with him to have a drink; I go. We end up at this hooker's apartment. Turns out that Tommy had picked up his stripper/hooker friend, with a huge rack, at Aqua-Fest, right in front of the girls while they were working the counter. Poor Mattie humiliated in front of her sister and cousin. She got dumped way worse than me, at least I got a proper F— off. We have a drink. Tommy is all grin like the Cheshire cat, he has a new toy and is proud of what he's got. The hooker wants to come over to our place. Tommy is a great photographer; he's the assistant manager of Steadman's Photo on the corner of 19th and Guadalupe. Evidently Tommy has developed some film she wants to see. Tommy took a bunch of pictures of them having sex last night. Tommy has a two seat Chevy Monza coupe. The hooker has to sit on my lap for the trip. She squirms her ass on me and tells me that we ought to have a threesome. I decline; they laugh at me.

I can see it happening. Just after I moved in Tommy bought an old boat and motor. In order to get it in the carport to work on it, we completely disassemble my blue midget and my two Penton 125 dirt bikes, one given to me by the guys in the shop, and one is Mark Walker's. I intend to make one good bike out of the two. We then shoved all the parts in the attic. This becomes a running joke with anyone who knows me. Tommy is a perfectionist who has a temper. I saw it while we were re-glassing his 13 foot Glasspar boat and rebuilding the Johnson 75. He would be working along steadily getting frustrated, then instead of walking away or taking a deep breath like I do, he would get steadily more impatient then explode, sometimes destroying what he's working on. Mattie told me that they were waterskiing once, out on Decker Lake when he scared her. The motor quit and Tommy couldn't get it started, he lost it. Mattie and Sam's golden Retriever Zlote went over the side and swam to shore and left. Tommy's Dad is an Army Col. Tommy knows Doreen and all the folks from Killeen, they went to high school together. Tommy is an adopted West German. His dad adopted him when he was stationed there. His mom and Dad still dote on him like he's ten-years-old. He acts like he's ten around them. Mattie says it's sick. Tommy must have got impatient, lost his temper and wrecked everything.

    Mattie and I pal around together after that, two wounded birds. She and Sam have a pact that they will never date each other's ex. It's safe, I love Sam and Mattie will never cross that line, even if I am totally kicked to the curb and run over by a bus Sam is driving. Sam tries to covertly set me up with one of her friends, a mousey little thing, on a waterskiing outing Mattie and I have planned. She is just there when I show up and tags along, doesn't even have a swimsuit on, she is wearing jeans and a long sleeved shirt. I'm in the dark, I say ok she can ride along if she wants to. Mattie and I are tearing it up on Lake Austin. Sam's friend the pale mouse doesn't ski or swim and is acting weird. I ask Mattie what's up with a whisper in her ear. Mattie doesn't like the girl or what Sam is doing, testing me to see if I will go for it, she makes me wise. I tell Mattie that we should discourage her; we should make out. We are both sopping wet standing in the stern, recovering the tow rope. The boat is drifting. We have a long sloppy loud smacking kiss. I sneak a peek at our guest and see that she is perfectly mortified. This is not the date she was sold. Mission accomplished. If I had socks on, Mattie would have knocked them off. I love my friend.

Kissing Mattie reminds me of happier times. The girls and I were at Denny's, in a booth, having breakfast, after I had spent the night. Sam said something sweet that I liked so I ask her for a kiss. Sam gives me one of her tight lipped kisses on the cheek, I mockingly complain about the kiss to her cousins, saying "that's all I ever get, just a little peck; Sam is a little pecker". Frankie loses it, she is drinking a Dr. Pepper, her drink flies out of every hole in her face spraying all of us when she puts her hand to her mouth; we bust out laughing hysterically. Everyone in the place turns and looks. Now that's a shared moment. I would like to be a fly on the wall when Sam gets her report.

Mattie had quit her administrative assistant job a while ago. She tried construction because she was interested in it that was after working as a cocktail waitress at the Magic Time Machine. Her costume was a Playboy Bunny outfit. Steve worked there at the same time, as a waiter, he was Benny the Beaver and Sean Sean the Leprechaun. Mattie and I have lots of long talks. I tell Mattie that she should go to school and get a teachers certificate. With her temperament, charm, easy laugh, and sweet voice, she could easily get a pack of kids to do what she wants and they would love her for it. I support her in whatever she wants to do.

Tommy meets and starts dating Beverly Allen. How they meet I don't know, she's just there. I like Bev, she's a sweet country farm girl from McGregor west of Waco. Bev looks like a sorority girl, a gorgeous blond with blue eyes who talks with a lazy drawl, quite different from the girls I have seen Tommy with lately. Beverly works at Westwood Country Club the most exclusive Tennis Club in town. She gets me some court time. How she got that job I never ask. Beverly doesn't like sports, play tennis or string racquets, but she looks really great in tennis togs. She works the counter in the Pro-shop, chatting up the members and scheduling court time. The ladies lover her, she looks like their daughters. Bev also runs the store selling whatever they have. She sells high end tennis clothes and shoes to the members, modeling what she sells. She gets outfits comp from vendors because she is a walking advertisement. Tommy is serious about this girl; he puts Bev on a pedestal. He talks her into moving in with us. Tommy buys Bev a fire engine red tan roof tan interior Fiat 124 spider convertible. They call each other sweet sick pet names. They got it going on so I think about getting going.

# 1979 Barrington Way

About this time Dad gets fed up with the renters in his Mother's house in LA. It's too far away to manage through friends. Dad works a slick deal where he gets the folks who want the house in California to buy a house that he wants in Austin then he gets a Lawyer to effect a trade for $250.00. He pays no reality fees or taxes on the deal. I am there for the signing. The house is a 3500sf 5bdr 3 bath mother in-law plan single story house with a two car garage on a corner lot off Barrington Way. It is located in north Austin near Spicewood and HW 183. Tim, Bill, and Maureen move in. Gary moves in too summer of 1980; I move in for my last semester. Jim is banned from ever entering the premises. Tim has transferred from San Marcos to study with the UT Geology department. Bill was going to San Angelo State majoring in Chemistry and Journalism, he was on the yearbook committee in High School. Bill had an incident where his roommate went nuts and held poor Bill hostage with a knife to his throat during the subsequent SWAT standoff. Bill has transferred to UT too. Maureen graduated from High School in "78 the year of the great" (she says). She spent a year at Texas Tech in Lubbock. Tech is the big school in west Texas. Reenie went up there with a bunch of her friends. Dad buys her a new Toyota Corolla, "Lola Corolla" so she could drive home. Gary graduated Spring 1980.

Maureen, Bill and Gary all played on the Abilene High Tennis Team and did the Rose Park thing with Dad. Maureen heads out to Lakeway World of Tennis to check it out with a recommendation from Rose Park. Cliff Drysdale is the Pro, he has a lot of South Africans like Billy Freer working there along with some great College level players. Maureen applies for a job at the pro shop, they like her and her game; she's hired. Maureen wrangles a job for Gary at the lonely remote satellite courts, used mainly for overflow and condo occupants; Gary is left to himself most of the time. Gary has game. He is a lefty. Gary plays and beats me at intermural courts. I always have trouble with lefties but that is not an excuse; Gary is really good.

I introduce Gary to Beverly, she tells him that he should come out to Westwood. Beverly introduces Gary to Dave Severtson the head Pro and his wife who runs the Pro shop and instructs. Gary does a stint working the desk and stringing rackets. Dave hits with Gary to check out his game and lets him help with the juniors program when he has a lot of students. When Dave sees how good he is with kids, Dave gives Gary a chance even though he has reservations. Gary is only eighteen-years-old. Gary quit the tennis team his Senior year over a dispute with his Coach about the challenge rules. He hasn't played the tournament circuit and he didn't play college tennis so he doesn't have a reputation; but there is no doubt that he can do the job and play the game. Gary is also handsome and very personable. The members like him. Gary, although he takes some courses at San Marcos and ACC, is not really interested in going to school. He has found his calling and a home. The babies dodge the storms that assailed me and dance between the raindrops.

Right after Dad has us all squared away and settled, he suffers a heart attack in October. He is flying and sets down in Brownwood. Tim Steve and I drive up to Brownwood to see him. Dad is medevacked to Brooks VA Hospital in San Antonio. He is then then transferred to Heart Hospital, the USAF Hospital at Lackland AFB. Dad gets a quadruple bypass. We all go in a group to see him in San Antonio. He said that they have to tie him in bed because he has bad dreams about wrestling bears. They are afraid of him falling out of bed. Dad pulls a medicine bottle with no label from his bedside table and shakes it like a baby rattle. It's a chunk of flack. Krupp iron, half the size of a 9mm round that he had buried in his sternum. You could feel it under his skin. He caught it the same time two of his crewmen were wounded. It had to have gone through his: May West, leather sheepskin electric heated suit, uniform and long johns plus more. Dad said he wore every stitch he had when he flew. Dad didn't say anything about it at the time. He didn't want his crew to finish their tour with another Pilot. Dad can't stand to watch 12'oclock High with those guys bombing from 30,000 feet, wearing their fifty mission crush caps, open collar type A uniforms and A-1 leather flying jackets with no oxygen masks on.

This is busy time for me with interviews exams and still holding down a job. I don't have a car that I can leave town in, so I don't get to spend time with Dad before he is transferred home. Dad grounds himself. He can no longer pass the physical. I tell him the ANG is looking for WSO's for the local F-4 Wild Weasels flying out of Bergstrom and I was thinking about it. They offer a decent signing bonus. He looks at me and says "Why would you want to pull all those G's when you're not even flying?" That was the end of that.

Before I graduate in the winter of 79 I get a call from Sam, she wants me to meet her to go CW dancing. When I show up at the dance hall I see Sam sitting on some guy's knee. Great, asks me out but she came with someone else. She spy's me, gets up, walks onto the dance floor, and greets me, before I can change my mind. I ask her to dance, it's a nice two step. She says "I didn't know you could dance", I tell her "I can do a lot of things". To myself I say you just never bothered to find out. When the music stops, I leave, as she heads for her table, I head for the door. She probably never even noticed I'd left.

I don't know it yet but that is the last time I ever see Samantha. This is just not healthy, my heart is in my throat and my stomach is in knots every time I see her, I'm pathetic. Sam is playing with me like a cat plays with a mouse, except the mouse doesn't love the cat, wounding and maiming it, holding it with its claws, swatting it away, pouncing when it dares to run, finally killing it. Sam has been wounding me for years. She calls me up a few days later. Sam and the girls want to throw me a birthday/graduation party next weekend. Problem is, it's not my birthday next weekend. My birthday is the day after Christmas three weeks away; no one goes to that party. I am in my MG and heading down the road to the pity party at Frankie's, down the road that I have driven for three years now, all through school; I pull over and turn around. Not going down that road anymore. Whenever I hear the song "Run Around Sue" I think of Samantha and can't help but feel a little sad. I am lucky to get away, withdrawal, like quitting a nasty habit. My good buddy Mattie later tells me that Sam said there was never anything there, she never really cared. I was just one of Sam's "guys" a sap who couldn't say no, a "Mike the butcher". I keep seeing Mattie for a while longer, we go sailing, boating, and water skiing after Zach, Tom and I buy a speed boat, and a 22 foot sailboat, ZABOTA, we keep in a slip on Lake Travis, ready to go. She comes to my Soccer games sometimes and the win or lose after parties. We don't talk about Sam; I don't want to know. I never see Mattie again after I start traveling, can't have Mattie w/o Sam...

When I graduate I don't walk, there's no ceremony for winter graduates. It's all so anticlimactic. Dad gives me an oriental rug that he bought in Taiwan. It's a 6 x 12 foot grade one carpet. It's gold with turquois navy blue and rose accents, pretty. It's moth eaten. Dad offers that I will have to have it cleaned and treated. Another thing to fix, they always give me things that need to be fixed for presents. I remember this rug. Dad had ordered a single color Air Force blue rug with a simple gold border and accents. When he goes to pick up the rug, Dad discovers that the weavers couldn't help themselves from weaving in a large ornate pineapple in one corner, ruining it. The weavers offered this much more expensive rug to compensate for their mistake. 400 hand tied knots per square inch. Little kids and women knot and trim the individual rug threads and yarn with scissors contouring the designs. The weavers get Mom's order right, about ten ivory colored rugs with roses with green stems branches and leaves in the border as accents. These rugs were ordered to match her custom teak and green marble inlay tables and chairs and couches. Dad later allows that I had it harder than anyone else.

I have offers from the Lab, Tracor (started by folks from the Lab), Rockwell International, Lockheed, Hughes Tool, and Lufkin Industries, thanks to a letter of recommendation from Dean Rylander. I also had an offer from Vector Torch immediately after presenting my senior project. Senior project was a lot of work, this real bright double Engineering EE/ME major and myself had to carry two guys who had flunked the course before. We wrote their parts. I designed and produced drawings for a multiple input constant velocity output transmission. A mechanical angular velocity summer. We all received an A. I am lucky to get all these offers. The NASA effort and the space race is winding down, so is the oil business. There were also a lot of people who went into engineering because of the space program further flooding the profession. There is not much work for MEs in town, that is more of a Houston thing. I tell Mattie that I am thinking of taking Rockwell's offer. I would be working on airplanes at Love Field I want to leave town, too much baggage here. I want to start over. I really want to burn this bridge in a bad way. Mattie tells me I'm just trying to make myself even more miserable than I already am, I shouldn't run away, if I'm unhappy here I will be unhappy there too. I accept the Lab's offer, not my best but six times what they were paying me. How about that, half a year's pay a month! I'm twenty-six-years-old. It has taken me seven years on a five year plan to get an Engineering degree, 130 hours. I have over 160 when I graduate.

## Roomies

I am Tom and Zach's roommate now. Tom got his Masters in Accounting he and Zach are working as Petroleum Landmen, they go to remote locations, research and buy up mineral rights from property owners for oil companies for a percentage. They work three weeks and take a week off. They work in Texas, Tennessee and later western Washington. I travel too, so we are rarely there at the same time, the boys make good money and need toys for that week off. We throw a party for everyone we know. When it gets rolling, Tim tells me Sam has shown up with a date. I leave my own party, don't want to see it, don't want to deal with it, and don't want to be sad. I drive to Fandango's for a very long late solo dinner. This is when I throw in the towel, I am just not good at this stuff, relationships. Not going to waste my time or thoughts on anyone anymore. I admit to myself that I'm about as lovable as a sticker burr. I can piss someone off at fifty yards and not even know it. All my life I have been trying not to be a Jim, have my brother's weakness, the failing or my Dad's hard ass judgmental hair trigger temper that my brother shares with him. Yet here I am. What they taught me is when you get mad and emotional, you just stop thinking. I adopt a strict no party policy it's just work, cars, boats, scuba diving and sports for me. For years I'm not looking; if anyone is interested in me I'm oblivious, don't care to care. I need to get out of the nice guy business; never volunteer, should have known that. I start a city League Soccer team and join a city league tennis team. I also play on the Lab Tennis team with Nancy and Mark Bedford, Pat Pit, Clark Penrod, Karl Folk, and Stewart Rohre. The lab also has a softball team. We are all sponsored by the kitchen fund. That all goes away when Campus finds out we are running a cafeteria and making a profit. Too bad, those little social gatherings and activities brought a lot to the Lab...Pit and Rohre are real bad about calling and asking me to split early to play a set. Always thought they should have courts at the commons, at least the director would know where to find us!

 I want to fix my MG, old Annabelle. I used to say a little prayer each day before I started her up, promising that if she would just start and run I would fix her up when I graduated. I want to keep my promise to the old girl but I need another car to drive first. I buy Evelyn's TR-6. The car is a beautiful chocolate metal flake brown with tan roof and interior. Evelyn can't keep it running. She is chauffeured around everywhere anyway. The car has no reverse gear so she is selling it. I ask Evelyn how she deals with that. She says in a matter of fact voice that she just stands by her car and pretty soon some guys will come by and push it back for her. Of course they do, Evelyn is very pretty. She's in Dallas selling cosmetics.

I have to pull and rebuild the tranny on the TR6, it also has a clutch yoke problem that I fix along with a problem with the independent rear suspension. I give it a brake job and a tune, then I balance the three carbonators. She runs like a top but she is all Bondo and rust under all that pretty paint. Next I tackle the MG. Annabelle gets a rebuilt transmission I make one out of the two I have. I balance and blueprint the motor. I install a ¾ race cam and a supercharger that doubles her horsepower. The body gets a roll bar, a fire engine red paint job, a black interior and new convertible top at Tiny's on Lamar. She also gets new Pirellis on her wire wheels. She will turn 7 grand and go 110 mph with a 1 liter engine. I get to work and back real fast.

All my brothers and my good friends are on my soccer team. We are the Austin Hawks City League flight C. We are the first Verde, our colors are kelly-green and white. We practice at Murchison Elementary in North West Hills and play every Sunday on the Great Lawn at Zilker Park. Tim is our goalie Bill and Steve play defense, Gary is a midfielder, Tom Luckenbach plays Intimidation wherever he wants, I play striker and coach, we pick up two walk on ringers who play wing for me. One is a thirty-year-old named Wally who has skills and an Iranian kid who is fast as hell. Tommy Gairloff plays and Beverly comes to cheer him on along with everyone else's girl, Mattie is there for me usually with Kerry for Steve. Mattie gets into it, she starts her own team with Frankie. I've never been in better shape in my life. I can run forever. It's just a joy to run in cleats on grass. I have the best foot on the team and score most of the goals. There is nothing more satisfying then to hit the ball dead center and have it come off your foot like a cannon shot. We play for fun, not so our opponents they live for the weekend game. The refs want combat pay. I wear a splint on my left wrist that they check for weapons. I am afraid that one leg tackle would be enough to fold my hand back again. We are pretty resilient but that old bane of sports, injuries, rears its ugly head. First I take a round house kick to the chest as I score a goal and break three ribs. It leaves me writhing on the ground gasping for air in agony. Then it's a nasty sprain to my right ankle, as I go for another goal; my opponent "missed" the ball; followed by having all the toes on my right foot buckled. The final blow was. Timmy twisting on his left knee after planting his foot in the new cleats I bought him to clear a ball, dislocating his knee, tearing his ACL and MCL. The people we are playing are all out for blood; their girlfriends on the sidelines shout "kill him", they live all week just for this. We are out here just having fun with our friends, getting some exercise and the after game win or lose parties. We play two seasons winning half our games. I sell the Austin Hawks to Wally for $300.00, slots are very limited. We can't absorb the injuries, Tim and I are out of commission.

Tim has to have surgery. The initial orthopedic procedure turns into a major incision. He drops two classes to get down to 12 hours of school. Tim has to have s second surgery seven months later because of excessive scar tissue resultant from the prior incision, a large amount of scar tissue is removed.

## Kerry

Steve meets Kerry Rodgers in 78 while the both of them are working at The Magic Time Machine. She is one of the old gang that used to go to Tom Luckenbach's Lake weekends; she was dating Ray Weigum back then. Kerry is one of the girls, she knows Frankie, Sam, and Mattie. She is pals with Doreen too. Steve moves in with her. He graduated Magna Cum Laude, Phi Beta Cape with a degree in Government, and a teaching certificate. He's working on his Master's in Community and Regional planning. Steve's working on a Haz-Mat rout for the City. Steve and Kerry are very happy. Kerry is going to UT as well. She is majoring in anthropology and archeology. Kerry was working at the Indian art and Artifact Lab at Balcones Research Lab where ARL is located. She got bored of cataloging Mezzo Indian pottery shards after she went to Italy for six months on a dig. No comparison. Steve takes Kerry up to Abilene to meet the parents; they love her. Everyone loves Kerry, she has big brown eyes and a big smile to match; her laugh is infectious. Steve is working at the IRS, he works in Micro film, in the secure area, he handles 1040 X and problem files. Kerry takes a job at the Lab as an Admin. Asst. in my division. We go to lunch and chat in the patio when she goes on break. This starts the rumor mill turning. Kerry is so sweet, she wants to have a baby badly, she'd be a great mom, but Steve can't see it while he is poor and in school. Kerry's biological clock is ticking. She says "babies don't worry about money, they will always be taken care of" Kerry quits the Lab and takes a position as a nanny for the single child of an established Architect, Ernesto, who has lost his wife. If she is not going to have a child now, then being a mother to this poor little boy who lost his, is the next best thing. It is not the next best thing for Steve. Kerry leaves Steve for

Ernesto and his readymade family, money, and fancy home. She sold out. Found a better man. Breaking Stevie's heart. Turns out all those girls were like that, everyone except Doreen, another loyal Military Brat with a big heart… Kerry has a child of her own a little girl. She is the first of our little group to pass, didn't even know she had bone cancer.

# ZABOTA

Zach wants to buy a sailboat. Lake Travis is famous for sailing and is sixty miles long. He found one he likes the price of at the Sailboat Shop. Zach knows nothing about sailboats but in a rare stroke of luck he has found a real jewel, a Chrysler 22. The boat comes with a trailer. It has a cuddy cabin with a Pop top; 6' 4" Tom can stand up in the cabin with the top popped. It has a swing keel so you can easily refloat it if it runs aground. It also comes with an outboard motor. It is also fast, it can easily out sail a Catalina 22, a very popular hull on the lake. The boat has a double hull and deck cored with Balsa wood, it is unsinkable. The pitch is it will only cost us 2.5K each, we can keep it in a slip, ready to go and it doesn't use gas. Every time we go water skiing we burn a couple of hours launching and recovering the boat, and fifty bucks in gas. Its Memorial Day weekend. We buy the boat with the provision that it goes in the water no later than Friday afternoon. Zach names the boat ZABOTA, after our names, Zach, Bob and Tom. The salesman tows the boat to Mansfield boat launch on the main pool of Lake Travis, located near the dam. We help the salesman rig and launch the boat, then climb aboard. The wind is 15kts and steady, from the South. The salesman takes the three of us and Brother Tim on one tack, a broad reach, the easiest tack, across the lake and back, then he dives overboard and swims to shore, telling us "you've got it" and that we are on our own. We don't know how to trim sails and leave them how they are. Sailing for us is going as fast as possible. Our plan was to head up the lake and look for a cove where we can tie off to shore and spend the night. We are not making any headway up the lake. We don't know how to set a course to get from point A to point B. We have no anchor. It's starting to get dark, we are running out of beer, we are also getting hungry. The only other provisions we have are glass bottles of whiskey, coke and 7-up in the shelves above the port and starboard berths. We drop the sails, haul in on the main sheet to keep the boom from flopping around and start the 7.5hp Evenrude outboard mounted to the stern then head for the U-Float-Em a store and burger joint attached to Lake Travis Marina. We remember the salesman saying that we could cruise faster on the motor if we crank up the keel to reduce drag, another nice feature of the boat. We crank it up and take off at 6kts. In under an hour we are at the U-Float Em. We tie up to the fuel pier, head in, buy beer, stock our cooler and order burgers and fries. There is only one guy working the kitchen and orders in front of ours so it is going to be awhile. We drink beer and play pool while we wait; which is hilarious because the pool table is on a floating barge. The wakes of passing boats make the balls roll all over the table. We are also animated by our first adventure on the boat. This is great fun! Our order is up so we grab our sacks of food and jump on board. We put our burgers on the dinette table, cast off and start the motor. The wind has freshened blowing 20kts. We haven't had enough of sailing so we raise the mainsail without breaking loose the mainsail sheet. The boat immediately wind-vanes and heads up into the wind overpowering the rudder. We don't have the keel down so we get knocked down burying the rail. The whiskey flies across the boat, hits the coke bottle, they shatter. The shards with the contents fly into the bilge to mix with our dinner. We are lucky none of us went over the side in the dark. After the boat turns into the wind we gain control and drop sail. We are barefoot. We scramble over the berths and dinette and gather up whatever food didn't go into the bilge. That, beer, and a bottle of scotch is now our dinner. We manage to nose into a ravine. Tom jumps overboard and ties us off fore and aft to trees on shore. We are done. Then comes the dawn. Learned quite a lot last night. The next day we lease a slip at Commanders Point. Tom hauls the trailer to his Dad's Place in Marble Falls.

This is the first of many weekends on the lake. One of our favorite things to do is "keelhauling" we tie a knotted 75 foot line with a loop on the end to the stern and drag behind the boat. You can "Porpoise" down 15 feet pretty quick using a free arm, if you wear a snorkel and mask. I made a crude underwater sled with elevators and rudder to enhance this, but deemed it unsafe due to low visibility when I almost hit a stump going at a good clip. You have to trust your skipper too.

The boat turns out to be a great dive platform; plenty of room for gear, some privacy to get suited up and a self-draining cockpit. You can also leave your gear on the boat, if you hang it up to dry. Tim, Gary and I make a lot of dives in all kinds of weather off ZABOTA all over the lake. I take Mom and Dad out, Mom still can't swim and Dad thinks I am out of my mind to dive when I blow a high pressure hose on my regulator. I just pulled a spare out of my bag, installed it and went over the side. He thinks what I do is dangerous!!!

Zach wants to buy a condo; he schools me on cash flow and leverage, he's a UT Macomb's Business school grad. He buys in just off 2222 and Mesa, a real fancy place. Zach also purchases an Audi

5 cylinder coupe. He has a steady girl now Lisa, that he eventually marries. They are the first to have a kid, his name is Logan.

Tom and I get a new place too, a big luxury creek-side duplex off Heart lane and a new roommate, Frank, he used to date Evelyn. Frank is a Dallas boy, a good guy, another Mechanical Engineer, a classmate. He looks like he could be my brother. Frank is big help keeping all the toys running, he's a sailing enthusiast too, we sail ZABOTA at every opportunity. We go out during small craft advisories to get heavy weather experience. Frank and I get caught out during the great Memorial Day flood. Tom has friends in from Midland and Frank has friends in from Dallas. We have over a dozen people with us doing our usual thing, towing the ski boat with ZABOTA under sail to where we want to ski to save gas, use it to carry the ice chests people and food. Most folks stay on the sailboat while the skier, a driver, and a spotter, put on a show around the boat. We are just off Starnes Island after skiing all afternoon when we see an absolutely black wall coming in from the northwest. We are just drifting. There's no wind, yet. The Carlson unties and takes off for the boat ramp with about five people. I crank up the outboard and head for port. It runs out of gas after going 100 yards. Then a huge down draft with buckets of rain hit; we are off to the races under bare poles and a lashed flagging jib. I jump in the cuddy cabin with seven other folks and install three hatch boards leaving the top one out so I can reach the tiller. I manage to steer a course and not broach by looking at the Texas flag streaming from the backstay. I coast into Commanders Point where we have our slip and tie off to the nearest pier. It rains for a couple hours more. We get over 10 inches of rain. The folks in the ski boat were trapped in their cars when Bull Creek over flowed and closed the highway. Shoal Creek which runs through town floods and does major damage downtown. We are in my little sailboat snug and dry, laughing about all the excitement and with all the provisions… We wait it out. We have no idea. Epic floods seem to follow me around. I vow that if I ever buy a place it will be sited on a high piece of ground and on solid rock.

Frank is a part time roommate, he's working in San Antonio for Baker Tool. He stays at his married big Sisters house during the week and comes in town on the weekends; he puts a lot of miles on his 260 Z. It's Friday, I've just showered and shaved after a long day's work, outside today. I have the place to myself. Tom is out of town and Frank hasn't shown up. I go downstairs and crash on the couch. I wake up to find a beautiful slim young girl with blue eyes and long blond hair held back with a head band looking at me. One of the he prettiest girls I have ever seen. This girl has style and taste dripping off of her. I ask her how long had she been sitting there looking at me; "quite a while" she said laughing. I ask her if I drooled or was snoring, she says "you're cute". I love her voice, the pitch and the cadence. She's Jennifer, Frank's sister Marie's best friend from Dallas. Jen is going to school in Galesburg Indiana. She's down to see her parents and Marie during spring break. Jen was supposed to meet Frank and Marie here, the door was unlocked so she came on in. It takes me about ten minutes to see that this girl is not a shallow, empty headed girl; she's very bright, engaging, and ambitious; she's special. I like her. After an hour or so Marie shows up and I take them both out to dinner, no sign of Frank. I ask the girls if they want to go sailing the next day. We agree to meet at the marina.

The wind is 5-10 the next day, slow sailing. We mainly drift and drag on a rope hung from the stern, one of my favorite things to do. The benign conditions give me chance to know Jen better. After a while we drop sail and are just lying on the deck and benches soaking up the sun. I keep sneaking a peek at Jen, she is absolutely beautiful, I catch her checking me out and we smile at each other. We head on in, we're out of drinks and Marie wants to try to call Frank again plus get us some sodas. We tie up and Marie starts the long climb up to Commanders Point Club House from the docks Jen and I wait on the boat. We're alone, we slip into the water because the heat is oppressive in the cove where there is not a breath of wind. We immediately start making out in the warm water like teenagers rolling around under the bleachers, can't keep or hands off each other. Marie walks up on us and her jaw drops, Marie's plan was to fix up Jen with Frank. Jen spends the rest of her time in town staying with me. Frank literally missed the boat, he has been with Mark Andres sister, at her place, the whole time.

We like the same music and we're reading each other's mind; the conversation is very engaging, this girl is way scary, another Samantha. I find myself fighting to resist her, it's like there are three people in the relationship. Pretty girl like her, she is just going to make me fall hard for her then dump me to date other guys. She is probably dating someone else now, I don't ask. I fly her in a few times for the weekend, but I can't seem to let myself go, I struggle not get my hopes up. I sabotage at least one visit flying her home early, even though we never argue or fight. She knows that she has me when I take her to the airport, she gives me a knowing Mona Lisa smile when I kiss her goodbye. We write each other for years. I love to get Jen's letters. Jen wants to be a curator. She considers coming to Austin to continue her studies, and I consider asking her to marry me if she does, I never tell her that, she goes on to study at the Met. She used to live in the City. A good decision, She does what's best for her, like I once did. I fly to New York to see her and we do the town. I wander around the Met during the day while she's working. She's specializing in Oriental Art and Artifacts. Jen learns Chinese and goes to Taiwan where she works to become fluent while teaching English and studying at the same National Museum I once told her about. She visits the Mainland after she travels all over South East Asia, India and Nepal. I get postcards that I pin up in my office.

This pen-pal, occasional visit, relationship is the only one I have for years. I am constantly traveling myself, spending not a few months at sea. My instincts were right. It was never going to work out, what bankers call surety I don't have it. It's unfair to make someone limit their potential for your own selfish reasons; I never try to persuade Jen one way or another, we would have been great together. Long distance suits me fine. I love my brilliant Pen-Pal. I should have never read all those Sir Walter Scott Quintin Derwood and Ivanhoe books Mr. French encouraged me to read at RFA, you know, worship your true love from afar stuff; I'm a hopeless romantic with no romance. Don't trust, don't verify. I also read War and Peace, same thing.

# 1981 San Francisco

If you live in Texas and work for the Navy you have to travel. There are no Navy bases in Austin. I am away working for months at Charleston and Portsmouth Naval Shipyards, Ballast Point San Diego, Kittery Maine and Mare Island Naval Shipyard in San Francisco Bay. Working at Mare Island gives me a chance to see Kathleen. Kathleen graduated from Boulder with a degree in Anthropology and French in 1973. Anthropology enlightened her to the fact that everything her Mother and the Catholic Church had taught her was a lie, all fairytales, magic, old wives tales and guilt. To finish off her French minor she needs immersion. She took a position as a Nanny, an Au Per, to a French couple in Strasburg through CU 1974-1975. Dad funds this excursion and the round trip Airline tickets. The couple turns out to be horrid. The husband has taken a lover. And the wife enlists Kathleen to find out who it is, asking her to do things that she is uncomfortable with. She is often left with the child, who is sweet when they are gone and acts up when the parents are there. Obviously begging for attention. Kathleen is unhappy and isn't making any money. She wants to cut her commitment short and come home. Mother writes her letters advising her to stick it out. Kathleen is able to travel some and it makes a difference. She travels to Berlin with stops along the way staying at Hostels, traveling by train. Kathleen also attends free classes at the University.

Kathleen finds that she likes the Germans better than the French who are narcissistic and driven by emotions. She also misses her boyfriend Detliv Doring. When Kathleen and Steve were at CU they lived in the Language Dorm. Kathleen was on the French floor Steve was on the Spanish floor and Detliv is on the German floor. Detliv and his little brother escaped East Germany and an abusive Nazi Army father with their mother crammed in the trunk of a car before the wall went up. His mom married an American serviceman who adopted him and brought him to the States. Kathleen and Detliv have been dating for two years. When Kathleen comes home for Christmas 1975 she tells me of her new found sexuality. Kathleen still sleeps in the same bed with Maureen when she is home. Kathleen shares things with her that no thirteen-year-old should hear. Mom and Dad are in the dark; Reenie and I are their substitute, Kathleen has to tell someone. Steve of course is in the know, he fills me in on some of the details.

Detliv studied Structural Engineering at CU but didn't graduate, he's moved to Berkley to be a draftsman and finish his degree at Berkley. He is working part time for a firm in The City. The two have been writing each other. At first Kathleen is not so sure she wants to marry him, he is a mama's boy and there is another fellow that she is interested in. Loneliness finally gets to her. They both want to get married. Kathleen flies home, sews her wedding dress, and flies out to Los Angles where she is going to meet Detliv. Her plan is to get married at her Uncle Bill's house; her cousin Tony is getting married, she is going to horn in and make it a dual ceremony. The only one there for her is her mother who flies out. Detliv is staying at Bill's too. Kathleen and Detliv haven't seen each other for a long time and want to spend the night together. Bill is appalled and practically calls her a whore. Mom defends Kathleen but makes the mistake of attacking Irene, Bill's second wife calling her "used goods" since she had been married and divorced. She always liked Shirley, Bill's deceased first wife better. Shirley was a saint. This sends Bill and Irene in a tizzy after she throws that in their face. Kathleen ends up defending Bill and Irene and attacking her mother for defending her. Somehow everything goes off in spite of the row. They are married.

When I see Kathleen on a trip to Mare in 1980 she seems quite happy. She has her little apartment in Berkley filled with things she loves. Kathleen has inherited Mom's taste for fine things. She has landed a job as personal assistant to the Chancellor of U C Berkley. She is taking class and performing with Berkley Ballet. Kathleen has always loved the Ballet; she got her first tutu and had her first recital in Japan; she was mom's living doll. Mom made a lot of clothes for her, sewing little outfits that she kept in a miniature armoire. Kathleen kept it up in Utah getting more tutu's and performing in more recitals we have 8mm movies of Daddy's little princess on stage. Dad's camera requires an obnoxious high power two lamp light bar that blinds everyone in the darkened theater; it's brighter than the spotlight, that's a lot hutzpah on the part of the proud Poppa.

Berkley is an open campus. There are crazy people wandering all over the place. California has changed the law and closed the institutions. Kathleen mans the desk in the Chancellor's outer office. She screens his mail and all media for subjects that might interest him. Kathleen also has a security button under her desk for the nuts who occasionally wander in. Kathleen is easy on the eyes, she just chats them up till security gets there and hauls them off.

I am out at Mare Island for months sometime. I play soccer with Detliv and catch a few Bear Football games with him. Kathleen isn't into sports. I escort Kathleen to a fancy Jewish wedding in Sausalito. One of her good friends is getting married. I have a date with another one of her friends, we all have a picnic in Golden Gate Park. I help cook a fresh salmon that my date caught, on a grill, it's huge, longer than the grill, more than enough to feed everybody at the picnic. My date and I exchange a few letters when I fly home. But it goes nowhere.

I'm the only one who goes out to visit Kathleen besides her Mother; I worry about her. I see Kathleen after one of Mom's visits. She has new issues with her Mother, like she needs more. Mom came up ostensibly to visit her daughter but turned it into a tour of all her friends who reside in Sacramento and around the Bay area, making it all about her, having Kathleen chauffer her around in her car. Kathleen is really burned. They went to Sausalito. Mom bought some artwork there that Kathleen had picked out for her place, two paintings; a pair by a famous Chinese artist of Chinese children. I have seen them both, they are very moving pieces. Mom bought them for Kathleen but then decided to keep one of them just before she left and took it home with her, promising to give it to Kathleen "later", which never happened. Mom had it mounted, framed and hung it on her wall, Now Kathleen looks at the painting of the little girl and only sees the missing painting of the little boy.

I am worried, Kathleen is thoroughly infected by the California state of mind. She is really into herself, self-analysis psychobabble, trying to dive into why she feels the way she feels. The search for someone to blame in our blameless society. Just a bunch of circular logic. I say nothing. I know that it would be unhealthy, at least for me, to engage in that sort of thing. She and Detliv don't seem to have anything in common. They don't do much together. They even have their own set of friends, the Ballet and Soccer hooligans. Detliv has long days with his commute to the City, class and homework. A boring Engineer like me. On weekends he likes to kick a soccer ball around with his buds and drink beer. Kathleen is always at the ballet studio taking class, or rehearsing. The next thing I know Kathleen is divorced and in Austin; I am never clear why, and I never ask, but I think I know.

I purchase a house just the other side of 183 from the Barrington Way house. It's close enough to come home for lunch if I want to. I ask Kathleen to decorate it for me and give her my Foley's card. She does a real good job. It's a quiet neighborhood. I can walk in and just flop down, no one will bother me. I have a place to myself for the first time in my life and I savor it.

Kathleen, Jim, Steve, and I have all failed to fly, we have all crashed and burned on takeoff. No hometown, no long term friends, no career waiting for us out of high school, no rich Dad to pay our way, no romantic fairytales here, no high school sweetheart, no extended family, no rendezvous in Paris, no love, nothing lasts long, just bent airframes and burned bridges.

## Kathleen is Back

Its 1982 Kathleen moves in with her sister Maureen. Maureen is rooming with her two best friends that she has known since Taylor Elementary, Debbie and Melody. Debbie and Melody are attending Southwest Texas so their place is on the south side of town. Kathleen moves in when Melody moves out. I like these girls. Maureen is following my example and going into Mechanical Engineering, working her way through doing tennis at Lakeway.

It doesn't take long before Kathleen lands a job with the UT Health Center. She starts taking class with ABT Austin Ballet Theater; Stanly Hall is the Artistic Director. He is also the Artistic Director for the UT dance department. Stanly puts on performances from the Armadillo to the Paramount. Stanly, who I meet, is a real character, an ex-British Royal Navy Signalman Sailor. He served in the Atlantic during WWII. This same man trained with the Royal Ballet, where he performed as a soloist, Ballet de Paris, danced in Diamonds are a Girl's Best Friend with Marilyn plus many other movies. He has a very good troupe with lots of boys. Greg Easley inherits Stanly's estate and goes on to be the artistic director of the National ballet theater in DC. Kathleen has got her groove back. She's with likeminded individuals.

Kathleen doesn't get along with Melody or Debbie she's not acclimated; these girls are as West Texas as they come. Debbie is a tiny slight cleaver girl with big brown eyes brunette hair and bangs that accentuate her eyes; she's studying to be a programmer. Melody is a slim blond girl with blue eyes, she's dating Jim Snow who is working on his PhD in Chemical Engineering. Neither of these girls are very athletic, which Kathleen has no respect for, plus they've nothing in common. Kathleen is older and wiser, she also hogs the bathroom, which can create tension. Kathleen tanks the setup. The girls have been covering for Debbie should her parents call for her when she is staying with her boyfriend Charles Ford, her future husband, another programmer. Debbie's Dad calls one evening and Kathleen informs her Father that Debbie has been shacking up with Charles, ruining her relationship with her father

## 1981 SAN FRANCISCO

who refuses to attend her eventual wedding. After that Kathleen moves in with Arleta Howard, a principal dancer, and her sister Alesia. Arleta has leased a house off Lakeshore on Pruitt close to the ABT Studio off Deep Eddy.

Arleta is dating John Logan, an Electrical Engineer and one of the dancers. Kathleen fixes me up with some of the dancers. They are very fine women; we are worlds apart. I take out one of the principals, a very classy lady. She takes me to her home in Tarrytown where she lives with her parents. Her brother plays tennis too. We go sailing. She has a dance studio tan. She shows up in a long shear dress and a big floppy straw hat with a picnic basket containing a bottle of white wine, a beget, cheese, peal and eat shrimp with all the fixings. I enjoy her company but we totally fail to connect, it's most likely me. How could I not go for this elegant graceful creature? I just don't have it in me anymore my brain doesn't trend in that direction.

John and Arleta move in together, Alecia moves out too. Maureen moves in with Kathleen, Jim moves in with Melody. Kathleen asks Cindie Hardee, a stipend ABT principal dancer and instructor, to move in with her. Cindie is a quiet, serious, nineteen-year-old, a freshman at UT, one of the original Armadillo Ballet Dancers; she's been instructing since she was thirteen. The Armadillo used to sponsor ABT performances once a month to crowds over 1500. Cindie has been bouncing around house sitting ballet patrons houses; she left her home in South Austin when still attending Crocket High School. Her brother assaulted her and her parents took his side. I don't hear about any of this till much later. Yep I had to ask, she is the great, great, great, granddaughter of General William Joseph Hardee CSA. The family has a ranch by Camp Swift in Bastrop. I ask why her great, great, great grandpa settled in Bastrop County. Cindie says that her Ma-ma told her "he shot a man in Florida".

Kathleen and Maureen like Cindie and sort of adopt her like a sister. Cindie has a dancer's body, all legs. Her upper body is so thin you can see her ribs. Cindie doesn't smoke cigarettes like the other dancers to stay thin, to kill their appetite, she's been surviving on Saigon Egg Rolls and Rahman. She has big brown eyes, long brown hair that she wears in a single braid that reaches to her kidneys. Cindie is a very modest, low maintenance girl, nothing fancy, fancy is for the stage. She is happy to be barefoot, in shorts and a tee shirt or wearing one of the shifts her sister Jackie has sewn for her. I help my sisters out with whatever they need, fix their cars and appliances etc. I take them out to dinner once in a while and give them a few bucks now and then. I go to the ABT performances, I like the ones that they put on at Zilker Park, just the thing for a warm summer night.

Tim has a party at the Barrington way house, he is managing the place now. It's a Mamba party; you fill a baby pool with ice and Mamba Beer. I am on travel for this one. The guests are a bunch of tennis players, ballet people and geologists plus anyone else we know. Traci Baker, another Principal, lives right across the street. Kathleen gets into a friendly argument with some tennis players as to who's the better athlete. When, according to Tim, Kathleen jumps on a table and tries to unscrew a lightbulb with her toes to demonstrate her prowess and blows out her knee. You should always warm up and stretch before you unscrew lightbulbs with your toes. Kathleen needs major invasive surgery, arthroscopic won't do. Dad helps her out with the bill. Dr. Malone Hill performs the operation. Dr. Malone Hill is the same Surgeon that operated on Tim's knee. His Dad General Jim Dan Hill lives across the street from Dad.

## 1981 SAN FRANCISCO

The General and his wife are some of the parent's favorite people. General Hill was the president of the Wisconsin State Teachers College and commander of the 190th field Artillery regiment during world war two, rising to the grade of Major General in the National Guard. I have many of the books he authored; among my favorites is *The Texas Navy*. Some toughs tried to mug the general on his walk, now an old man with a cane. They were surprised to see the General pull a 30 inch blade from the cane. They ran away. Malone hill is something else too. He was in a horrendous plane wreck and broke nearly every bone in his body. He was pieced together, rehabilitated himself and is now back practicing.

In rehabbing her knee Kathleen becomes the queen of Hyde Park Gym. I'm at work, in the cafeteria and I overhear a conversation that goes sort of like this: "there's this girl, at the gym, man… she is so hot…you should see some of the things she does…" I have to say "hey man, that's my sister you're talking about". The trainers tell her that she has the legs, that's the hardest muscle set to build, all she needs to do is build her upper body and she could be a competitive body builder. She goes for it, gets ripped, gets spray on tanned, gets sponsored by Eddie Wilson and Threadgills. Kathleen is working the loading dock at Whole Foods now just to get a workout. She wins trophies taller than she is in the State 98 pound class. When they start pushing supplements and steroids she quits but keeps up her gym regimen. Now it's get up at 0400 to bake bread, hit the gym, go to work, and go to ballet class. A busy girl.

Around this time Kathleen accepts a position as personal assistant to Congresswoman Barbra Jordan. Rep Jordan is retired from congress but has a teaching position at the LBJ Library; she's to speak at West Point and will continue on to France from there. With Kathleen's PA experience at Berkley, her Health Center experience and her fluency in French she is a shoe in. Rep Jordan is in poor health and wheel chair bound so Kathleen is a great help to her.

Kathleen starts dating Robert King an ex-Marine Lance Corporal at least ten years her junior. They meet at the gym, he has a great body...They make big plans, both want to be EMTs. Robert's plan is to first be a Corpsman in the Marines, trouble is Corpsmen are all Navy personnel, Robert would have to enlist in the Navy to do that.

Kathleen and Robert are married at Steve and Kathy Newhouse Scanlon's house. This time we are all in attendance. Steve was the first of us boys to get married; their house is a few blocks away from the Barrington Way house. We meet Kathy through Jim, he had ingratiated himself with Steve Newhouse, her husband at the time; he's their daughter Kelly's babysitter, if you can believe that. He was living on their couch. Steve is an A/C technician Jim is his helper but he thinks he is the boss. Steve N is the son of a Sargent not that there is anything wrong with that. Kathy, valedictorian of her High School class and Magna Cum Laude Accounting Grad from UT, is the daughter of Lt. General Ayers who is heading up central command in Germany. Kathy is a CPA and has her own firm, she's set Steve N up in business several times and he always manages to fail. Steve N and Jim are in a close race as to who is the laziest. Kathy finally got fed up with Steve N and divorced him, but not before having another child, Travis. When Steve and Kathy got together I had my reservations. Steve was vulnerable, Kerry had just left him for Ernesto. I thought that Kathy was just looking for a replacement, a better one, the bar was pretty low. Then again, Steve is a tax examiner and Kathy has an accounting firm. Steve starts working with her and gets his CPA license too. They have a lot in common.

# 1981 SAN FRANCISCO

Robert's mother hates Kathleen, calls her a cradle robber; his sisters are little delinquents who steal stuff from Kathleen's room and purse. Robert wants to change his last name to Scanlon but I tell him there's already one of those. He is seriously considering changing it to Atridies after the character in Dune, he hates his family so much. My sister is not marrying into money or class.

Robert enlists in the Navy and goes through Boot Camp again; he says that he wants the exercise. This make me question Robert's intelligence. The only reason I can see for dong that, is that he would be at the top of his class. He goes to Great Lakes Naval Training Center, in Chicago, in the winter...Kathleen flies up to see him midway through Boot. When Robert finishes up with something he could have skipped, he receives orders for Camp Pendleton and off they go, with two dogs that Kathleen ends up shipping to people who adopt them, one back to Texas and one to New York. Kathleen likes Pendleton; they live in Fallbrook, the Avocado Capital of California; there's no on Post family housing for E3s.

Kathleen enjoys working out at the base gym and riding her bike around driving all the Marines crazy. She makes friends with Robert's Company Commander's wife, a likeminded individual. They become even better friends when Robert is deployed to Guam, where Robert runs up the credit card buying dive equipment, much to Kathleen's annoyance. And later for Desert Storm. I talk to Robert about Desert Storm. His only comment that stuck was that he thought that a lot of his fellow Marines had a death wish. After two years enlistment's up its back to Texas and Nursing School. The plan is for Robert to go first using his benefits then it would be Kathleen's turn. Kathleen should have gone into nursing in the first place. She has a talent for it. Life is a like a single pass compiler. You had better think it out before you run your program.

Robert falls asleep at the wheel heading home to Texas; he rolls the car and U-Haul trailer somewhere in nowhere New Mexico. Dad drives out and rescues them, I am at sea… they get an apartment. Robert enrolls in Nursing School using his benefits and Kathleen gets a job as a sterile surgical nurse for an oral surgeon. Things are going well until Robert cheats on Kathleen and to make it worse it's with another nursing student, "a fat ugly Jewish girl". Kathleen is furious, she throws a television at him. Robert calls the Cops and files assault charges on her! Some Marine. They are done. Robert is later rumored to have been seen at Oil Can Harry's in a wet underwear contest.

# 1983 Cindie

I start dating Cindie a little before Kathleen gets married to Robert. She moves in with me when Kathleen moves on; Maureen moves back into the Barrington Way house, now Tim's place, at the same time. Cindie has no idea of how little she has to work with and I never tell. I think that she just likes my cooking. I don't believe she ever really loves me. I am just her better man. I know this. Cindie is a good person. I think that she will be good for me. We start off real slow. Cindie is nineteen, and I'm thirty one, she has absolutely no interest in me at first. I had just bought a house. I took her over to see it when I met her, at the Barrington Way house for the first time, she wasn't that impressed with it or me. I take my sisters out every once in a while to see movies and for dinner when I'm in town, Cindie tags along too if she isn't busy. Cindie drives a Ford Torino with a 427 big block and a 4bbl Rochester that she starts having problems with. I rescue her a couple of times, one weird thing, when you turn on the wipers the radio comes on there's a short somewhere killing her battery, I disconnect the radio after jumping her several times. I'm a hero. I rub, Hi Cindie, in the oxidized blue paint on her door with my thumb. Her car still runs like crap so I offer to go over it. I rebuild the carb and give it a full tune. When Cindie comes to pick it up, she lights up the tires leaving; she is very pleased.

A week or so later Kathleen tells me that ABT is putting on a performance at Ziliker Park, Ballet in the Park and that I should come see it. It's a nice warm Texas summer night. She and Cindie will be performing. When I show up I see Reenie, Cindie's mom, Lavern, and her sister Vicki, they've staked out a good spot, stage left, up front, spreading out a quilt and pillows. Both Lavern and Vicki look a little uncomfortable to see me, Reenie puts me wise. Cindie has asked a fellow she is seeing to come to the performance; he's just been by. She has been dating some Star-Flight ET. He's sitting somewhere stage right. Just what I thought, good thing I didn't get my hopes up. I start to leave, my evening is just another disappointment; I'll add it to the list. As I get up I see Cindie comes running up in costume, she looks very pretty. I tell her that I was just about to leave, I have to meet someone, somewhere…about something… She asks if I can wait awhile, they are just about to go on. The company is doing a 40's dance hall, jitterbug routine that Stanly has choreographed to show off all his dancers. She doesn't know that I know. I decide to stay and see how it plays out, its sick, but I'm curious to see how she juggles us; my bet is on the other guy. After the performance, which is very good, I'm amused to see Cindie with her face all shiny going from stage right to left keeping us apart. I leave to spare her and myself anymore embarrassment.

The next time I drop by the Pruitt house, for movie/dinner night, as I'm getting ready to go out the door, Maureen and Kathleen start grabbing their purses; Cindie tells them that it's going to be just us tonight. We get wry smiles. She called me after the park performance and said that we should get together some time.

We go see *Dune* after dinner; I've read the book in fact the whole series, and comment on the movie re. the book, Cindie is disgusted by the whole thing; my inner nerd is really showing. I take Cindie home to the Pruit house. We sit in my car and chat in the driveway. I have a navy blue 79 Corvette with an L-82 350 and a navy interior with leather seats. I sold my TR-6 to a friend of Maureen's who was really jonesing for it. I bought the Vette from this Air Force Officer who dolled it all up for his wife but couldn't get it to run for a song. I found a manifold leak in ten minutes, now it looks and runs great. Despite my best efforts Cindie seems to be enjoying my company. I try to kiss her, hard to do with the tranny tunnel in a C3 Vette. Just as I am leaning over there is a loud thump; we knock heads. Fido, Cindie's nineteen pound black cat is on my hood looking through the windshield at us bobbing and nodding his grapefruit sized head. This cracks Cindie up no end. She has a very loud uninhibited laugh. He makes me laugh too as he paces back and forth peering in. climbing the roof and looking at us upside down. Fido breaks the mood. I am allergic to cats, Cindie comes with one. He's a funny animal, he rules Kathleen's big black lab and her little terrier. He will jump on the lab and ride her if she gets out of line. When Fido is hungry he will climb the chain-link backyard gate and rock back and forth making it clang like a dinner bell. He walks Cindie to the bus stop in the morning and spends the day waiting under the bushes in front of the house, in ambush, for her to come back. The only thing you see is a paw darting out of the bushes, a Ninja cat. She tells me to wait as she opens the door, jumps out and runs into the house. She come back clutching a small bag. We go to my place. We live together for seven years along with Fido until we are married, I needed to be sure that I could trust her to not use, ghost, or dump me for a better prospect someday.

We barely see each other. I am traveling a lot and staying a long time when I do. Cindie is busy with: school, homework, rehearsals, performances, teaching class, taking class, and volunteer work for Suicide Hotline, she's studying to be a Social Worker and minoring in French. We can usually only manage Wednesday evenings together, performances are on the weekends of course. I make all of her performances. It makes me proud to see her picture in ads and posters, or in the lobby of the Paramount and Bass Hall Theaters, plus several other venues downtown.

When Cindie graduates she goes to France as a present to herself. The girl she was going to travel with backed out and the people she was supposed to stay with left for the south on vacation. She finds a note on their door. Undeterred she goes anyway, she is in Paris for Bastille Day. She rides Euro Rail from Paris to Marseilles to Brussels and points in between on a chocolate and beer tour, staying in Hostels and making friends, some who later visit us. Coming home after a month of fun, she misses a charter connecting flight, has no money so sleeps on the floor at JFK until her sister Jackie buys her a ticket home. I worry about Cindie but I'm incommunicado. I don't hear any of this till I get back from two months at sea.

## Boom

While at sea, working on a prototype mine hunting sonar for Dr. John Huckabay, I have one of several sobering events working for the Lab that I make it through, unscathed. We were still in port, I had just installed the system which consisted of cabling into the ships generator and welding down a dog house with electronics and a 10,000 pound line pull 40hp marine winch, I designed and built, to deploy and capture the 1000 pound sensor. The winch had a control station near the stern so the operator could observe the deploy/capture evolutions. The control station had controls enclosed in a two foot square six inch deep sealed cast aluminum water proof box with a bolted on O ring sealed 3/8 thick aluminum lid. My good friend Charlie Jaecks thought that he had let some rainwater get in the box. He was working on it, while I was in the galley getting some chow. He flushed the electronics with pure reagent grade alcohol sloshed it around and dumped it out on the deck, bolted the lid back on and reattached it to the control station.

After chow we want to do some pier side testing. It comes time to deploy the sensor. I climb up and punch the power on button and the thing explodes in my face. The lid curls up popping the ¼-20 hex bolt heads off like a zipper. Everything, the lid, bolt heads, my arm, hits me in the chest pretty hard and would have knocked me overboard without the rails I built into the operator station. I roll on to the deck wishing I had a PFD on, the sea was calm, I was in a cage, it would have taken some of the blow. I immediately start tearing my clothes off looking for holes. I have three or four layers on, its cold, it's November. No holes. Arm chest and hand blunt trauma. Everything hit me just right. Folks are asking me if I'm ok. I can see their mouths moving but I can't hear what they are saying, My ears are ringing. We call it a day; head back to the hotel, clean up and have a "thank God Bob is alive party" at the best restaurant in Ft Lauderdale. The box is rebuilt with a new lid and vented; I'm functional, we go to sea. I'm glad that I decided to move back and perch on the railing instead of standing and leaning over the control panel at the last moment, everything would have hit me square in the face. Talk about choices. My boss and good friend John Brady keeps the bent up control box lid on top a filing cabinet in his office for years. Cindie may not have wanted what was left of me. This is the first in a series of accidents. To date: I've had two diving accidents, two shark attacks, Paul Eisenman punched one of them in the snout for me when it was going for my leg, for which I am truly grateful, got trapped in the Okefenokee Swamp at low tide towing a RIB African Queen style while the swamp burned, contracted amebic dysentery in Bahrain and was on bananas and rice for a year loosing thirty pounds, nearly caved in my head in Bahrain when I fell and hit a deck pad-eye, but the most disturbing incident was being flushed like a rat in a sewer pipe. I was doing a pier side Ship Alt at Charleston Naval Shipyard. I was on my belly, army crawling under the deck in the free flood spaces under the sail of a 637 class sub. I was accessing a hull penetrator. Someone executed a low pressure blow. It was supposed to be secured and tagged out but they had been shooting water slugs all morning. I nearly drowned backing out. I had wrenches in my overall pockets that hung up and I had to back up to empty my pockets and start out again. Good thing I can hold my breath for a little over a minute. This was supposed to be safer than walking top plate or the oil field. I am really pissed. John Brady intercepts me; he will do the protesting I will take the afternoon off. Two Engineers, EEs, who were helping me quit.

 Cindie and I are the last ones to get married. Gary makes a name for himself playing tennis, he marries Tanya who he has been dating since he moved into the Barrington Way house. Maureen Marries Peter Ellis an architect she works with at Texas Parks and Recreation. Kathleen finds Scott a real rock for her stormy seas.

None of my sisters have children, they have pets, fur babies. Cindie and I have three little girls. Suzanne, Katie and Emilie. Gary and Tanya have a girl, Shelby. Steve and Kathy have two boys Cory and Cameron. I don't know why things turn out like they do. When asked why they got married people responded in the main that it was time, or it was expected of me, or I wanted children. Most admitted to seeking, security, someone steady, dependable, someone like their parents or unlike their parents, someone that looked like them or that they could identify with, rarely for passion. I had passion aplenty but to what end? Decades later, during a rough patch, I ask Cindie why she married me. She replied "because you're smart". We think that we are in control making all these decisions that guide our future. I pose here: if you believe in the time space continuum, it's the souls of our unborn children wanting to be, that determines what happens, because in the end it's all about the children. Time is something that man came up with. An empty nest is a real test of a marriage. Did you stay together just for the children? All I really know for sure is what Debbie Schmidt once told me," In the end, the most important thing is to be able to reach out at night and find your someone there".

## Bill

Bill is another story. He tried to follow me and Maureen into Mechanical Engineering but he couldn't pass Statics, never got Newton's first. I tried to tutor him but he flunked twice. If you can't get Statics you won't get Dynamics, Newton's second. Bill works for a while at Catfish Parlor during the great Hush Puppy scandal where they were loading up customers on laced hush puppies that would swell, during all you can eat catfish night. After that Bill took up making sandwiches when not playing Dungeons and Dragons. He later progresses to pizza delivery, plying hacky-sack on the South Mall or sand volleyball at Barton Springs. To assist in his Pizza delivery Mom gives Bill her powder blue four door 75 Ford Comet, apply named. Bill overheats it twice. The first time he blows the head gasket the second time he cracks the head and I get him a rebuilt one. I fix it for him both times. He later pulls into the Barrington Way driveway, goes in the house and asks for help because his car is on fire. Everyone jumps up and pulls their cars onto the lawn and street away from the flaming Comet while Tim puts out the fire with a chemical extinguisher pulled from the kitchen. Bill had a gas leak he probably smelled but didn't attend to. I tell him he has a nonfunctional external combustion engine. What a mess. I blow off the extinguisher powder, fix the fuel leak, re-hose and rewire all the burned stuff under the hood. Bill is back in the Pizza business. I tell him there is no future in Pizza, this I know. Bill is becoming sensitive about other people telling him what to do, Dad is on his ass writing letters once a week and requiring updates from him. He tells him to stop drinking, to get a haircut, clean himself up, bathe and shave every day, dress like a gentleman, have some self-respect. Mom buys Bill a polyester suit and he get a job selling timeshares out at Lakeway. Tim tells him that it's a looser but Bill is nonplused, he knows what he's doing. I don't think he ever made a dime in the six months he worked at it. He kept saying that he was going to get a spiff, whatever that was. Dad footed his bills.

 The fire in the Comet's engine bay ruined the A frame bushings, they soon turned to dust. Bill needs to get back and forth to Lakeway for his new commission only sales job. To replace the bushings I had to pull the suspension's coil spring. In order to do this, you have to compress it with two long bolts with hooks to catch the spring's coils. Very dangerous with all that stored energy. I'm doing this out in the street with the front of the car jacked\up, sitting on the curb, in front of the Barrington Way house. I am compressing the spring using a large open end wrench, alternating a half turn for each bolt. I carefully lift the assembly a few times to see if it will come out setting it back down to compress it some more. On one of my try's Bill, who is looking over my shoulder, decides to "help". He reaches over my shoulder jerks on the spring and collapses the bolts on my right hand. It's still not ready to come out. I have to tighten it a little more before I can lift it free from the A frame. I lay it on the lawn with my hand trapped and loosen one of the bolt's nut a turn at a time with my left hand while looking at my mangled right hand. My fingers look like squid tentacles they are blue, purple and bloody red, going every which way. Bill is absolutely no help, he's hopping on one foot and then the other and making faces. I must be in shock but I'm not puking. The pain is so overwhelming I no longer feel it. After I get it off I have to hit the yellow pages to find a nearby clinic. I call them and get directions, then I start losing it. Tim drives me there. They X-ray my hand, the bones aren't broken, they say, but the joints on three of my finger are crushed there are screw threads cut in the top of my knuckles and the first inside joints of my index, middle, and ring fingers. They can't do anything for me they won't even attempt to straighten them. I pull my fingers straight gauze and tape them to tongue depressors myself. I am really bummed; I make my living drawing. Everything I do is hand written, and I am in the middle of rebuilding Dad's truck engine.

In a stroke of good luck, UT in cooperation with Apple, is offering first flight 128k production Macintosh computers for a reasonable price. I buy one and get a printer too. I bring the first computer into the lab to do Mechanical Engineering, It's my own.

I can manage to keep working, make purchase requests, memos, work requests, documentation, Parts Lists, and do rough drawings by turning on and off pixels. I even get a Basic software package for it to do some simple programming. I can click a mouse and type, its great exercise for my hand. I can work left handed while kneading a piece of clay with my right hand. When decent drafting packages like AutoCAD come out I start lobbying for one. The response is "can't you use the Main Frame, the IBM 6600" and "ME's don't need computers". Little Scotty soldiers on shared with my students for five years before I get a 2D system. In just a few years I now I draw in 3D and the detail drawings are automatically generated and updated. My projects are so complicated, with thousands of parts in an assembly now that I could not do without such a system. One person can do the work of ten these days. An engineer can have an entire project on his desk top. Parts linked to cost, availability, reliability, and production schedules.

Bill lands a Tech Job with Motorola, he works in a clean room making wafers. I think, good for him. He is getting in on bleeding edge technology, his future is bright. After a couple of years, Bill takes an early buyout for 10K during a RIF in 2001. He didn't need to take it, his position wasn't threatened, he said he just didn't like the work, wearing a bunny suit, or the stupid engineers he had disagreements with… Back to Pizza. The Comet finally flames out. Cindie and I give Bill the Torino when Cindie's sister gives Cindie her old Corolla. He leaves it on the side of the road. We then give him the Corolla when Cindie buys an Olds, her first new car. Dad gives him a little truck then a Sebring, same outcome. Bill is finally on a bike and in the staging business, a Union man. He tells me that he prefers repetitive tasks. We have lots of events in Austin from Dino's Alive to SXSW film festival and ACL Austin City Limits music festival.

## Jim

I was rebuilding Dad's truck engine at the Barrington way house because I had orders to repo Dad's Airstream trailer. It's 1981. Dad had always wanted an Airstream, he subscribed to their magazine. He bought this real nice used thirty six footer that came with a F250 that had a 5 speed manual tranny to tow it. He moved and sold the FAN to the Normans, an Air Force couple that had retired in Kingsland January 1977 after Jim got busted at the Brentwood house. The Normans are on one of those runway communities. They lived in the FAN while they built their house. The FAN was parked in their professionally built commercial grade metal hanger with their Mooney. They built their home themselves being short of funds. They used hollow Styrofoam blocks that you fill with concrete; it went real slow, but got it done. The masons were over sixty. They have their priorities straight though.

Right after Dad bought the Airstream rig, Mom made him loan it to Jim, who had fallen on hard times again. When Jim first conned the trailer, he parked it in a real nice trailer park off Barton Springs. Jim failed to pay his rent. The trailer was impounded, bonded out and towed to Col Tredie's place off Manchaca in South Austin. He and Mrs. Tredie have ten acers there. Jim had been living in the trailer since that time. Mom persuaded Mrs. Tredie to let Jim stay on her property. The Col is not living at his place. He had a couple of tours to Nam and retired out of Bergstrom. When he got to Austin he started taking art classes and hanging with Hippies. He changed; I don't know what happened after that. He is not living at his place anymore, neither is Jim. Jim has moved on and abandoned the Airstream.

The truck was left at the Barrington Way house for everyone to use so no one took care of it. The engine wouldn't turn, there was very little coolant in the radiator when I checked it out, I discovered that the coolant overflow bottle was missing. I pulled the motor intending to rebuild it; Ford parts are cheap. I assemble it left handed because my right hand is still dodgy and tender after mangling it working on Bill's Comet. Larry Overfeld helps me stab it back in, after that the rest is slow but easy. I eventually go get the trailer, it's a mess inside. I tow it to the Barrington Way house and park it in the street by the side of the house, it's on a corner lot. While I'm getting my hair cut I tell my friend Mari my sad story. Mari is a Fin, an artist, a self-employed business woman and one of my good friends. I meet her Mom and Dad; they sometimes hang out at her salon. Her Dad is a retired Engineer. I like it when they get into it in their native language very unique. She wants to buy the Airstream it appeals to her sense of style. She can see it as a mobile salon. I give her Dad's number, and they make a deal less the truck. I deliver the trailer to Mari at her salon, The Headrest, a place she built herself, off Jollyville Road. The next time I get my hair cut Mari has a sad story for me. She cleaned up the trailer detailing everything but when she hooked up the water she flooded the rig. Jim had let the pipes freeze and burst, that's why he wasn't living in it. Of course he couldn't fess up to that. I tell Dad; he's angry and embarrassed. Dad calls Mari and insists on giving her some of her money back. Dad then sells the truck and with his remaining trailer money he buys a little conversion camper van with a bathroom and kitchenette. He's much better off with the smaller rig which he takes on many hunting and fishing trips. One trip is with Tim to Wyoming to fish with some of Dad's old crew members, September through October 1989. Tim gets to hear a lot of war stories on this trip. Dad, on impulse, sells this unit and immediately regrets it, he's trapped.

## 1983 CINDIE

He has always told Mom that's all he needs to get by, he doesn't need all the stuff, now he doesn't have that option to lever Mom.

I lost track of Jim after he was busted, he stayed at a friend's house in south Austin and became a regular at the Continental Club. Kathy Newhouse said that Joe Ely and folks like him were no stranger to Jim's couch. He has a room with a married couple, he's an offshore platform oil field worker and is gone a month at a time. This is too much temptation for Jim; he fools around with his friend's wife and gets the boot when they are found out. He moved to a tiny apartment on North Burnet after that. He put most everything he had that wouldn't fit in a storage facility. Jim failed to pay the rent for his apartment. Then he was evicted. He was kicked out and left his remaining stuff there. He didn't pay the bill at the storage facility either, so he lost all his household goods and everything that was valuable to him, including: Kathleen's bow, Dad's 20ga and hunting knife, my hockey gear, carpentry tools and harmonicas, all his records and his stereo, all the tools and chain out of the war wagon, also Tim's silver flute that he stole in 76, because Tim "owed him for the phone bill". Jim then bums around crashing at the homes of anyone who would have him. This is when Mom makes Dad lend Jim his Airstream, before Dad ever had a chance to execute any of the plans he had for it.

While living in the Airstream Jim meets Steve Newhouse and starts working for him. Steve N lives a few block from the Barrington way house. It's on the other side of town from the Airstream. From 1980 to 1984 Jim worked for Steve Newhouse, sleeping on his couch and babysitting.

I see Jim at a 1981 New Year's Eve party Bobby Ayers is throwing. Bobby is one of the old Killeen kids a good friend of Tommy G since high school. He has two kids now, a boy and a girl, Travis and Austin of course. It's late and my evening has been pretty much a bust. Bill fixed me up with a blind date, a friend of the sandwich girls he has been working with. She is working at a charity casino as a Black Jack dealer. Bill and I go pick her up at the City Colosseum in the Comet since I don't know what she looks like and there is only room for two in my TR-6. She is not impressed with Bill's car or her date. She puts on a sour face. We go downtown to Ester's Follies on Sixth Street where the backdrop to the stage is the front show case window and the gawkers passing by. Bill has snagged tickets for the early show, not the midnight bash. We get one free glass of bad champagne. I tried to be pleasant but my date is not having it. The show is over at 2200 to make time for the second show. I ask Bill to drop me off at my car and I ditch my date, to be with some friends for midnight. I hadn't been to Bobby's place yet, a lot of folks I know are there. I'm chatting with Bobby and Tommy when Jim walks up and hands me a drink in a glass and says happy New Year. I think nothing of it and drink the drink. I start feeling pretty weird. I want to lie down all of a sudden. I can't just lie down here, so I stagger out the door to my car and try to drive home. I'm rooming with Chip Becker, he and his wife Dandy are old sailing buddies. Chip could be Tom Lukenbach's twin brother. Unfortunately Chip and Dandy have split up, so I'm helping out Chip with his mortgage until he can sell his house off north Lamar. I get pulled over on North Lamar about 500 yards from the turn into the neighborhood. The cops, Hwy Patrol, say that they were first attracted to me because I had the top down, it's January 1, 1981, and it's real cold out. I had put it down to snap me out of whatever was making me dull. The nice officer says that I was moving at a good clip but I failed to signal that last lane change. He says the he smells beer. I tell him that I spilled a beer that I didn't get to drink, on the passenger side floor that afternoon. Stacy, Chip's pretty new girlfriend goosed me when I was cleaning my car, and I knocked it over. I drove the TR-6 because the Midget wasn't running. If you are going to run British cars you got 'a run two or better three. They ask me to blow but there are Lawyer ads on TV that say you shouldn't do that. I get a ride downtown to be handed over to the County Sherriff. They must be busy, it is New Year's Eve, The Deputies take me to a large room that looks like a scene out of Barney Miller. They make me walk; I don't seem to have any trouble with that. They just let me sit and veg. they stick me in a cell when the shift changes. My sisters come and bail me out in the morning for $500 bucks.

I should be mad at Jim, He obviously slipped me something in that drink. But this one is on me. Jim had done this to me before. Did it to Tim too. Jim slipped him some acid in 1975. I found him the crotch of the tree in the front yard of the Brentwood house at 0400, when I was headed to work at Monkey Wards, back when we were all living together. We were all going to see a Springsteen concert a City Hall Auditorium after a big feed at Spaghetti warehouse on 4$^{th}$. After we take our seats Jim asks me if I want a Quaalude, a depressant. I said hell no. He tells me that he is going to get a beer and do I want one. I say yes to that. Jim comes back with some tall 20oz beers in wax paper cups. I drink some of the beer. Next thing I know is I am waking up and puking spaghetti on the head of a girl seated in front of me. She turns around and gives me a look that I will never forget. I jump up, scramble down the aisle and run for the bathroom where I puke some more in the nearest sink. Jim followed me to the bathroom. I see him pointing and laughing at me. I don't know what's wrong with me, all I can think of is going home. I drove my Midget to the venue, I find it and take off down

I-35 and somehow make it back to Brentwood. After that I swore that I would never again accept an open container from him.

I plead no contest. I get a fine with probation, suffer personal and public humiliation, a smear on my record, and gain another reason not to have anything to do with Jim, like I needed another one.

Kathy Newhouse finally gets fed up with Steve N and divorces him. Jim gets the boot with Steve N. Before Jim lost that gig he bought an old work pickup truck from Tim in the summer of 1985 for $400.00. Tim bought the truck from his employer. Tim is working as a Geologist for a small oil company, Oiltex/Osiris Petroleum. Oiltex bought Tim a new one. The truck is Tim's personal ride, a short narrow bed step-side Ford with a V8 and a 4 speed std. transmission in it that Jim just had to have. Tim sells his truck to Jim for a song since Jim has no wheels. Jim goes on to get 36 tickets from 1985-1986 in it, never transferring the title. The first time Jim pulled this scam was in 1980, while driving Tim's car, it worked like a charm. When the Cops pull Jim over and ask him if he is Tim Scanlon he says yes, when they ask him for a driver's license and current registration he tells them that he doesn't have it on him and he doesn't have insurance. Jim wrecks the truck in 1986 and walks away from it. The first thing Tim knows about this is when a cop pulls him over on I-20 West of Fort Worth in 1990. The Cop puts a pistol in his ear, then arrests him for multiple outstanding warrants. It takes a lot of explaining and three separate letters to the Judge to expunge his record before the law lets Tim go. The law is now wise to Jim.

After Jim wrecked the truck, he buys a black Lincoln, a real pimp car. He runs it out of gas with Maureen and some of her friends in the car. The Police see it on the side of the road with the hood up and call it in. They stake it out and nab Jim. Jim calls his

## 1983 CINDIE

Mom with his one phone call but she is in Abilene and there is no way Dad will help after giving us the, you end up in jail you are on your own brief a million times. Mom calls me and asks me to bail him out. I do what my mother asks, lord knows she has bailed me out of trouble. Jim calls me after magistrate and begs me to get him out. He needs to pony up $750.00 to walk. He swears he will change his ways and promises to pay me back. I don't believe any of it. When I pick Jim up the officer that lets him out says "I hope you think you're getting a bargain, ha ha" someone else is waiting to pick him up. Jim must have kept the deputies amused. Of course Jim never pays me back, he tells me that it's really me that owes him; if I do, which I doubt, it isn't $750.00. A couple of years later I get another get me out of jail call from Jim, I just laugh and hang up. Felt good, but I'd rather have my money back. What a jerk, never stiff the guy who bails you out of jail.

After that around 1985, Jim ends up living out at Dave Osser's place just east of Austin, in Garfield, Yea that Dave Osser, the guy who gave him his first break. He has bought some property on the Colorado. He's married now to a real nice woman he calls Moose. That is a misnomer, Elaine is a petite strawberry blond who obviously deeply cares for Dave. He puts Jim up in an out building and finds him something to do. Dave has a pile of Alabama hardwood from his family's place, black walnut, black cherry, pecan and oak in his barn with commercial woodworking tools. He employs Jim in making furniture. Moose and Dave tell me Jim is a full blown alcoholic. They only keep light beer around for him to drink so he won't drive to get his booze. They tell me this while Jim is out tending to his dogs, a pair of one year old Weimaraners, a male and a female, beautiful animals. Jim called me to come out and shoot some ducks and see his dogs and of course Dave who I haven't seen in years. He is still looks like a cowboy version of Mel Gibson, still has a mischievous sparkle in his blue eyes. I don't get any ducks, I'm carrying Dad's old Stevens 12ga double barrel which he has given to me, Jim is using Dave's gun.

Anyhow it's good to see that Jim has a home with folks that care for him. Unfortunately for Jim, within a year, Dave has a heart attack and dies, but not before the Weimaraners get snake bit down in the high grass by the river and die, he told me there were snakes down there when we were hunting so we made a lot of noise to run them off. He should have had the dogs inoculated. Doc Jim sure is careless with his dogs. He inherited the original Carin dame and sire Tolly and Toby, lost them and one of their pups Malcom. Drunk and careless of them they ran off. He doesn't have any more pups after the Weimaraners.

It's Wednesday March of 1990 a cold, cloudy, drizzly, gray spring day common this time of year. We are all getting along with our lives, working and raising our kids. I'm at work when I get the call, "you better get down to the hospital while there is still time." Jim has been involved in an auto accident. I jump in my 84 Vette and head to Breckinridge hospital down by campus, next to Custer's old Headquarters during reconstruction. I pull into the parking garage across the street. I head down the pedestrian walkway over the street connecting the two structures. As I enter the lobby I hear a banging on the metal roof of the walkway. I look up to see through the widows above the walkway a bunch of buzzards; not a good sign. The receptionist directs me to the emergency room where they let me into a small room where he is laying on a gurney by himself. He looks bad. They say his head trauma is the worst of his injuries, all of his limbs are banged up and his arm has a temporary splint on it. Jim's face is black blue and bloody. His long hair, his vanity is

# 1983 CINDIE

caked with blood. His prognosis is not good. He is indigent. I am supposed to be saying goodbye. I look at Jim, knowing he can't hear me, I say, well Jimmy you have finally gone and done it haven't you. I kiss his bloody brow and leave the room. There is nothing I can do for him it is up to the Doctors and God's mercy now.

Tim shows up around 1430 just after I leave to go back to work. I am a supervisor and have many responsibilities. Tim was at work at Mcully Frick and Gillman at their Balcones office. Debbie McDonald, Maureen's dear friend and one time Barrington way roomie, runs into Tim as he enters the lobby. She is the manager of the hospital's IT system. She had seen that Jim was admitted and looked up his file. His blood toxicology test show the presence of drugs and alcohol Jim was very drunk, no surprise. It was 11 o'clock in the morning when he wrecked. The preliminary accident report says that he was making a left turn on South Ben White Blvd, lost control and hit a light pickup truck head on with his full sized pickup. The other vehicle was in the left turn lane for the opposing traffic. Jim was ejected from his vehicle through the windshield. He wasn't wearing a seatbelt. The other driver was killed along with his cow dog, a blue heeler. He was a retired Army Lt Col with seven kids...

Jim survives. The next time I see him, the hospital is trying to decide what to do with him. If he had insurance they would try to do more with his mangled right arm. The hospital wants him out, now that he can stagger to the head to relive himself. Jim has Herpes and an STD so the nurses are not too happy to help him with that, also dealing with his catheter. Jim can talk but his speech is slurred. He is somewhat cognate but is deemed unable to make decisions for himself. Mom is by his side with a young woman named Patsy. Dad is nowhere in sight Jim has been disowned since he got busted, like the Col promised, he is on his own. This event has not moved him. Dad is a man of his word. Jim got Patsy, pregnant. Patsy is a married woman. Jim's plan is for her to go back to her husband to have their child on her husband's insurance then leave him. They will get together after the baby is born. That will never happen. Jim is not going to get better, he knows it. Mom escorts Patsy out of the room to chat in the hall. With the girls out of the room, Jim beckons me and Tim close to his bed. Jim says, "Do me". We don't understand what he wants with his slur, which gets worse the harder he tries to communicate. He says "Do me…do me out the window". Jim wants us to throw him out the window and kill him. He may have suffered brain damaged but he knows that he has really fucked up this time. He wants a coward's way out and to burden us with it; classic Jim.

Mom has been in town all week and had done her homework. The hospital wants to discharge him but someone, some place has to take him. She has found a nursing home near me that has agreed to take him for his Social Security Disability about $250.00 a month. Mom wants me to assume Medical Power of Attorney. He will be transported to the home. All I have to do is sign some papers. Sounds so simple. I have a lot of reservations but I do what my mother asks of me.

It all goes down like Mom said. I start going by to see Jim after work on the way home. At first he is vegged out, the doctor told me to expect him to be childlike, a seven or eight-year-old. Jim is sleeping eating or watching TV three hot's and a cot, no worries. He's finally got it made. I come by one day to find that they have tied him to the bed. They tell me that he is getting up and wandering around. I am aghast. I ask them, isn't that a good thing? Jim has a roommate who is nearly comatose, he is bored and getting interested in his surroundings. They untie him and let him roam. Jim is cradling his left wing with his right arm and moves his curled left hand like a lobster claw. He didn't have a cast he is all plated and screwed together. There is no money for PT. I give him an old Samsonite briefcase of mine with some paper pens and pencils, a triangle and a triangular scale ruler. I tell him to carry the briefcase in his bum arm and practice writing. Jim has retained some higher learning. He can still read and write. It is a small thing but Jim appreciates it. When I come to visit, he always has the briefcase close at hand. He shows me how he can hold it now, using it like a dumbbell to do curls. I check out Jim and take him to church on Sundays he likes to go to our church "because", he tells me," they don't know what I've done". He takes Communion without Confession. I take Jim to my dentist, he hasn't seen one since he lost his dependent status. Jim has meth teeth, he has ground them down to stumps. My dentist fixes him up as best he can and I pay for it out of pocket. I ask Jim what he plans to do now that he is getting better. Jim raises his right hand and rubs his index finger on the tip of his thumb. He would go back to dealing. I just shake my head. The tiger never changes its stripes.

Jim is functioning at too high a level to be in a nursing home anymore, he is getting better but is faking it. What a deal three hots and a cot and he doesn't have to work for it. Should have done this sooner. The staff thinks that he still cannot be responsible for himself. Jim is sent to the Blind and Deaf School in San Antonio for speech therapy. Before he goes the staff and residents throw him a big party. A lot of people leave the home but hardly anyone walks out. Jim writes me plaintive letters from the facility to come get him. Jim does not think that he belongs there with the other residents he calls dummies. Jim thinks he is just fine. That opinion is not shared by anyone. The facility does what it can for him. The nursing home takes him back, offers him a job as a bus boy and sets him up in a title nine apartment complex next to the home; this is a successes story for them. A morale booster for the residents. Unfortunately Jim believes that he is the social director and his job is walking around chatting up the residents. Jim gets paid and discovers that there are two liquor stores around the corner. He shows up drunk to work and is fired. I am livid when I find out. I drag Jimmy over to the two liquor stores and tell them to not sell liquor to Jim or I will sue them. I ask them, "can't you tell he is brain damaged". Jim's speech is still not proper. It is obvious that there is something wrong with him. I give them a notice with his picture on it to post, too late to help.

I have a life, a wife and a little baby girl. I cannot supervise Jim. I check on Jim after a three week trip. His place is a trash heap. It smells like BO and booze. I chew him out. His clothes are laying on the floor. He has done no laundry since he moved in. I ask him what he has been doing. He shows me a book he ordered on government grants, free money. As shown on TV in the wee hours of the morning. Free, just send shipping and handling. He tells me that he intends to hire competent people who will work for him to get and manage the grant. I want to correct him but I can see that he is mystified by what he is holding in his hands. He has no Idea how to proceed. The SBA catalogue might keep him occupied for a while. Why shatter his illusions. I leave and come back with some laundry soap, a sock of quarters, and a spare vacuum cleaner that I had at my house. I tell him that I will be back for a surprise inspection so he had better get busy. Jim salutes me like Gunga Din and says Yes Sir.

Jim gets bored; he takes to wandering around the apartment complex. They have a play scape. Jim likes to watch the little kids play. He asks them if they want to come over to his apartment to play, creepy at best. Jim likes little kids. The moms do not understand and want him gone. He is given an eviction notice. Mom comes through for Jim again she finds a religious associated charity NGO that will provide a supervised room and board situation, a small apartment/bedroom and a common area he will share with five other people. They will take all his SSI. I sign more papers. Mom moves him in. I have quite a stack of papers on Jim now, including about 330K in hospital bills that Dad has told me to ignore. Jim was never prosecuted. No money in it. Dad wrote a letter to the family of the retired officer who died in the wreck apologizing for his son.

This situation lasts until, and I am fuzzy on this, he starts inviting hookers into the place followed by drinking and the whores stealing his checkbook and writing hot checks. Anyway he is kicked out. One of the church members, a carpenter, takes him in and lets him live in his trailer with him. His roommate brings him by for family gatherings, Christmas and Thanksgiving usually held at Steve's place. I am not quite sure what to think. Slim Jim is obese now. He still talks with a slur of sorts and favors his left wing. His roommate tells Tim that he only acts that way around us...Jim passes August 2003 his roommate finds him lying on his side, in bed, eyes open, looking out the window, a heart attack, a broken heart. He was forty nine years old. Jim is cremated in Abilene. Dad has a mass said for him, all his Abilene friends are there. Dad is the only one to speak. He says "my son Jim was a carpenter, he would have liked this box". Steve has a commemorative, a wake at his place in Austin where I see folks that I haven't seen in years, everyone had a favorite story. Jim loved to entertain. No matter what he thought Jim was loved. We are sorry that he is dead and gone.

# 2009 Abilene

Been through Brownwood, Early, Zephyr and Wiley, now Rising Star down HW 183, now it's down HW 36 to Colman and Cross Planes next stop is Abilene all a good day's ride on a horse apart. As you crest the cutoff for the road to Potosi you can see Abilene just past KRBC hill where the radio and cell towers are; it's the highest point in Taylor County. You can see the tower of Abilene Regional Airport, where Dad was an Instructor and Charter Pilot. It was from here that he flew his last charter, setting down in Brownwood after suffering his heart attack. Dad was never the same after that. It really took the wind out of his sails. He effected the feeble old man for years afterward, but it was a ruse; he would still show his old self when provoked. He actually improved quite a bit after receiving a pacemaker in 1994. He got that after he passed out on the putting green at Dyess, picking up his ball. He scared the crap out of his foursome, they said that they didn't want to play with him anymore after that. Like he said of Jim "the tiger never changes his stripes".

"Snake Scanlon" would come to the fore, he was not to be trusted. I am headed for the nursing home. It's "nothing fancy" in Dad's words. Mom is over every day it's her riason d'etre. They have lunch together, free for Mom, she likes that. Something to do, a reason to get up and get dressed in the morning. It smells of urine and ammonia in general. Dad can't understand why he's not free to wander around, after all Mom is free to do so. It's been that way since he took a tumble stepping off a porch, on Christmas Day 2007, while he was with Tim and Mom at the Sales Ranch Christmas party, out at Buffalo Gap. Dad broke four ribs in the fall and skinned his shin to the bone. He had already been diagnosed with Louie Body and was somewhat unsteady before the fall but still able to get around ok. He and Tim were going to go deer hunting the next day with Dad's Border Patrol buddy Art Torrez. Tim tried to get them to go to a real nice luxury assisted living retirement community for nearly two years, a place where they could be together; there was a long waiting list; they never followed through. The place was across the street from the church they attended; a lot of their friends lived there. They could well afford it. OBE. Kathleen who has had strong differences with Mom about Dad's care had just traveled with him to Nashville in October 2008 for the Aviation Class 43K reunion, without Mom, where he was honored for his service. His acceptance speech: "It is with a profound sense of unworthiness and deep humility that I accept your flawless good taste and unerring eye for excellence." That was exactly one year ago. Dad and Kathleen had ditched mom to go to Europe four years prior to visit all the places he had bombed after Sal and Gracie backed out on a tour of Europe. Kathleen has to tell him not to be so loud about that. He joined the CAF in 1994 where he is the local hero with his portrait in the clubhouse. Dad attended the 453$^{rd}$ reunion at old Buckingham in 1984 for the D-Day Anniversary Celebration. Mom got to get her picture taken with General Jimmy Stewart; they make quite a contrast side by side. Dad attended all the B-57, 43K and 2nd Air Div reunions. The change was fast. His doctor, Dad says canker mechanic, told him that he would not wish this on his worst enemy.

I am driving up because I received a call that Dad wasn't eating. They said that people do that when they are giving up. The last time I was up to see Dad he complained that his caretakers wouldn't let him drink anything that wasn't thickened. They're afraid of liquid going down the wrong pipe and giving him pneumonia, so everything is a goo that he has to swallow. He misses a simple cool drink of water or iced tea. He also wanted some PT, just to be walked around. When I asked the Nurse about it, they told me that he had been taken off of PT because it wouldn't help. He won't get better and is rapidly deteriorating. I didn't tell him that.

Every visit with Dad includes the same briefing on what he wants done with his ashes. Dad doesn't want to be in the mud, he has always said that they ship you home in a match box and that is the way he wants to go, reduced to ashes. He wants to be scattered at the end of the runways at Coleman and Dyess, half where he started and half where he ended his flying career. I tell him he can't die, because that is not Mom's plan for him. Her plan is to be buried in a proper grave in consecrated ground next to a Catholic Church like the Church proscribes. She is miffed that he wants to be cremated and doesn't want to lie next to her for eternity. James Francis is already going in the box with her and so will Dear Old Daddy Jim.

When I get there Mom and Sandra, his longtime caregiver, are in the room. They tell me that he has been flying in his dreams all morning. They have been sitting bedside and quietly chatting.

## 2009 ABILENE

Dad is just wakening and smiles at that. He is flat on his back in bed looking thin, grisly and haggard, they haven't cleaned him up yet. Men look much older with a growth and Dad is always clean shaven; it's unsettling. I give him a kiss on the forehead and take his hand in mine, with a boy scout grip, the one with the interlocking little finger, the one that won't slip, the one you use to pull a friend to safety, the one Dad insisted we use with each other.

I note the birds and the bird feeder in the window. Sandra says that she put it there for Dad. He had been doing a lot of bird watching before his troubles put a halt to that, wandering around in mesquite, osage and cane breaks. I can't help but try to get a rise out of him when Sandra and Mom tell me how much he loves the little critters. I say, "As a good Catholic he is only doing penance for all the birds he has shot". That makes him mad, God love him; the Tiger Never Changes its Stripes. Those steely blue eyes burn a hole in me one more time. The look I've seen a million times. Dad would always demand that you look him directly in the eyes when he was addressing you. Because of that, I am unable to lie about anything. Brother Jim could look you in the eye and lie away about anything, he had been intimidated by the best.

A couple of caregivers come in, they are going to bathe Dad, clean the room and change his bedding so Mom and I go to lunch. Sandra has stuff to do so she leaves too. When we come back Dad is back in bed, his room is dark, and the shades are pulled. Getting cleaned up has totally exhausted him. They tell me that he needs to rest even though he just woke up, but I can go in and see him alone for a little bit. With just the light coming through the partially open door to see, he looks much better, all cleaned up and shaved, his hair combed. He is resting and at peace. Dad sirs when I pull up a chair, he turns his head and opens his blue eyes to see who's there.

I see the depth of his exhaustion in his eyes, the fire is gone from when I teased him earlier. He is a fighter but he is tired of the fight, he has been fighting for five years now. To what end Moncrieff? I kiss him on the cheek and ask, "Colonel, are you ready to bail out of this flaming wreck?" Dad nods his head with his eyes closed in the affirmative. I kiss him again and leave him in peace as he drifts off. Mom wants me to stay but I have suppressed issues with her that I don't want to vent. I'm bitter that she robbed us of more time with Dad, or at a minimum failed to make accommodations so Dad could have stayed in their house, or in a nicer place in the time he had remaining to him, even though he said that he wanted nothing fancy. I want to remember Dad like I left him, so I head back to Austin. By the time I get home the Colonel has passed. When I come back to Abilene for the service I meet and shake hands with folks who tell me with a reflective look in their eye "I flew with your Father".

Mom gets her way, like always. I had her going not too long ago. I told her that we were all chipping in for some jewelry for her birthday. I told her that it would be, a gold chain with a gold pendant, this is when her eyes started shining and her breathing became rapid, engraved with the word "Stubborn". This was after a thirty minute argument a month earlier about whether her house coat was inside out or not, it was. The buttons were on the inside and the hems were outside. She looked when I pointed this out to her but she denied it. Arguing that it was not so and she was correct. The truth of what I was saying finally dawned and she said "damn". One of the only times that I have heard her curse.

Dad is divided up into eight brass urns, one each. Gary and I put the contents of ours at the end of the two runways as ordered. Tim has taken three urns back to Colorado with him; Dad said he wouldn't mind that when he went up to visit Tim's ranch in Saguache in 2005. Pretty Country, where he once hunted Elk and Deer. Tim has been living in Colorado since he took a job on an EPA Superfund site cleaning up Bonanza. One urn is interred in a wall, at the Military Texas State Veterans Cemetery, in Abilene, not two miles from the house, behind an engraved bronze plate. One urn goes to Cousin Pat, his Brother Bob's eldest son, Dan has passed, much too early in this life. One urn is with Mom, in the back of a Flag Case I had custom made for that purpose, on the shelf next to Jim.

# Commendations

| | |
|---|---|
| DFC | Distinguished Flying Cross |
| AM w/6 ofc | Air Medal with 6 Oak leaf clusters |
| AFCM w/ofc | Air Force Commendation Medal with Oak Leaf Cluster |
| AFOUA w/1 olc | Airforce Outstanding Unit Award with 1 Oak Leaf Cluster |
| ACM | American Campaign Medal |
| EAMECM w 4bs | Euro Africa Middle East Campaign Medal with 4 Battle Stars |
| WWIIVM | World War Two Victory Medal |
| AOM (Germany) | Army of Occupation (Germany) |
| MHA | Medal for Humane Action |
| NDSMW/1BSS | National Defense Service Medal with 1 Battle Star |
| AFEM | Armed Forces Expeditionary Medal |
| AFLSA w/5 olc | Air Force Longevity Service award with 5 oak leaf clusters |
| AFRESM | Air Force Korean Service Medal |
| SAEMR | Small Arms Expert Marksmanship Ribbon |

# Addendum

Dad wrote many articles and did some public speaking.

## Icing

Jim Gardner asked me to speak on the topic of Cold Weather and Icing. Cold weather, when temperatures are below freezing, really present no significant problems since we operate a great deal of the time in sub-freezing temperatures. For example depart Abilene at 60 degrees F and climb to 10,000feet and you are at 32F but then put some moisture into the picture, when the temperature is near freezing, now we have problems.

I would like to relate to you a few things I have seen concerning ice and cold weather then briefly review some precautions to be take regarding cold weather operations.

I have been flying since 1942 and have encountered innumerable icing conditions but I have only been in real danger four times. Once in a B-24 in England, once in a T-33 Jet fighter at Little Rock, once in a Convair in So Carolina and once in a Cessna 310 at Abilene Texas. Now I'm certainly not saying that icing is no real threat to flying, since I've only been caught in bad icing four times…what I'm telling you is: I am scared to death of any situation that even hints of icing…and I avoid icing conditions like the plague. I landed that T-33 at Little Rock at 135kts using 85% power and was sinking fast. Heavy clear ice- freezing rain caught me unawares in the instrument landing pattern. The normal landing speed for that little jet was 95kts.

The Convair in South Carolina came falling out of the sky from 16,000' down to 5,000' caused by heavy clear ice in a cumulus cloud build up. The Convair hot wing only cleared the leading edge and allowed a fence or ridge of ice to form aft of the leading edge which destroyed most all lift.

The Cessna 310 episode at Abilene was also during an instrument approach when I encountered freezing rain which was not forecast. When I came over the strobe lights and looked up it was like trying to see thru wax paper so I kicked it in a crab and looked out the side window to land.

Years ago while flying out of England I personally witnessed on different days, four B-24 Liberators crash on takeoff due to severe icing in freezing rain or freezing drizzle. We eventually got rid of the stupid commander who ordered missions aloft in such poor conditions but only after forty aircrew had died.

I remember one early November morning: we briefed at 0400 for a 0600 pre-dawn departure for a target deep in Germany, Magnaberge was primary target and Chenitz the secondary. We had a full bomb load and full fuel tanks we were over max gross weight by quite a bit. The weather was freezing drizzle and some freezing rain which is just bigger drops of super cooled H2O. The ground was coated with ice that was so slick you couldn't even stand still on the ice coated ground much less walk. We went to the mess hall and directly to our A/C in weapons carriers and 6 x 6 trucks with chains on both front and rear wheels. When we arrived at the hard stands our A/C were being run up by the ground crews. The A/C were held in place with the wheel chocks staked to the ground. After the run-up all surfaces and props were deiced by hand held spray rigs, then the aircraft were towed to the T.O. position with cleat track tractors. At the end of the runway, each aircraft was again sprayed all over with de-ice fluid. Engines were cranked on and off we went. I remember breaking out of the clouds at 19,000ft in a near stalled condition and loaded with ice. I was turning 2600 rpm and pulling 45 inches of Hg barely hanging in the sky at 140kts. We had no deicer boots, no prop alcohol deicers because the alcohol was highly explosive and flammable. The Brits removed them for some unknown reason. What is more flammable than 100 octane? Soon after getting on top, I saw a fellow B-24 in a flat spin go down over 20,000 feet and hit flat in six feet of sea water just east of Norwich England. This particular crash was one of the minor miracles of the war. All ten crew members survived. They were broken and battered but all survived. Ours was the only group to launch that day. We managed to get twelve Liberators aloft and eleven hit Ohlmitz and one hit the wash. Once clear of the cloud vapor the ice sublimated off. Ice turns from solid to vapor. it was sure lonely, only eleven aircraft of the mighty 8th air force hit Germany that day. Fortunately the weather over the continent was even worse than in England. We had no fighter opposition, just flack over target. We returned after seven plus hours and in a slightly warmer climate. There was almost a mutiny but it was avoided by the quick dismissal of our crappy commander.

Now what is the point of all this verbiage? Facts: #1 we were in a war. #2 the aircrews had no choice but to takeoff as ordered. I have personally observed General Aviation A/C C-310, C-320, MU-2's Beach 55s, Kingairs, C182s, 210s etc— you name it. Depart the airfield in freezing rain. Not all of them made it to destination, quite a few aborted after run-up when they found they couldn't see out of the windscreen and not one of them was ordered to T.O. not one pilot was threatened with courts martial; no one had a gun at their head, and I am sure they were not at war. So why would a pilot leap off in such conditions. Is it ignorance? It surly is gross stupidity. For instance I saw a 182 depart VFR under an overcast out of which freezing drizzle was falling. They were going to Dallas, Love field for dinner. They did not make it there and back. You should have seen that bird, coated with clear ice, all over, the pilot said that they almost bought the farm. He had to use full power to maintain level flight at 5000'and was losing altitude at 2 to 300 feet per min. He was very lucky that the carburetor/air filter didn't freeze over cause he would then have had an ice glider.

If the weather is below freezing and stays that way from the ground up, you have very few problems. It's when the weather oscillates near the freezing point when troubles multiply. The most dangerous situation is when you have freezing rain or drizzle in the landing pattern. You gotta land somewhere some tine so if there is even a hint of freezing rain, stay on the ground.

Cold weather preflights do require a bit more attention. If you're A/C is in a heated hanger the only additional caution is to follow the pilots handbook for engine start and catch it the first time around or your sparkplugs will frost over and you have about as much chance of getting a start as trying to fire up a sack of potatoes. After start let the engines warm up gradually, you have cold thick oil, don't be in a hurry, and remember that you used a lot of prime to start. If you are parked in a heated hanger, preflight in the hanger except for fuel checks. Wheel it out drain the sumps, check the tanks and bug out. A warm A/C moved into damp cold air can frost over where as an already frozen bird will not attract moisture. If you have to put your A/C in a heated hanger to remove frost, ice or snow, make sure you leave it in long enough to dry out. If you tow it out dripping wet don't be surprised if you can't open the door that is frozen shut or your controls don't work quite right. Ice in all the cracks and crevices is dangerous.

There is no easy mechanical method of removing frost or ice, de-ice fluid is the only easy fluid means of removing the stuff. Never, never, never use water!! I have seen people weld their A/C to the ground by attempting to remove frost and ice with a water hose. The chance of cracking wind screens and windows is very high especially if one choses to dump a bucket of warm H2O on the wind screens.

If your fuel tank and sump drains won't drain, you must apply heat by using a portable hot air type heater or put the beast in the warm air hanger until it is thawed out, you will not know the extent of the water or ice in the fuel tanks unless you drain them and you can't drain ice.

Portable hot-air-heaters are usually available at F'B.Os if there is any doubt, use the telephone and find out what services are available at your destination. On a trip from Houston to Amarillo in January you will encounter freezing temperatures aloft and on the ground at Amarillo. You will defiantly have to encounter any freeing weather with knowledge of not only the exact procedures to use in preflight, engine start, taxi, run-up etc., but you must also be certain you know the effects freezing temperatures have after you are airborne. Often pilots attempt to use combustion heaters on the ground and the heater quits. Why? Insufficient air flow, so the combustion heater overheats and the circuit breaker pops and no heat. How do you check a combustion heater? You check and see if you have blown a fuse. You check to see if the blower fan works and that it is kind of a catch 22. Granted some combustion heaters will operate on the ground but most will overheat and pop the circuit breaker if you have a remote circuit breaker in the cockpit you are in business allow a cooling period reset and light up. If you have the normal set up once the circuit breaker pops, you can only reset after landing. This is a good system since it keeps people from setting an aircraft on fire. It forces a mechanic to check it and see why it overheats. The exhaust shroud heater common to single engine A/C is almost fool proof. Almost since an exhaust leak of CO is a killer. I would recommend that every aircraft have a detector installed, a small button that would turn color at the first hint of carbon monoxide. One should activate all anti ice and deice systems during preflight for proper service and operation.

Now let's talk about surface ice in flight. Most twins have ice lights, a small spotlight on the outboard side of the engine nacelle that shines on a dark area on the wing leading edge. In clouds it is difficult sometimes to determine what you are looking at thru the windshield it could be clouds and it could be ice, use your flashlight and your side windows. If you are accumulating ice, get out of the ice zone. You will either have to climb or descend. Look at your temperature gage and altimeter. You will soon know which way to go. If you are west of Denver, I'd suggest you climb but don't waste time. Declare an emergency if you can't get a quick clearance.

A standard rule of thumb is to allow ice to build so it completely covers the leading edge or the entire pneumatic ice boot to ¼ to ½ inch thick and activate the deicer boots long enough to crack off the ice then turn the deicer boot off.

Of course the anti-ice systems of alternate air for engine prop electric heater, prop alcohol systems, windshield heater or windshield alcohol systems must be used before you enter the icing zone. So keep a close eye out during the entire trip for any icing possibility. Know you're A/C and its system's operation limitations. I know of a young fellow that lost both engines in a 55 Barron at 16,000 feet over Silver City New Mexico in a heavy snowstorm. He forgot to go to alternate engine air. When the two engines quit he did two fast laps around the rosary, hit every handle he could reach and luckily got the manual alternate air handles in the right position and got an engine back in operation. Now some Beech Barons have automatic alternate air for engines…the point here is to know you're A/C real well before you run into the need for it in an emergency situation.

Pitot heat should be turned on any time there is visible moisture and should be used during every decent if you have been cursing in cold temperatures. Jets turn on their pitot heater for takeoff and turn it off after landing. Get in the habit of using pitot heat, its real cheap insurance.

Slush is a real menace. It can pack up around wheels, in wheel pants and in wheel wells to such an extent that you can end up with wheels frozen solid in the up position in a solid block of ice. If you have enough gas to fly to Florida or Panama from wherever you picked up the slush then you have no problem. Again know the A/C limitations regarding snow, slush, ice and frost then you will stay out of trouble. Get a through weather briefing, use common sense, stay out of icy conditions and stay alive.

## Additional Articles

Dad and GW Ford researched, contacted, compiled and published an article in: Air Power History Summer 1995 titled Double Duty of the Eighth. They identified 60 pilots who flew a tour in bombers and then volunteered for fighters. They were able to contact 16 pilots and get their stories and photos. Another work in itself too voluminous to present here.

# References

***B-57 Canberra at War 1964-1972***
Robert C Mikesh
(Dad was a contributor)

**History of the 453rd Bomb Group**
(Dad was a contributor)

***Forbidden Nation a History of Taiwan***
Jonathan Manthorpe

***Truckbusters From Dogpatch The combat Diary of the 18th Fighter-Bomber Wing in the Korean War 1950-1953***
Tracy D. Connors

***Mao The Unknown Story***
Jung Chang and Jon Halliday

www.ingramcontent.com/pod-product-compliance
Lightning Source LLC
Chambersburg PA
CBHW020656060526
44119CB00090B/398/J